Hartwig Steusloff (Ed.)

Integrated Water Resources Management

International Conference Karlsruhe
24 - 25 November 2010

Fraunhofer Institute of Optronics, System Technologies and Image Exploitation
Karlsruhe

Integrated Water Resources Management Karlsruhe 2010

International Conference
24 - 25 November 2010

Conference Proceedings
Hartwig Steusloff (Ed.)

KIT Scientific Publishing

Cover image by Filipebvarela – Fotolia.com

Imprint

Karlsruher Institut für Technologie (KIT)
KIT Scientific Publishing
Straße am Forum 2
D-76131 Karlsruhe
www.ksp.kit.edu

KIT – University of the State of Baden-Wuerttemberg and
National Research Center of the Helmholtz Association

KIT Scientific Publishing 2010
Print on Demand

ISBN 978-3-86644-545-1

Editor's Preface

Fighting water scarcity means employing the full range of countermeasures, from possibly increasing the water resources to making optimal use of the water resources available. During the first decade of the 21st century, the limitations of the global water resources, however, became evident in a way that demanded a new quality of water resource management. Taking into account the multiplicity of water grades and water utilization requirements, we face a complex real-time management problem of matching current water demands with frequently very limited water supplies of appropriate quality and quantity. The IWRM Karlsruhe 2010 conference, including an accompanying exhibition, sets out to establish a bi-annual platform for an international exchange of methods for and application experiences on up-to-date and future Integrated Water Resource Management (IWRM).

IWRM Karlsruhe 2010 will bring together a global water management community from approximately 20 countries worldwide. The topic areas ranging from rainwater retention and storage via waste water management, water ingredients management, water resource management and modeling, up to water management under extreme conditions. Within the three water consuming communities (agriculture, households, industry) IWRM Karlsruhe 2010 draws attention to agriculture and to the use of waste water for irrigation. A main area of the conference programme deals with mathematical modeling of water processes including seepage. Further highlights are the influences of climate change and capacity building in water resource management.

The conference proceedings in hand comprise four keynotes and 60 papers which are arranged in three so called "tracks": Water Utilization, Water Processes and Water Management Methods. These tracks will provide guidance for our visitors through the broad thematic offer of IWRM Karlsruhe 2010.

The scope of the IWRM Karlsruhe 2010 conference has been developed by its advisory board, listed overleaf, which also served as the programme committee. The editor wants to express his sincere thanks to all members of the advisory board as well as to many highly motivated and competent supporters from Messe Karlsruhe and from the Fraunhofer Institute of Optronics, System Technologies and Image Exploitation (Fraunhofer IOSB, Karlsruhe). The Editor and the Advisory Board would be delighted to start, based on the IWRM Karlsruhe 2010 conference proceedings, a bi-annual publication platform on the status and the future of integrated water management worldwide.

Prof. Dr. Hartwig Steusloff
Fraunhofer-IOSB,
Karlsruhe (Germany)

Advisory Board

Table of Contents

IWRM – GETTING GROUNDWATER INTO THE PROCESS

Prof Dr Stephen Foster
World Bank Groundwater Management Advisory Team (GW-MATE) – Director
International Association of Hydrogeologists (IAH) – Immediate Past President
E-mail: GWMATEfoster@aol.com or IAHfoster@aol.com

ABSTRACT

A brief, but systematic, overview of the promotion of an IWRM approach to groundwater management and protection in the developing world is presented, including consideration of the conceptual and practical difficulties sometimes encountered. The importance of the nexus between the groundwater resource base and certain facets of agricultural policy and urban infrastructure development is stressed, since without integrated vision and action at these interfaces serious sustainability concerns are likely to arise. In this paper the topic is treated generically, but specific examples can be found through the key bibliographic references provided.

THE BACKGROUND

IWRM – a Balanced Concept

- Integrated Water Resources Management (IWRM) is a process to manage water resources holistically. It promotes the coordinated consideration of water, land and related natural resources during developmental activity through recognition that:
  - freshwater is a finite and vulnerable resource, essential to sustain human livelihoods and the natural environment
  - freshwater is a public good, and has a social and economic value in all its competing uses
  - water resource development and management should be based on a participatory approach – involving users, planners and policy makers at all levels (and specifically women, given their often central role in water-supply provision and protection) .
- IWRM is thus not an end in itself, but a means of ensuring that the development of water resources is efficient, and maximises resultant socioeconomic welfare in an equitable manner without compromising vital ecosystem sustainability.

Groundwater – a Critical Role in Development

- In the final decades of the last century advances in waterwell technology and hydrogeological knowledge facilitated a massive expansion in groundwater use across the developing world – especially in Asia. Groundwater became of major importance for :
    - supplying the growing population (some 2.0+ billion in urban areas alone) and the expansion of industrial enterprises
    - providing irrigation for both staple and cash crops (globally to more than 40% of irrigated land).
- The provision of low-cost, drought-reliable and (mainly) high-quality water supplies has produced enormous social benefits, with many countries developing large groundwater-dependent economies. And further expansion of groundwater resource development will be instrumental for achieving progress towards the 'UN Millennium Development Goals'. Moreover, with climate-change issues looming larger each year, groundwater resources have come under increasing focus with realisation that natural and enhanced aquifer storage can play an important role in adaptation strategies.

Groundwater Management & Protection – a Pressing Need

- In the developing world investment in overall governance and practical management of the groundwater resource base has been seriously neglected. Groundwater stocks in many aquifers are vast but their replenishment is finite. Indiscriminate resource exploitation has widely led to serious water-table decline, in some cases resulting locally in irreversible degradation associated with aquifer salinisation and/or land subsidence, and serious impacts on dependent down-gradient interests in stream baseflow and/or aquatic ecosystems.
- Concomitantly there has been increasing pollution of shallow groundwater – due mainly to uncontrolled urbanization and to some intensive agricultural production regimes, and sometimes to natural contaminant mobilisation. The flow dynamics of many lower-lying aquifers in particular means that they are often the 'final sink' for pollution from the land surface, and the nature of groundwater systems often means that their clean-up is either technically impractical or extremely costly. In other aquifers with more pronounced upland recharge areas, the ecological protection of these recharge areas is a critical, but achievable, concern.
- All of the above is beginning to impact human livelihoods and health – and thus improving the management and protection of groundwater (a classic 'common-pool' resource) represents a pressing need and, in many senses, one of the greatest challenges in stewardship of the natural environment.

The IWRM Approach to Groundwater – Relevance & Impediments

- Since groundwater is a very widely distributed resource, it is affected by a plethora of local users and polluters, whose behavior in turn is influenced by national policy affecting land and water use. Thus mobilisation on improved management and protection needs to be strongly participatory, integrated across sectors and at a wide range of scales.
- Although IWRM has been widely adopted in the 'developed world' and its axioms are gaining increased acceptance in 'developing nations' – incorporation of groundwater resources into National IWM Plans has encountered some professional impediments given that:
    - many senior water resource managers who have a strong IWRM vision, and who are trying to put principles into practice, have a rather limited grasp of groundwater system scales, dynamics and vulnerabilities
    - most hydrogeologists, who have a sound understanding of groundwater dynamics, 'up-gradient linkages' and 'down-gradient dependencies', tend not to focus on the socioeconomic drivers of resource use and pollution load nor on the institutional framework for addressing land-use and water management.

PROMOTING INTEGRATED GROUNDWATER MANAGEMENT

Addressing Some Important Conceptual Challenges

- Groundwater and surface water are intimately linked, being part of the same overall hydrological cycle. Aquifers discharge to surface water bodies, and can be recharged by them, depending on local conditions. However, while river systems are naturally flow-dominated, most aquifers are characterized by large storage (stocks) and much lower flux (flow rates) with the implication that :
    - groundwater systems have capacity to buffer surface water variability
    - for groundwater 'upstream-downstream considerations' neither predominate nor are necessarily fixed.

  However, despite natural hydrogeological complexity and given adequate monitoring, groundwater system behaviour can be predicted with equal confidence to that of surface water and the 'storage buffer' makes it is easier to accommodate uncertainty in management decision-making.
- The 'river basin' is the fundamental spatial unit for application of the IWRM process, and this has to be reconciled with the fact that 'groundwater bodies' (defined on hydrogeological criteria) are a much more appropriate spatial framework within which to address the needs of groundwater management and protection – with specific hydrogeological settings requiring a different approach (Table 1).
- While groundwater resource management needs do not normally relate to 'upstream-downstream' conflicts, successful management requires a much more 'integrated approach' to the :

- land-water management interface in the interest of conserving groundwater recharge and quality
- spatial allocation of resources to different uses (including ecosystems)

than is usually attempted in river basin management. The Global Water Partnership has issued guidelines on the conceptual framework for National IWRM Plans and on the practical steps needed for implementation. These constitute the basis of what is recommended here for groundwater, essentially developing and aggregating specific groundwater body management plans as the hydrological realities dictate.

Table 1: Hydrologically-consistent approach to reconciling river basin catchments with groundwater bodies for water resource management

important groundwater bodies of limited extension compared to river basin	independent local groundwater management plans required, but these should recognise that aquifer recharge may result from upstream riverflow and downstream baseflow will often be dependent on aquifer discharge
river basin underlain by extensive shallow Quaternary aquifer	surface water-groundwater relations (and their management) require fully integrated appraisal to avoid double resource-accounting and various problems (including salt mobilisation on land clearance, soil water logging and salinisation from irrigated agriculture, etc)
extensive deep aquifer systems of arid regions	groundwater flow system dominates – there is little permanent surface water and thus not helpful to adopt a river-basin approach
minor aquifers of shallow depth and patchy distribution predominate	limited groundwater interaction with river basin and (despite socioeconomic importance of minor aquifers for rural water-supply) integrated groundwater/surface water planning and management not really essential

- In the promotion of integrated groundwater resource management certain misconceptions are often encountered and need to be addressed :
  - resource deficits can be met by supply-side measures alone – when in reality measures to control demand will always be necessary because aquifer recharge enhancement is likely otherwise to stimulate increased groundwater abstraction
  - reluctance to recognise non-renewable resource exploitation – whereas in reality the exploitation of non-renewable or weakly recharged groundwater resources should only be undertaken with a high-level of resource planning, enhanced management and fully-justified socioeconomic returns.

A Pragmatic Framework for Action

- When trying to promote improved groundwater resource sustainability, 'high-level reform' may not be the best (or a necessary) entry-point for action. Experience demonstrates there is no 'simple blueprint' for action due to intrinsic variability of:
  - the hydrogeologic setting of the resource, which tends both to define the nature and scale of the problem and to constrain the solution

  - the socioeconomic context of groundwater resource utilisation – it being very different to manage a few large abstractions for municipal water- supply or industrial development than to control of a very large number of individually-small waterwells used for subsistence irrigated agriculture.

- Thus a 'pragmatic framework' for the definition of 'groundwater management plans' has been devised (Table 2), using local hydrogeological realities and resource use dynamics to identify an appropriate mix of the following :
  - local institutional arrangements – with an empowered government agency facilitating community awareness and participation, and where appropriate self-regulation
  - finance and implementation of technical demand-side and supply-side measures
  - macro-policy modifications – to constrain groundwater demand
  - enhanced resource administration and targeted use regulation.

  Progress will require a balance between 'bottom-up' and 'top-down' action, with 'political support' for prioritized, sequenced, practical and patient interventions.

- Groundwater quality protection should follow a comparable strategy comprising the following steps :
  - systematic assessment of groundwater pollution hazard – based on mapping of aquifer pollution vulnerability and subsurface contaminant loads
  - definition of a 'groundwater protection plan' – to reduce this hazard in priority areas through differential land-use management (involving an appropriate mix of local technical measures, stakeholder mobilization and regulatory control) and where necessary national restrictions on the sale of certain agrochemicals.

Table 2: The GW-MATE Pragmatic Framework for elaboration of groundwater resource management plans

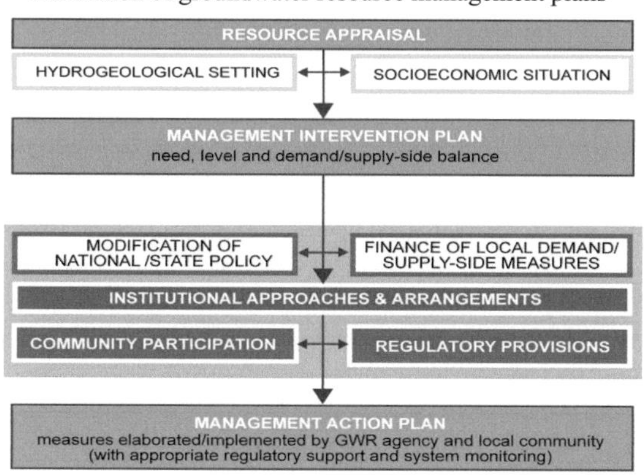

- Groundwater is a 'highly decentralised resource' often developed by private initiative, thus its management and protection will not be effective without social (user and polluter) participation – and investments to strengthen institutional capacity in this regard are especially critical. Government usually has to play a central role as 'resource guardian' (usually via a 'local-level agency') and 'to make the first move' by:

- profiling groundwater users and uses – as a basis for understanding and communicating the socioeconomic importance of the resource and the consequences of 'non-action' on its management and protection
- selecting 'pilot areas' to try out participatory resource management and quality protection — the boundaries of such areas should be scientifically-consistent 'groundwater bodies'.
- working along a 'parallel-track approach' to achieve incremental improvements in resource management/protection whilst continuing investigation and monitoring so as to advance aquifer characterisation and to consolidate conceptual and numerical modelling of the groundwater system.

GROUNDWATER & AGRICULTURE – SPECIFIC CROSS-SECTOR ISSUES

- In many parts of the developing world irrigated agriculture has become the major user and predominant consumer of groundwater – and serious questions of resource sustainability (and even irreversible degradation) are arising. A number of key cross-sector policy issues arise when attempting to promote the more sustainable use of groundwater in irrigated agriculture and to reduce the impact of intensive agriculture on groundwater quality.

Role of Irrigation Technology Improvements

- Investments to improve (so-called) 'irrigation efficiency', whilst highly desirable from the standpoints of energy saving and water productivity, do not equate to equivalent groundwater resource saving, since in many cases (and especially on permeable soils) their introduction mainly reduces irrigation water returns (infiltration) to groundwater – and as such is not a 'real water resource saving'. Thus where canal water is the source of irrigation such investments usually result in a marked reduction in groundwater recharge, and where groundwater itself is used for crop irrigation they usually result in an increase in net consumptive use.
- If they are to constitute an effective groundwater management measure, it is essential that these improvements are accompanied by parallel interventions in terms of reduction in the volume of groundwater rights (or licensed abstraction) and/or constraints on the total irrigated area.

Influence of Rural Energy Pricing Policy

- Electricity subsidies for groundwater pumping are often argued to be the key factor provoking excessive resource exploitation in areas of irrigated agriculture. And certainly the adoption of 'flat-rate tariffs' (which has occurred in some countries and provinces) must always be regarded as perverse, since it provides absolutely no incentive for constraining groundwater pumping nor improving the efficiency of waterwell pumps – resulting in very high energy consumption on irrigated land (in terms of kWhr/ha/a) and enormous pressures on electricity generation.
- But on more detailed consideration the influence of normal energy subsidies may prove less significant (and in political terms is often justified in terms of helping to level the cost of irrigation water between those outside and inside major irrigation canal commands). Other macroeconomic interventions, such as waterwell drilling/deepening finance and crop guarantee-prices, may thus exert more influence on the evolution of groundwater-based agriculture and defer the transition to less 'water-use intensive' livelihoods.

Optimising Conjunctive Use of Groundwater & Surface Water

- Conjunctive use of groundwater and surface water resources represents a great opportunity for improving irrigation-water availability/security and for sustainable expansion of agricultural production – especially on major alluvial plains. However, in most parts of the developing world the practice it is still largely spontaneous and sub-optimized.
- Widely a concerted effort is needed to tune conjunctive use practice to the dynamics of groundwater/surface water interaction and to confront the socio-economic and institutional impediments – rigid historic surface water-rights, ineffective irrigation canal management and split institutional responsibility.

Impacts of Intensive Agricultural Cropping

- The intensification of agricultural cropping practices on permeable soils results in a negative diffuse impact on groundwater recharge quality associated with the leaching of nutrients and pesticides – and the introduction of such practices in arid terrains can also mobilise salts held in the vadose zone to groundwater.
- In certain vulnerable hydrogeological conditions, some intensive agricultural cropping practices and monocultures generate a level of diffuse pollution which is incompatible with the use of groundwater for potable water-supply and/or threatens the sustainability of local groundwater-dependent ecosystems. This situation tends to result in a conflict of land-use interests which is not easy to address – but whose

solution lies in negotiations and trade-offs which reduce the aerial extension of land under intensive cultivation and/or the use of certain types of agrochemicals in hydrogeologically-specified zones.

URBAN GROUNDWATER-RELATED POLICY ISSUES

- City development results in a continually-evolving relationship with groundwater in underlying aquifers – urbanisation processes tend to augment groundwater recharge and degrade groundwater quality, and in turn groundwater system changes can impact urban infrastructure. Without improved policy and planning, based on an 'integrated vision', the problems arising often turn out to be persistent and costly. In the urban environment all too often 'one persons infrastructure solution becoming another persons problem' !
- All too often degradation of groundwater resources in urban areas is not a result of insufficient scientific understanding nor failure to mobilise capital investment – but more due to an 'institutional vacuum' whch leads to lack of policy coordination. Groundwater considerations need to be inserted into metropolitan area and municipal infrastructure development plans through better communication of resource presence and protection needs (including dissemination of land-surface zoning maps).
- In discussing the policy issues arising it is convenient to distinguish cities/towns where the municipal water-supply utility or the autonomous company/ concessionaire has major dependence on groundwater abstraction from those where private in-situ self-supply from groundwater by residential, commercial and/or industrial users predominates – although in some situations they co-exist.

Managing the Groundwater-Sanitation Nexus

- In-situ sanitation of major urban areas often presents a significant groundwater quality hazard, which needs to be recognised and managed. In most hydrogeological conditions, except the extremely vulnerable, there will be sufficient natural groundwater protection to eliminate faecal pathogens in percolating wastewater. But the hazard can increase markedly with sub-standard waterwell construction and/or certain types of sub-standard in-situ sanitation practices. However, troublesome concentrations of N compounds (usually nitrate) and DOC in groundwater also arise to varying degree with population density served, and can penetrate to considerable depth in some aquifers and persist for many years after the source of contamination is removed.
- Groundwater pollution can be reduced substantially by deploying so-called dry or eco-sanitation units (in which urine is separated and not discharged to the ground), although this solution is not always socially-acceptable and retro-installation in existing housing presents problems. The alternatives in terms of protecting urban

groundwater quality (where considered high priority) are either prioritising the installation of main sewerage systems or controlling the housing/population density for urban development in areas of vulnerable groundwater.

Planning & Managing Major Municipal Abstraction

- Quite commonly municipal waterwells have been constrained within the built-up area or engulfed by rapid urban expansion. In such cases more effort generally needs to be put into assessing potential pollution risks and the surveillance of groundwater supply quality for the range of potential pollutants identified.
- The construction of 'external' municipal wellfields (or capture of other water-supply sources) is the rational policy response to loss of (or threat to) urban waterwells. But this will also require much improved coordination and 'appropriate compensation' between municipalities within and neighbouring 'metropolitan districts', to facilitate the implementation of equally-necessary protection measures for the associated investment.

Developing Policy on Private In-Situ Groundwater Use

- Private waterwell drilling by urban residential and other users is usually initiated as a 'coping strategy' at times of inadequate mains water-supply service – in-situ self-supply from groundwater representing an economical alternative to purchase from water tankers. Very large numbers of waterwells have been (and are being) constructed on this basis, especially in urban areas underlain by shallow aquifers.
- For the mains water-supply service-level to improve usually requires higher average domestic water-tariffs – and private waterwells are then often deployed as a 'cost-reduction strategy'. It is too simplistic to talk in terms of banning residential self-supply from groundwater, except where the risk of negative impacts from aquifer development or epidemic disease transmission are very high. The phenomenon has benefits and risks which need to carefully balanced and managed.

CONCLUDING REMARKS

An Important Message

- Groundwater resource managers need to voice more strongly their concerns about resource degradation in national debates and through public communication – where appropriate challenging macro-policies which are highly counterproductive in terms of groundwater sustainability and drawing intention to inadequate institutional provisions and capacity for resource management and protection.

Bibliographic Materials

- GW-MATE Briefing Note Series : Sustainable Groundwater Management–Concepts and Tools (2002-09) World Bank, Washington DC-USA www.worldbank.org/gwmate
- GW-MATE Case Profile Collection : Sustainable Groundwater Management–Lessons from Practice (2003-09) World Bank, Washington DC-USA www.worldbank.org/gwmate
- GW-MATE Strategic Overview Series : Sustainable Groundwater Management – Contributions to Policy Promotion (2009-10) World Bank, Washington DC-USA www.worldbank.org/gwmate

  1- Groundwater Governance–conceptual framework for assessment of provisions and needs.

  2- Conjunctive Use of Groundwater and Surface Water–from spontaneous coping strategy to adaptive resource management.

  3- Urban Groundwater Use Policy – balancing the benefits and risks in developing nations.

  4- Sustainable Groundwater Irrigation – approaches to reconciling demand with resources.
- GWP TEC Background Paper 10 (2004) Integrated Water Resources Management (IWRM) and Water Efficiency Plans by 2005 – Why, What and How ? Global Water Partnership, Stockholm-Sweden www.gwpforum.org

How to develop capacity for IWRM: Examples from UNW-DPC

Reza Ardakanian
Teresa Walter
UN-Water Decade Programme on Capacity Development (UNW-DPC),
United Nations University, 53113 Bonn, Germany,
e-mail: ardakanian@unwater.unu.edu, tel.: +49 2288150651

Abstract

As a response to growing water demands and increasing water scarcity the concept of Integrated Water Resources Management (IWRM) has evolved during the last decades, to propose and implement instruments to reverse the trend of declining water resources. During the last years several IWRM projects have been implemented and it became obvious that besides the establishment of appropriate frameworks and tools, capacity development especially on the individual and the institutional levels is required to promote and implement the IWRM concept. The UN-Water Decade Programme on Capacity Development (UNW-DPC) supports awareness creation of water use by strengthening global policy and by advocating enhanced funding and local action towards sustainable water use and management. As an organizer of several IWRM workshops, UNW-DPC pays special attention to the improvement of management and capacity development activities and plays a crucial role in the identification and assessment of capacity gaps and needs.

Keywords

Capacity development, IWRM, water scarcity, UNW-DPC

1 Introduction

Water is one of the most indispensable of all natural resources; it is essential for human beings, economic development and biological diversity. However, many countries have to face the challenge of rapidly growing water demands, driven by an increased population and economic growth, linked to growing urbanization, industrialization and mechanization (King 2004). Consequently water demand is stimulated and the worlds' freshwater resources are decreasing.

The concept of Integrated Water Resources Management (IWRM) as a response to water scarcity and water quality problems has emerged during the last decades. Although the issues of water resources assessment, water use and water efficiency were key components of the outcome document of the UN conference on water in 1977, the fundamental principles of the IWRM concept were discussed much later at the International Conference on Water and the Environment in Dublin in 1992 and adapted at the UN conference on the Environment and Development in Rio de Janeiro in the same year. Ten years later in 2002 water has been accorded high priority in the global development agenda and at the World Summit on Sustainable Development in Johannesburg 193 countries agreed to develop an IWRM and Water Efficiency Plan by 2005. It shall propose and implement instruments to reverse the trend of a decline of water resources through the integration of land, water and living resources management, while strengthening national capacities (Snellen and Schrevel 2004).

This requires water management institutions which are able to monitor water availability and water use and to adjust accordingly. The UN-Water Decade Programme on Capacity Development (UNW-DPC) provides important support to promote and implement the IWRM concept by creating awareness, strengthening global policy and advocating enhanced funding and local action towards sustainable use and management of water resources.

2 Capacity Development for IWRM

2.1 Capacity Development

"Making public sector organisations work better is one of the most persistent and difficult challenges in development and development cooperation. At the same time, nothing is more crucial for achieving sustained progress, growth and poverty reduction" (European Commission 2005). Hence assessing individual and institutional capacity development is an essential element of any kind of support.

Bolger (2000) defines capacity as the "abilities, skills, understanding, attitudes, values, relationships, behaviours, motivations, resources and conditions that enable individuals, organisations, networks/sectors and broader social systems to carry out functions and achieve their development objectives over time." The European Commission (2005) notes that “capacity can be defined as the ability to perform tasks and produce outputs to define and solve problems and make informed choices”.

The term capacity development extends the definition of capacity to an approach or process towards poverty reduction and as a development objective to develop individual and organisational capacity. Furthermore “capacity development refers to the approaches, strategies and methodologies used by developing countries, and/or stakeholders, to improve performance at the individual, organisational, network/sector or broader system level” (Bolger 2000).

Although capacity development has been an issue for many years, it is only today where it is receiving special attention in the issue of water management. Many irrigation management transfer programmes were implemented without raising necessary capacity of local people and it was recognised that there are serious shortcomings in development assistance (Kay and Renault 2004). Nowadays it is widely accepted that capacity development does not only include the strengthening of individual skills and abilities but also needs appropriate environment, opportunities and incentives, taking into account the individual, organisational and institutional levels to have a long term impact (Alaerts and Kaspersma 2009). The new approach is a rather locally driven process, which includes local stakeholders and motivates local capacity development by considering and analysing respective capacity factors. New capacity development should build on local capacities and different activities at various levels should be integrated to address complex problems (see Fig. 1). Capacity development on the individual level includes the transfer of knowledge to stakeholders, farmers, professionals and others, while the organisational level refers to groups of people as water user organisations, research groups and private companies. The sectoral level implies that irrigation is just one part of water management and that also other aspects like water supply and the environment should be considered. Enabling environment represents the broad national and international context and refers to the political and legal framework (Bolger 2000).

The objectives of capacity development include the enhancement of skills, abilities, and resources, the strengthening of understanding and relationships, and the assessment of issues of values, attitudes, motivations, and condition in order to support sustainable development. The development of indicators to monitor interventions and tracking progress on capacity development can contribute to the success of different programmes (Otoo, et al. 2009).

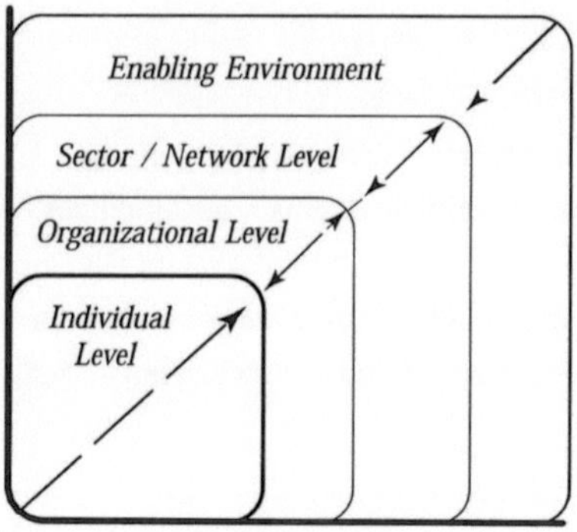

Fig. 1: Capacity Development: Conceptual Framework

As the water sector has particular complexity with each individual having impact on the water management, it strongly depends on institutional and individual capacities.

2.2 Key elements and objectives of IWRM

The concept of IWRM integrates water resources as a systematic process for the sustainable development, allocation and monitoring of water resources. This promotes more coordinated management of land and water, the river basin and upstream and downstream interests. IWRM comprises four core principles which were agreed upon at the Dublin Ministerial Conference that preceded the first World Summit on Sustainable Development in Rio de Janeiro in 1992 (UN 1992): (1) Fresh water is a finite and vulnerable resource, which is essential to life and sustainable development, (2) Water management is based on a participatory approach, where users, planners and decision makers are involved at all levels, (3) Women play a central role in the provision, management and safeguarding of water, and (4) Water should be recognized as an economic good that has an economic value.

Since IWRM is still an evolving concept with high complexity, there are various definitions (IWMI 2001). Although they differ in their terms of formulation, they all include the three key concepts of efficiency, social equity and sustainability shown in Fig. 2. Possible management instruments include the water resources assessment, and regulatory and economic instruments, while the environment is enabled through policies, legislations as well as financing. The according necessary institutional framework is a necessary prerequisite.

One possible and widely accepted definition of IWRM is given by the Global Water Partnership, which defines IWRM as the "process which promotes the coordinated development and management of water, land and related resources, in order to maximise the resultant economic and social welfare in an equitable manner without compromising the sustainability of vital ecosystems" (Global Water Partnership 2000).

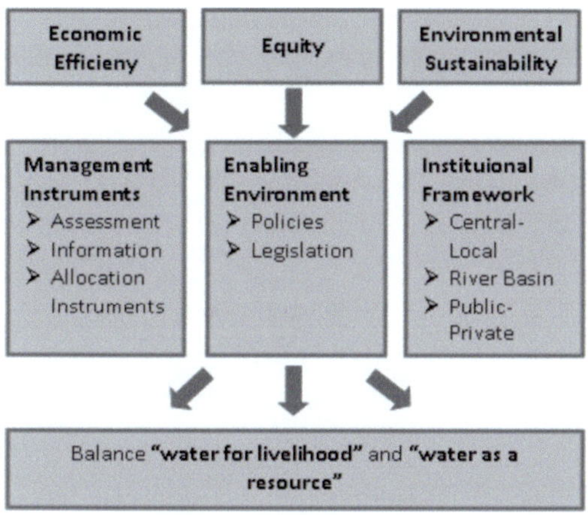

Fig. 2: Structure of IWRM

IWRM is based on management that accounts for the inter-related nature of different sources and that considers surface and groundwater as one entity. It seeks the balance of interests between competing water users through an integrated management approach, taking into account the available amount of surface and groundwater, its use and quality as well as environmental and social considerations. Hence, IWRM depicts the whole physical, social and economic complexity of the water cycle with supplies and demands and shall provide possibilities to optimally distribute water resources.

The shift towards a more integrated and participatory management of water resources requires adaptive management approaches and is a logical consequence of the need to fully take into account the complexity of the systems and uncertainties.

2.3 Examples of IWRM projects and lessons learned

During the last few years several IWRM projects have been established and implemented. They differ in their objectives, their approaches as well as in their target group and people involved. All of the projects have the main objective to develop a transferable approach for IWRM. However, as the concept of IWRM is very complex different issues are emphasised. While the *CuveWaters project* has a long time horizon and can therefore deal with many aspects of IWRM, the *LIFE project* in Egypt mainly considers water use efficiency improvement possibilities.

IWRM Project	**Duration**	**Main Objectives**
IWRM-Vietnam	2006-2010	Develop methods and tools for IWRM and for a better use of available data for problem, analysis, planning and decision making
Smart Project in Jordan	2006-2009	Develop a transferable approach for IWRM and analyze innovative technologies, decision support systems and management strategies
IWRM for Central Asia: Model Region Mongolia (MoMo)	2006-2009	Develop and implement IWRM through an integrated view of all water uses and protection of freshwater resources using a trans-disciplinary management approach
IWRM in the Cuvelai-Etosha Basin (Central Northern Namibia) – CuveWaters	2006-2012	Conceptually develop and practically implement IWRM in order to secure the livelihood of the resident population in the long term and thus make a substantial contribution toward poverty reduction and crisis prevention in the region
LIFE (Livelihood and Income from the Environment) IWRM in Egypt	2004-2008	Increase water use efficiency and productivity through providing technical assistance, training, commodities, and small grants to support decentralization of water management
IWRM pilot project in South Africa (Olifants project)	2006-2010	Develop an IWRM concept, covering all relevant issues and easy to implement. Economic aspects are explicitly considered by estimating water demand functions.

Tab. 1: Examples of IWRM Projects

These examples show the efforts that are done in the topic of IWRM but they also reveal that implementation of IWRM is still in its early stages. The river basin organizations are still looking for their role and responsibilities and struggle with limited human and financial resources. There are examples of well performing river basin organizations but common problems exist. Very often there is a lack of autonomy, poor recognition among stakeholders, a lack of human and financial resources and a lack of adaptive management. Furthermore the coordination between institutions and sectors is very often inadequate (CAP-Net, UNDP 2008).

3 The UNW-DPC and Capacity Development in IWRM

3.1 Activities of UNW-DPC related to IWRM

The UN-Water Decade Programme on Capacity Development (UNW-DPC) is one of the three programme offices of the inter-agency mechanism UN-Water that has been created to follow up on the World Summit on Sustainable Development's (2002) water related decisions and the Millennium Development Goals concerning water. The main objective of UNW-DPC includes the enhancement of concerted action and coherence of UN-Water Members and Partners providing support in assessing capacity development needs.

UNW-DPC pays special attention to the support of the improvement of management and dissemination of capacity development activities, including IWRM. Therefore UNW-DPC co-organised several workshops on IWRM during the last two years, mostly focusing on the individual and organisational level of capacity development (see Fig. 1). Therefore they co-organised a workshop on Adaptive Water Management (AWM) and IWRM in India in 2008. The workshop aimed at building capacity of trainers by providing support in curriculum development to be able to teach water and sustainability issues at universities with available facilities and possibilities. An earlier workshop held in New York in the same year focused on virtual learning approaches to capacity development for IWRM.

An important activity of UNW-DPC on IWRM was the "Training programme on Integrated Water Resources Management for decision makers from the Middle East and North Africa (MENA). UNW-DPC was one of the three partners in order to implement a training programme on IWRM for decision makers and professional managers from the water as well as complementarily relevant sectors for three identical eight day units from October through December 2009. The objective of the training programme was to contribute to increased cooperation and mutual learning among decision makers and professional managers in the MENA region. Theoretical and practical sessions were developed in the topics of network, team role, project management and multi-stakeholder cooperation. Participants concluded that the workshop was successful with its content and organisation and especially with its interactive modules.

A Training of Trainers course on Conflict Resolution and Negotiation Skills for Integrated Water Resources Management (IWRM) for South East Asia, which was held in March/April 2010 in Hanoi and co-organised by UNW-DPC aimed at capacity development of extensive cooperative activities concerning the management of water resources. Given the importance of water resources to all human communities, it is natural that conflicts arise with regard to access, allocation, development and management of the resource. Therefore, professionals from ministries, agencies and NGO's of the region participated for five days to be trained in conflict resolution, negotiation and management for IWRM.

Regarding the improvement of water use efficiency in the context of IWRM, UNW-DPC was co-organiser of a Training Event on "How to Improve Water Efficiency in Water Utilities" in Rio de Janeiro in March 2010. This training course provided technical knowledge and practical experiences so as to enable water utilities to improve their water efficiency. Other five workshops considering water use efficiency were done on "Capacity Development for Farm Management Strategies to Improve Crop-Water productivity, using AquaCrop" as a joint initiative of FAO and UNW-DPC in collaboration with local partners. The objective of the five day workshops was to train participants from various regions in the practical applications of AquaCrop, in order to improve participants' skills in strategic management, leading to improved water use efficiency as part of an IWRM.

3.2 Further needs in IWRM capacity development

Knowledge gaps in water management exist especially in the prediction of future global changes of water supply and demand and how they are going to affect people. It is not clear how responses to changes should be and how the water service delivery and the resource water should be managed more effectively (Alaerts and Kaspersma 2009). UNW-DPC can play a crucial role in capacity development in the topic of IWRM including identification and assessment of capacity needs to be pursued on a permanent basis. UNW-DPC supports full participation of stakeholders in the public and private sectors as part of the decision-making process and specially focuses on the organizational level, where most countries suffer weaknesses.

As limited knowledge and a limited number of professionals and institutions, laws and regulations are hindering effective action, capacity development is required to identify, address and communicate the challenge and to implement according action (Alaerts and Kaspersma 2009).

In cooperation with UN-Water Members and Partners, UNW-DPC creates awareness, influences global policy and facilitates information exchange towards sustainable use and management of water resources. They play an important role to monitor and report on how water resources management is advancing in member states, to identify barriers to progress and to suggest ways to overcome them. UNW-DPC supports and facilitates coordinated action both between UN agencies, UN-Water Members and Partners, bi-lateral donors, development banks, NGO's and civil society in a way that will enhance coherence and impact capacity development amongst others in water resources management.

4 Conclusion and Outlook

The numerous projects and workshops that have been done or are still ongoing show the worldwide recognition of the importance of scientific research in the area of IWRM and at the same time of concrete action that must follow on the individual and institutional levels.

Many scientific based research projects are facing the challenge of the development of complex models which are difficult to implement, while international organizations, which are rather trying to provide technical and advisory support, have to deal with the short term impact of their support. UNW-DPC tries to overcome this problem by organizing training of trainers workshops were local people shall be trained amongst others in IWRM topics so that they are able to use and disseminate their gained knowledge. In the future through a better cooperation of research institutes and international as well as national and local organizations, research and capacity development projects could be implemented in a sustainable way as gained scientific knowledge and knowledge transfer strategies could be applied simultaneously.

List of Illustrations

List of Tables

List of Literature

[1] Alaerts, G. J., and J. M. Kaspersma. "Progress and challenges in knowledge and capacity development." In Capacity development for improved water management, by M. W. Blokland, G. J. Alaerts and J. M. Kaspersma, 3-30. London: CRC Press Taylor & Francis Group, 2009.

[2] Bolger, J. Capacity development: why, what and how? Capacity development occassional papers Vol. 1 No. 1, CIDA, Policy branch, 2000.

[3] CAP-Net, UNDP. "Integrated water resources management for river basin organisations." Training Manual. 2008.

[4] European Commission. Institutional assessment and capacity development. Luxembourg: European Communities, 2005.

[5] Global Water Partnership. Integrated Water Resources Management. Stockholm, Sweden: Technical Advisory Committee (TAC), 2000.

[6] IWMI. "Strategic Plan 2001-2005." Colombo, 2001.

[7] Kay, M., and D. Renault. "Capcacity development engineering - a way forward for capacity building in irrigation and drainage." Capacity development in irrigation and drainage. Rome: FAO, 2004.

[8] King, N. "The economic value of water in SA." In Sustainable options: development lessons from applied environmental economics, by J. Blignaut and M. de Wit, 207-238. Cape Town: KCT Press, 2004.

[9] Otoo, S., N. Agapitova, J. Gold, and S. Fisher. "The need for a conceptual and results-oriented framework in capacity development: discussion of a new approach." April 2009.

[10] Snellen, W. B., and A. Schrevel. "IWRM: for sustainable use of water - 50 years of international experience with the concept of integrated water management." Background document to the FAO/Netherlands Conference on Water for Food and Ecoszstems. Wageningen: ministry of Agriculture, Nature and Food Quality, 2004.

[11] UN. The Dublin Statement on Water and Sustainable Development. 1992. http://www.un-documents.net/h2o-dub.htm (accessed July 17, 2010).

Taking care of the world's water

Aspects of Reuse, Desalination, Water Transport and Non-Revenue Water

Stefan Kraus

Siemens AG
Industry Sector; Industrial Solutions Division; Industrial Technologies
Water Distribution & Controls
91052 Erlangen, Germany
E-mail: stefan.ks.kraus@siemens.com, Telefon: +49(0)9131-729368

Water is our most precious resource

1.7 billion people around the world are lacking safe water. Lack of access to clean water kills about 5 million people annually and causes 3.3. billion illnesses. Every 8 seconds a child dies of a water-related disease. At least 1 billion people must walk three hours or more to obtain drinking water. Nearly 2 % of U.S. homes have no running water. In Mexico, 15% of the population must haul or carry water.

If present consumption patterns continue, two out of every three persons on Earth will live in water-stressed conditions by the year 2025.

Catchwords

Safe water, world water consumption, World Water Day, water reuse, sustainability, water transport, seawater desalination, water shortage, leak detection, leak control, non-revenue water, SIWA Solutions, SIWA Systems.

1 We can't do anything without water

Water is a resource which is needed in all kinds of production processes. No car, no industry product, no food can be produced without water.

The amount of water necessary for production of all-day-products is both, amazing as well as shocking. It is a condition sine qua non, that we use this precious resource most efficiently .

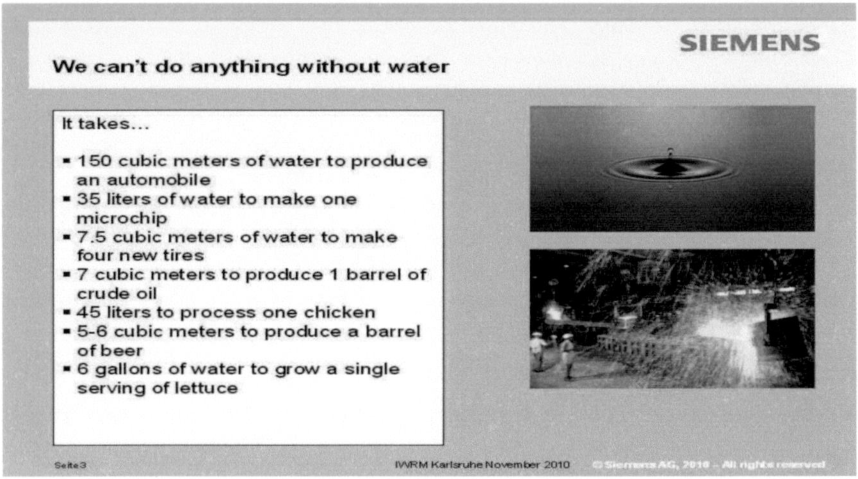

2 Aspects of Reuse, Desalination, Water Transport and Non-Revenue Water

2.1 World Water Consumption

World Water Consumption varies significantly regarding regions as well as consumer groups. It is notable that first of all the agricultural use of water resources is the predominant type of consumption in all parts of the world except Europe.

It is furthermore most significant that the daily consumption of water per capita for all types of consumption have a spread of up to 500%.

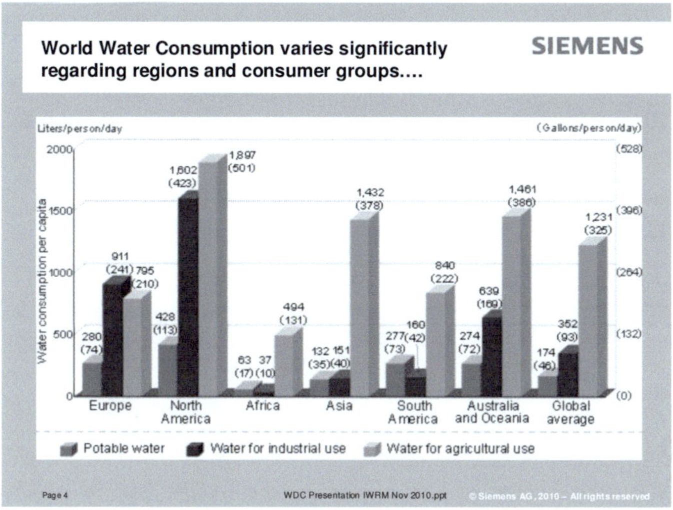

Over the last decades the awareness for clean water and the limitation of water resources has risen and just recently on the occasion of the World Water Day in March 2010 the WHO made a clear statement for the protection of the quality of water.

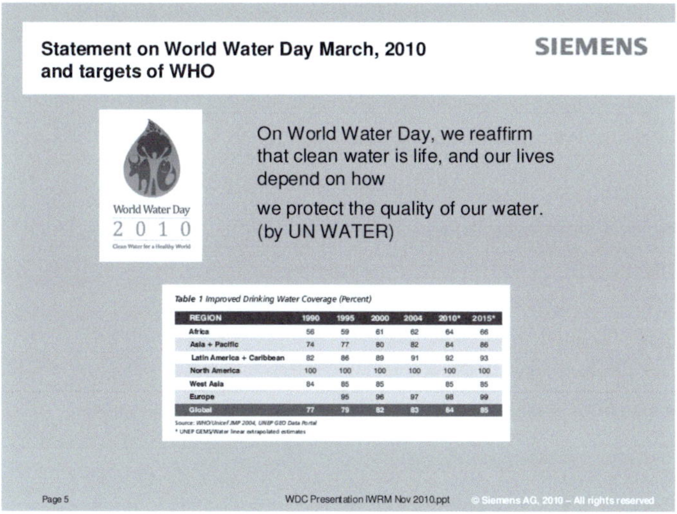

Table 1 Improved Drinking Water Coverage (Percent)

REGION	1990	1995	2000	2004	2010*	2015*
Africa	56	59	61	62	64	66
Asia + Pacific	74	77	80	82	84	86
Latin America + Caribbean	82	86	89	91	92	93
North America	100	100	100	100	100	100
West Asia	84	85	85		85	85
Europe		95	96	97	98	99
Global	77	79	82	83	84	85

Source: WHO/Unicef JMP 2004, UNEP GEO Data Portal
* UNEP GEMS/Water linear extrapolated estimates

2.2 Siemens History in the field of water

Siemens has a company history of more than 160 years in the field of electro – technology.

The fact, however, that Siemens has a history in the water branch of more than 100 years, is not widely known. Brands such as, but not limited to Wallace & Tiernan, Envirex and Zimpro stand for a long history of striving for higher efficiency and trendsetting technologies in water and wastewater applications.

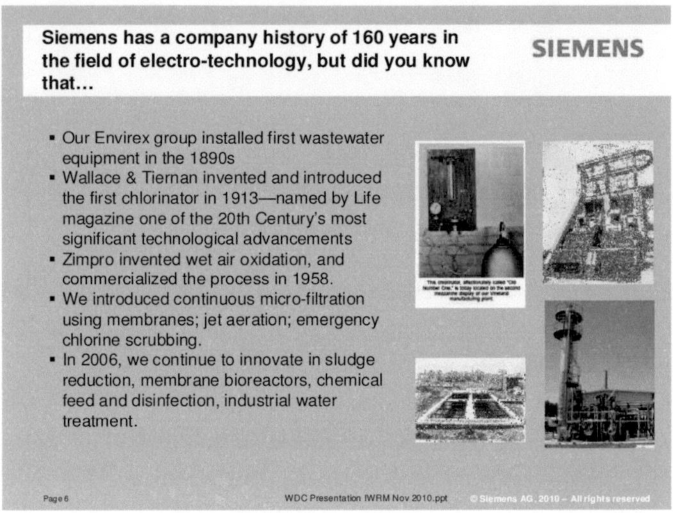

2.3 Siemens – the Water Reuse company

As stated before water is a limited resource.

Efficient use of natural water resources alone will not cover the enormous demand of the future. Water Reuse is the answer to this challenge.

To mention just some of the most important applications of water reuse:

- reuse of wastewater for irrigation
- Reuse of treated wastewater for industrial cooling, boiler feed water applications

- Recharge of groundwater, replenishment of seawater barrier with treated wastewater
- Desalination of all kind
- Reuse of industrial process water
- Zero liquid discharge

And today… we are the leader in helping customers achieve sustainability

- over 1.3 billion gallons of water a day are purified for reuse by our membranes, helping reduce the demand for water
- We provide ultra-pure water (purest in the world) for critical medical applications
- Customers call on us to clean up contaminated groundwater
- With our help, industries recover metals and other valuable materials from wastewater
- We make sludge safe for reuse and enable plants to reduce sludge volumes

We help cities and towns treat and reuse water

- thousands of communities on six continents use Siemens Water Technologies to provide safe drinking water to hundreds of millions of people
- as well, we help them treat and reuse wastewater
- Over 1,000 treatment plants use our odor control technology to make them “good neighbors”
- We provided emergency water for the tsunami in SE-Asia and the Katrina and Rita hurricane disasters

2.4 Water Transport Solutions – an important part in the puzzle of Reuse

Since decades natural resources such as geological water have been exploited for the sake of satisfying the demand for water of urban and agricultural consumers.

One example is the Great Man Made River Project in Libya, which taps geological water resources below the Sahara and transports the water over 600 kilometers to Tripoli.

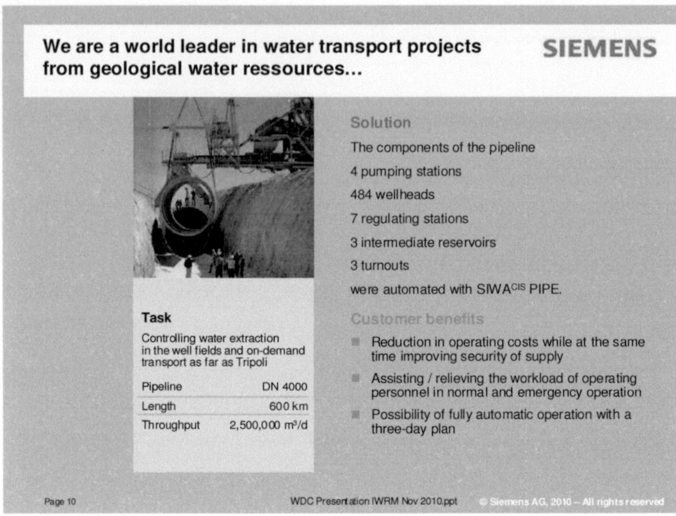

Another source for huge amounts of water to be transported over long distances is desalinated seawater. The task actually is the same, however in this case we really can talk about sustainability, as natural resources are not being used but reused.

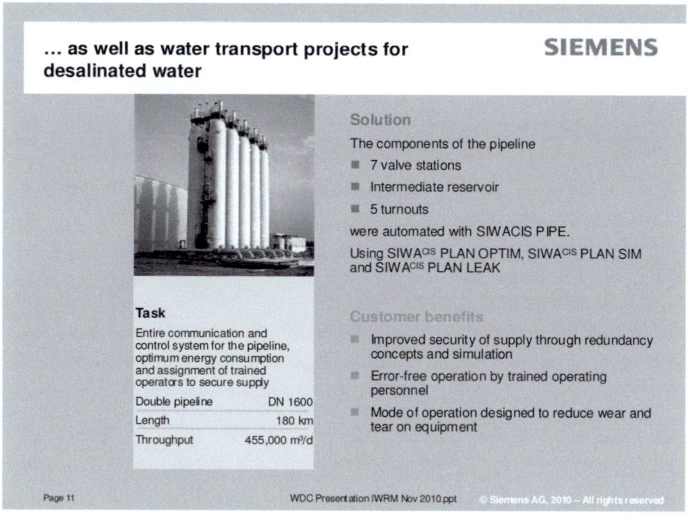

2.5 Desalination – the answer to global water shortages ?

Geological water – resources, as we mentioned before, won't last forever.

A rapidly increasing world population leads to significant growth of freshwater consumption. Ambitious irrigation projects all over the arid zones of the world are or have been realized.

Cost of membrane – based desalination decreases following economies of scale as well as quantum leaps in membrane innovation.

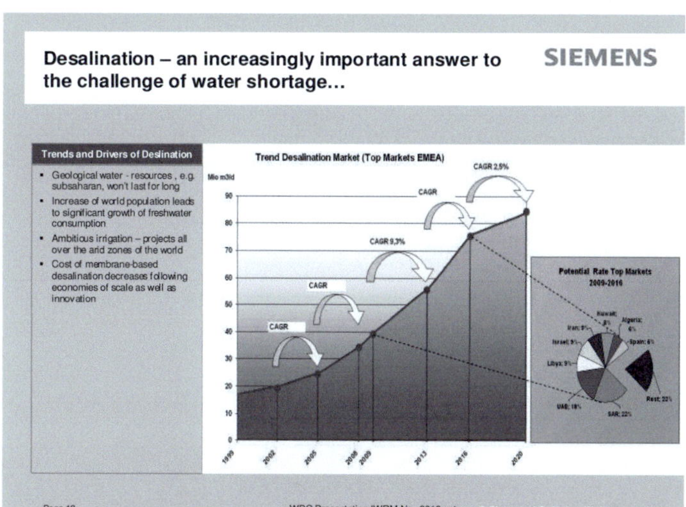

2.6 Leaks in water networks / Non – Revenue Water

In the sections before we've learnt, that other than optimizing the consumption of water, we can reuse it.

Another way of reducing water consumption is the reduction of Non – Revenue Water.

What are we actually talking about ?

Non – Revenue Water (NRW) is the difference between the water volume fed into the water network and the billed water volume consisting either of:

- **real losses**, resulting from inefficient operation/maintenance, missing proactive Leak – Management and bad quality of the water pipes in a water network
- **commercial losses**, resulting from imprecise, faulty data or water theft
- **non-revenue legal consumption** e.g. for the fire department

2.7 The solutions for water and wastewater networks in a nutshell...

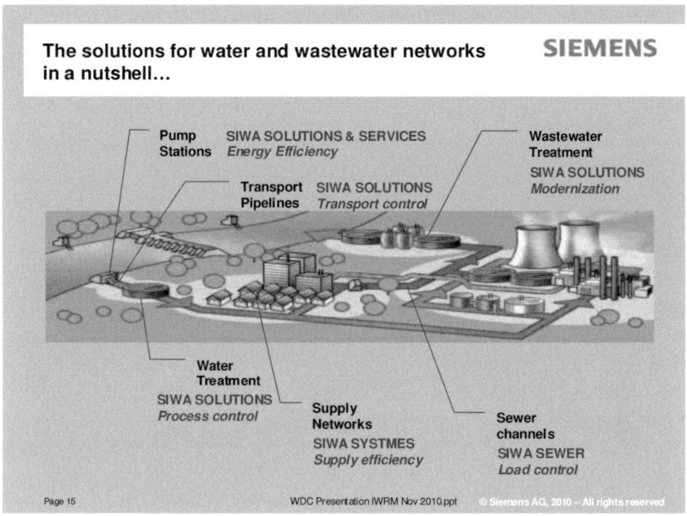

3 Conclusion and Forecast

By applying methods of efficient water usage, water reuse including desalination and reduction of Non-Revenue Water the natural water resources can be used also by our children and grandchildren.

These are proven technologies and just have to be applied in order to achieve the Millenium goals for 2015

List of Literature

[1] WHO/UNICEF/LMP 2004, UNEP Geo Data Portal

[2] UN WATER 2010

The development and application of a catchment surveillance methodology, employing an interaction matrix, for water resources vulnerability assessment and management

Prof. B J Lloyd, *DelAgua Ltd and Centre for Environmental Health Engineering (CEHE), World Health Organisation (WHO) Collaborating Centre for the protection of water and health,*
University of Surrey, England, GU2 7XH.
Email: b.lloyd@surrey.ac.uk Tel : 0044 (O) 1483 689532

Abstract

This paper focuses on the severe impact of changing land use patterns on surface water supply reliability, and how this can be reversed to provide sustainable supplies in the Caribbean island of St Lucia. It is concerned with the development of a novel methodology, the *interaction matrix*, to assess the vulnerability of river catchments to diffuse and point sources of pollution. This was applied in a GIS-based catchment surveillance overlay methodology to locate risk hot spots to guide land use planning, regulation and remediation. The overall *objective of catchment surveillance* is the conservation and protection of surface and groundwater sources to ensure that they may be used safely and sustainably.

Where terrain is increasingly permeable the risk of groundwater pollution is increased. Conversely where the land is impermeable, run-off and hence the likelihood of surface water pollution is increased. However, the probability of pollution occurring in either case is controlled by many different factors. The investigation of all the factors which influence the risk to which a water resource is exposed requires a multi-disciplinary approach. The interaction matrix provides the means to assess the interactions between those factors, and to calibrate their relative contribution to the vulnerability of the river system. Research projects in Southern England, in temperate low-land river catchments (Matthews & Lloyd, 1998), and in the Caribbean, sub-tropical mountainous river catchments (Lloyd & Thorpe, 1997) provide the context for developing and testing this methodology for risk assessment and risk management.

Keywords

Water resources protection, catchment vulnerability analysis, interaction matrix, GIS.

1 Introduction

The adverse impact of deforestation on water resources management was established more than 80 years ago as a result of the Wagon-Wheel Gap experiment in Southern Colorado (Bates & Henry,1928). Forestry confers many benefits including reducing the magnitude of ordinary seasonal floods, maintaining stream-flow in dry weather, and preventing erosion. In their natural state the volcanic islands of the Caribbean were densely forested with tropical rain forest in their steeper middle and upper catchments, but during the past 500 years deforestation has occurred to make way for various crops. These were typically grown intensively in the lower and less steep zones of these islands.

Within the last 50 years the increasing pressure on land use has resulted in many marginal farming activities extending to steeper, middle to upper catchments. Today, increasing population and anthropogenic pressure are imposing increasing demands on land and water resources in the Caribbean Islands. In the Windward Islands the use of land for agriculture (principally banana cultivation) in upper catchments is impacting adversely on the reliability and quality of water resources.

2 Project justification

Coupled with climate change, it is asserted that the principal effects of inappropriate land use on river systems in the Windward Islands are :

- greater variation in flows (increased flooding & drought periods)
- more erosion and land slides
- declining water quality (increased sediment loads and loss of biodiversity)
- increased river silt causing drinking water treatment plants to be out of service for extended periods

Thus forests are essential for reducing the risks imposed on water resources, so *land use management is fundamental to sustainable water resources management,* but we do not know what level of deforestation in any particular location will result in serious loss of reliability of water for domestic supply and other uses. The point at which a water resource becomes unreliable depends upon many environmental factors including the proportion of it which remains forested. There is therefore a practical need to examine the consequences of vegetation changes in different climatic conditions, topography, geology, soil type, etc., and define their effects. We need to know the magnitude and distribution of the effects of forest and different types of land uses on stream-flow and quality in order to manage. In the specific case of. St Lucia and the other Windward Islands a catchment surveillance studied was required because:

- the population is dependent on surface water streams for 95% of its demand
- water supply is regularly interrupted, sometimes for several weeks, because…
- water resources are becoming less reliable, due to
    - Longer drought periods and more flash flooding
    - Higher stream turbidities causing water treatment plant failures

Additionally it was believed that these problems are caused by unregulated banana cultivation extending to steeper upland forested areas. Unlike forest trees the banana root system is shallow and fibrous providing less soil stability. This problem is compounded by intensive, continuous cropping which is only possible with repeated herbicide application. This removes all other vegetation and roots and organic mulch. The net effect is that soil structure is degraded & erosion increases.

3 Catchment surveillance methodology

The 'bottom up' outline strategy and methodology applied in this project is summarised in Figure 1. Two principal types of environmental data gathering activities dominated the time required for the project:- field work (ground sensing) and satellite images (remote sensing). Following data validation the data were entered into a data base for the ArcView (ESRI) geographic information system (GIS). An interaction matrix approach was used to define and calibrate the key parameters which control catchment vulnerability to run-off. The GIS provided for the integration of raster and vector data sets, and the analytical environment to produce multiple data overlays and vulnerability maps. Vulnerable areas and hot spots provide the evidence for regulating land use and hence catchment management and protection. However, interventions to regulate and protect land use require appropriate legislation which was under consideration by the St Lucia government whilst the project was in progress.

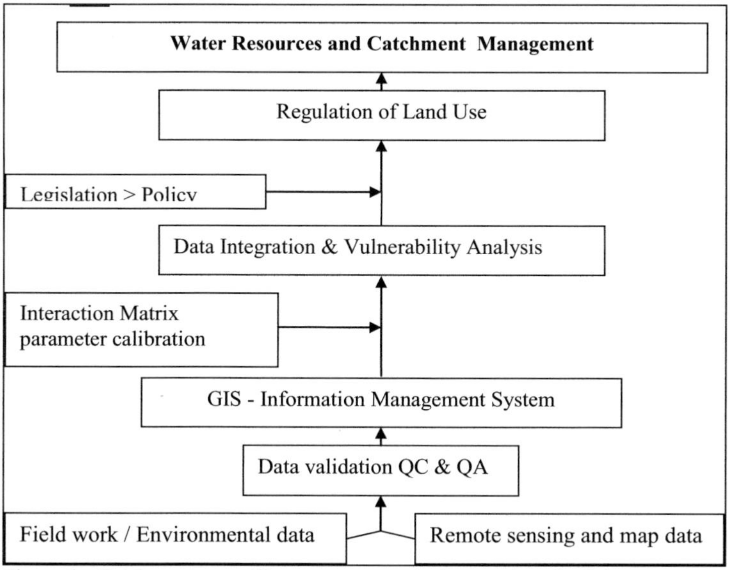

Figure 1. Catchment surveillance summary strategy and methodology

The methodology may be further broken down into the following activities:
a) selection of priority river catchments required for water supply (Figure 2a)
b) definition of river quality and hydrological monitoring network
c) execution of stream quality monitoring programme
d) acquisition of remote sensing land use data and conventional map data
e) data validation and input into databases and geographic information system
f) integration of geographic and environmental quality data
g) identification of hazard factors which contribute to the vulnerability of the rivers
h) application of interaction matrix principles to scale the hazards
i) GIS overlay of geographically referenced hazards and classification of risks
j) identification of high risk micro-catchment zones for intervention
k) pixel level identification of high risk areas
l) multi-sectoral consultation for catchment level land use policy and regulation
m) consultation with community groups and land owners for changes in land use
n) dissemination of advisory reports to community groups and regulatory departments
o) application of policy and regulations to conserve water authority forest reserves
p) practical interventions for conservation and management of precisely defined micro-catchment areas

Three examples of the map data coverages digitized for the project are shown in Figure 2; a) is a digitised map of St Lucia showing only those (14) river catchments used for water supply. These catchments are further subdivided to show where there are sub-catchments at lower, mid and upper altitudes zones specific for water supply.
The adjacent figure 2 b), shows all surface water streams and their catchment boundaries. Hydrological monitoring stations are also pinpointed. Figure 2 c) is the same map but with only rainfall isohyets plotted.

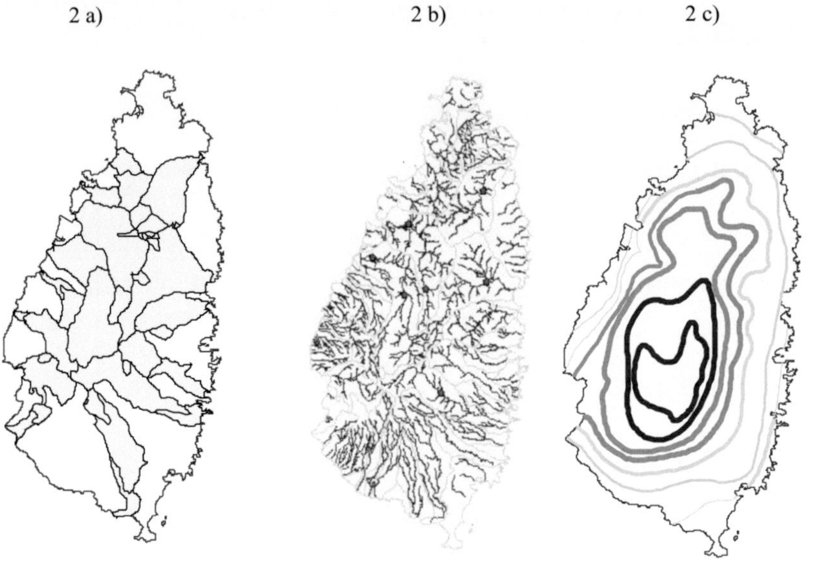

Figure 2. Examples of digitised coverages for St Lucia showing a) river catchments and water supply sub-catchments, b) streams and hydrological gauging stations, and c) annual rainfall isohyets plotted from rain gauging station data

More than 20 additional coverages were digitised or satellite images processed and entered into the GID data base. They included: Erosion, Geology, Stream sample stations (50), Stream gauging stations, Stream abstraction points, Prawn farms, Rain gauge/weather stations, Administrative boundaries, Roads, Urban areas, Training and Truth areas(1km squares) to identify and verify the land use and land cover types, Contour maps, Spot and Landsat images.

4. Preliminary sub-catchment risk assessment

The stimulus for the project was the increasing difficulty reported by the St Lucia Water Authority in maintaining water treatment plants operating due to the siltation of river intakes and blocking of filters. A preliminary study was therefore undertaken to identify and calibrate the relative vulnerability of all (13) treatment plants by a combination of desk study and walk-over surveys. Table 1 summarises the results for four water supply sub-catchments to demonstrate the criteria used in scoring and classifying their vulnerability to, and their actual, deterioration. The Low vulnerability sub-catchments, such as the upper Troumasses, are typically those in the uninhabited, upper forested areas with high aquatic biodiversity [A,B quality, Lloyd & Thorpe (1997a)], which provide uninterrupted water supplies throughout the year. The Extreme (worst class) example given is for the middle Troumassee which equates to a treatment plant which is out of service for many weeks of the year. Figure 2 uses the criteria in Table 2 to classify the vulnerability of all the water supply sub-catchments and shows two in the extreme condition and other 6 which are in the High vulnerability class.

Table 1: Preliminary vulnerability assessment of four sub-catchments for water supply

Subcatchment examples	**Human settle-ment**	**Farming above intake**	**Land slides**	**Siltation of water intakes**	**Inverteb Biol Quality**	**Vulnerability Hazard score & Risk Class**
Troumassee (upper)	None	* <5%	* Few	* Rare	A,B	**0-3 Low**
Dennery	None	** 5-10%	*	**	A,B	**4-6 Medium**
Roseau (upper)	* Little	** 5-10%	**	*	** C,D	**7-9 High**
Troumassee (middle)	* Little	*** >40%	*** Severe	*** Frequent	*** D E	**>9 Extreme**

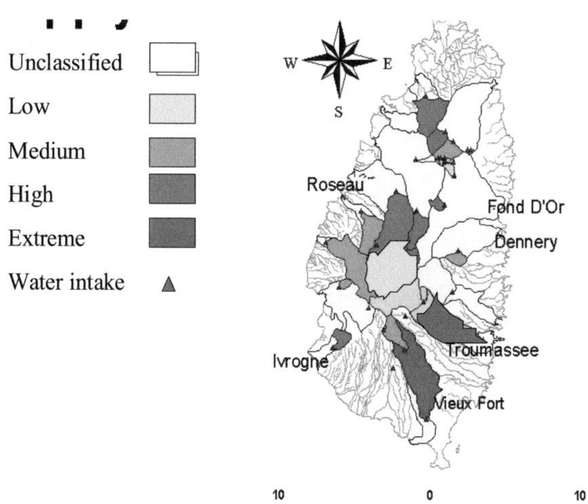

Figure 2. Preliminary classification of vulnerability of sub-catchments for water supply

5. Interaction Matrix development

To reduce the arbitrary nature of identifying and scoring the parameters (elements or thematic coverages) which influence catchment vulnerability a systems approach borrowed from rock engineering was used (Hudson, 1992). This starts with a 'top down' analysis of the total system rather than the traditional 'bottom up' approach of studying each element in isolation. In order to achieve this an interaction matrix (IM) for the catchments was developed as a means of representing total system behaviour. Once the key elements have been identified, the IM permits a systematic review and calibration of the interaction of each parameter with all other parameters by an expert multi-disciplinary panel. A simplified version using only four (diagonal) elements is presented in Figure 3 to show how a panel would define interactions. This starts at the top left with SOIL affecting SLOPE, and SLOPE's effect on SOIL.

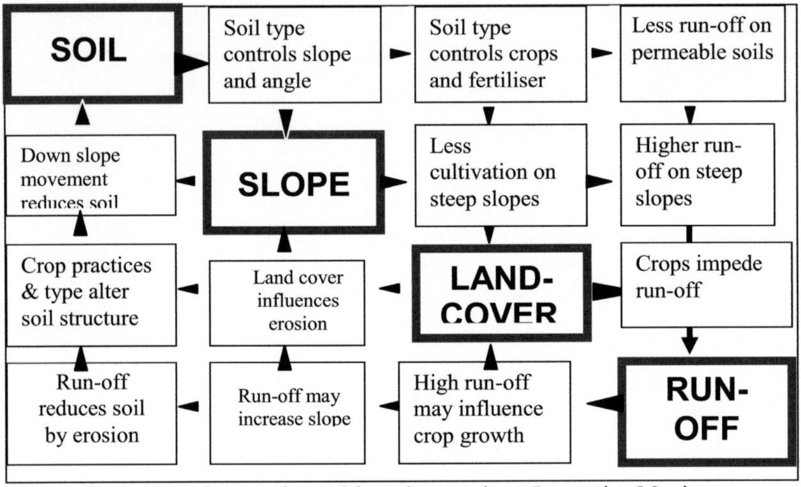

Figure 3. Definition of interactions of four elements in an Interaction Matrix.

After every interaction has been defined, the next stage is to semi-quantitatively score the interactions between all pairs of parameters using the degree of interaction agreed by the panel. Interactions are coded from 0 (no interaction) to 4 (critical interaction). The significance of the interactions is investigated by summing the scores for the rows of the matrix, to establish the influence of an individual parameter on the system (Cause), and the influence of the system on the associated parameter from the individual columns (Effect). A weighting system for GIS coverages is thus developed from the degree of interaction for a given parameter in terms of Cause + Effect. The contribution that each parameter makes to the vulnerability of a catchment can thus be quantified and listed (weighted) in order of significance. In the case of this St Lucia study the assessment is aimed at identifying the impact of catchment elements on RUN-OFF and also on WATER QUALITY. LAND COVER including land use, together with SLOPE (derived from a digitised contour map), were critically interactive elements and came out top of the weighting system. Full details of this method were reported in Matthews & Lloyd (1998).

6. Results of application of GIS rasterised weighted thematic overlays

The contribution which each element (coverage) makes to the vulnerability of a catchment can be quantified, plotted and identified in individual GIS coverage maps. By adding together the weighted values in each spatial coverage and combining them in a single GIS overlay a vulnerabilty map is produced and hot spots identified.

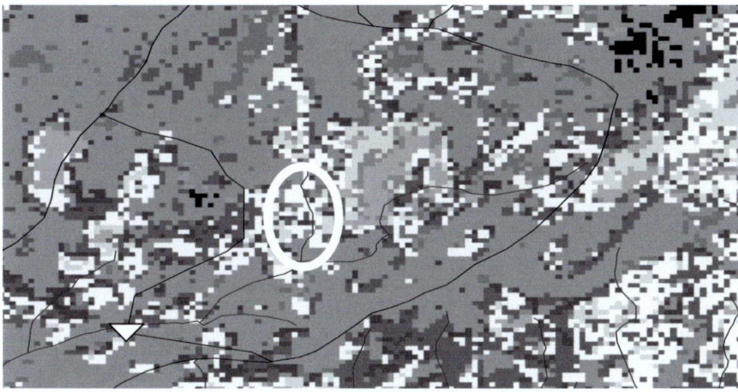

Figure 4 a). LAND USE analysis of the Ivrogne water supply sub-catchment.
The oval circle highlights banana cultivation (yellow pixels) adjacent to the stream.

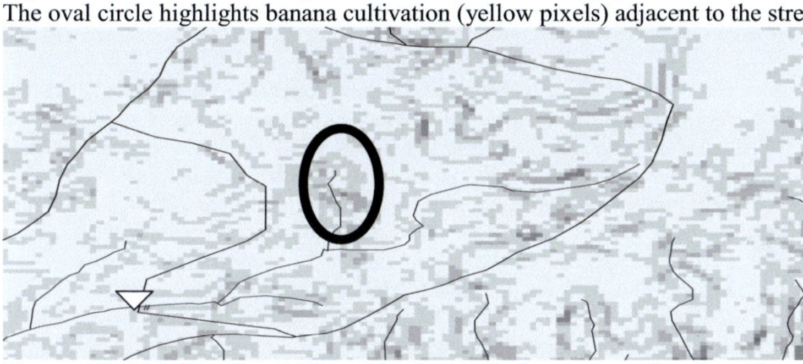

Figure 4b). SLOPE vulnerability calibration in the Ivrogne supply sub-catchment.

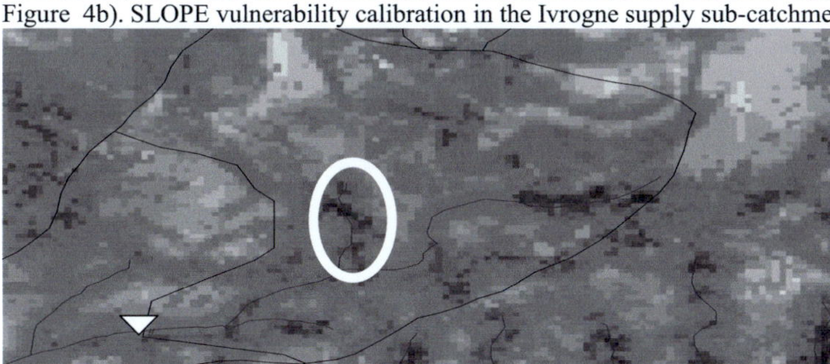

Figure 4 c). Multi overlay vulnerability analysis in the Ivrogne water supply sub-catchment, oval highlighting a hot spot for remedial action.

An example of the application of multiple thematic overlays is given in figures 4 a) b) c), to show how vulnerable hot spots can be identified for remedial action or protection. Figure 4 a) is part of a processed SPOT satellite image of LAND USE for St Lucia. The circled area of pixels is where bananas are cultivated adjacent to the Ivrogne stream and upstream of a water supply abstraction point (triangle). Figure 4 b) shows the same catchment area as in 4 a), but with SLOPE analysis from a digital elevation model, calibrated and colour shaded for increasing slope angle. Figure 4 c) shows the same area with multiple calibrated overlays identifying the vulnerable hot spots (darkest pixels) where change of land use is advisable.

7. Policy and regulation of land use

St Lucia has identified and defined upland forest reserve areas specifically for the protection of water resources for water supply. The catchment surveillance method provides a precise definition of large areas and small parcels of land (as little as one pixel, 20m x 20m) used for banana cultivation on steep inland slopes. These areas currently represent the greatest threat to the water resources. Remedial action i.e. change of land use to forest trees or perennial fruit trees, can thus be advised by agricultural extension officers equipped with GPS. In more general terms, Government Departments, in consultation with local communities and individual farmers, are able to make provision for the appropriate, sustainable use of such areas, and where necessary, the purchase and reforestation of selected, degraded upper catchment areas.

Discussion and Conclusions

During the past 30 years, in spite of increases in drinking water supply coverage, the continuity and physico-chemical quality of supplies have become increasingly severe problems in a number of countries bordering the Caribbean. Ironically this was largely due to economic success in agriculture, particularly the banana industry, as it has extended its area of activity into unsuitable terrain. Present land use practices, particularly the intensification and spread of agriculture into upland rain forest catchment areas, are causing drinking water supply shortages, discontinuity in supply and treatment plant failures which in turn have led to attempts to develop more complicated multistage treatment processes. This scenario of conflict of use and reducing resources is repeated in the great majority of catchments used for water supply.

This paper has outlined a catchment surveillance methodology and strategy based on the need to develop and manage land use and water resources in a sustainable way. A general comparison and individual analysis of the vulnerability status of 13 catchment areas and sub-catchments upstream of 31 water sources was undertaken in St Lucia. The criticality of the situation regarding the vulnerability of these sources both in terms of the quality and reliability of flow, and the hazards in the catchment areas which impact on flow and quality was defined. The GIS analysis allowed the identification of which stream reaches within rivers had poor water quality, and why this was the case. The results indicated that, generally, poorer water quality was recorded at the downstream sites with improvements in biological scores (diversity) at the upstream sites. Poor-fair water quality was generally recorded in the lower-middle catchments, whereas upper catchment sites, particularly those with >50% natural forest cover and little or no agricultural activity, displayed the best water quality.

An assessment of each individual catchment gave a clear picture of which river reaches have been affected by engineering works, current land use practices, severe weather events and other disturbances (Lloyd, & Thorpe, 1997a). Those upland sub-catchment areas upon which the towns and villages depend for their water supply are progressively becoming smaller as the Water Authority is having to move its river intakes ever further upstream to avoid excessive siltation of intakes and blockage of treatment plant filters. A number of abstraction points have been moved from <100 ft amsl to >500 ft amsl, and even some of the high intakes are now considered at risk.

Scarce financial and natural resources have been wasted e.g. by the World Bank in St Lucia, as a result of incorrect decisions on lower catchment management when the problems to be addressed are mainly in the upper catchment (Lloyd & Thorpe, 1997b). It is logical to use the best information on which to make management and protection decisions. It is argued that the only reliable means of making and taking these policy decisions is to use GIS information management systems.

The ArcView GIS technology applied in this project has the capability to link large geographically referenced environmental data sets, to provide new insights into the relationships between hazards and their effects on water resources. The application of the general method was aimed at the assessment of the relative risk to which a collection of river basins and micro-catchments are subjected. This will permit planners and regulators to take cost-effective, knowledge based and hence prioritised interventions, and manage increasingly scarce forest land and hence also, to protect water resources.

References

Bates, C G and Henry, A J (1928) *Forest and stream flow at Wagon Wheel Gap, Colorado.* Final report on completion of the second phase of the experiment. Monthly Weather Review, Supplement No. 30, 47pp.

ECLAC (1989), Report on *Integrated Water Resources Management* by the Economic Commission for Latin America and the Caribbean, Natural Resources and Energy Division.

ESRI (1999), *Getting to know ArcView GIS* . Published by Environmental Systems Research Institute, Redlands, California.

Hudson, J. A. (1992). *Rock Engineering Systems: Theory and Practice.* Published by. Ellis Horwood, London, 185pp.

Lloyd, B & Thorpe, T (1997a), Final report on *The development and integration of biotic and chemical monitoring with land use assessment for tropical river resource management.* ODA Environment Research Programme, Research contract R5936, 1994-1997. University of Surrey Pp 1-40 + 4 Appendices.

Lloyd, B & Thorpe, T (1997b) Catchment surveillance for water resources management. *Water Lines*, Vol. 16. No.1, July 1997 pp 23-25.

Matthews, M & Lloyd, B (1998) Final report on *The River Test Catchment Surveillance Project* Research contract CBR619. April, 1998.University of Surrey Pp 1-100

Watershed Management Under Terms of Depletive Water Balance and High Vulnerability in Wadi Fuqeen: West Bank-Palestine

Jawad Hasan[1], *Amer Marie*[1], *Haneen Froukh*[1],
Department of Applied Earth and Environmental Sciences, Al-Quds University, Jerusalem, POB 20002, Palestine: telephone: ++972222799502
Nader Al-Khateeb[2],
Water and Environmental Development Organization, Bethlehem, Palestine

(Corresponding author: Jawad Ali Hasan: jhassan@ccba.alquds.edu)

Abstract

Sustainable water supply problem becomes "strategic" when inelastic demand levels are overhauling maximum available supplies. The situation is more acute when the groundwater recharge area is heavily populated, consist of urban, industrial and agricultural areas and above all have typical karstic morphology and extensive regions of thin or null soil cover.

The springs of the Sera Al-Janobi and its sub catchments especially the springs within Wadi Fuqeen belong to perched karstic aquifers. Wadi Fuqeen basin encompasses 9 springs with total discharge of <1E6 m^3/yr, that are fed by local, perched, unconfined aquifers, which are completely separated from regional aquifers. These springs are often vulnerable to anthropogenic contamination because of the short residence time of water in discrete fast conduits, and many residential areas in the recharge zones.

During winter season the infiltrated water mixed with the wastewater leaking from the poorly designed cesspits and wastewater overflow from the treatment plants of the adjacent settlements, infiltration leachates from the piles of animal dung and agricultural return flow are all responsible for spring's water contamination.

The village faces environmental hazards caused by the unsustainable development of the settlements of Tzur Hadassah and Betar Illit. Currently, most of the recharge area is disturbed due to the ongoing urban development in Beitar Elite and Tzur Hadassah in conjunction with the planned Security Fence (apartheid wall) threatens to extend over ~70 % of the aquifer recharge area. Such massive destruction in a small watershed leads to considerable decrease in springs discharge and could completely dry-out the springs at the upper part of the valley.

Keywords:

Wastewater, Perched karstic aquifers, Apartheid wall, watershed, Mountain Aquifer.

1 Introduction

Groundwater is a natural drinking water resource often subjected to severe human impact. Strategies are required to preserve optimum groundwater quality, and so management of this vital natural resource has become a worldwide priority (Drew and Hötzl, 1999). Groundwater from karst aquifers is among the most important resources of drinking water supply of the worldwide population. In Europe, carbonate terrains occupy 35 % of the land surface, and in some countries, karst water contributes 50 % to the total drinking water supply. In some regions, it is the only available freshwater resource (COST 65, 1995).

In Palestine, the carbonate Mountain Aquifer is the largest (quantity) and best (quality) fresh water resource. This carbonate Mountain Aquifer is providing an average of 600 million cubic meters (MCM) of fresh water per year (Hydrology Service 2005). Springs draining smaller perched basins in higher elevations in the mountain aquifer outcrop area (Sarar Al-janobi catchment in Bethlehem) produce smaller quantities of water relatively to down gradient springs that drain the main aquifer, therefore in most cases they are not connected to modern water distribution system, nevertheless due to the scarcity of water in recent years these quantities can be significant. These springs are often vulnerable to anthropogenic contamination because of the short residence time of water in discrete fast conduits, and many residential areas in the recharge zones.

The fresh groundwater in the aquifers of Serar Al-janobi catchment area is the main source of drinking water and agricultural activity for the Palestinians, and therefore understanding the hydrology and the geochemistry of this system is of major interest. In addition, this research is necessary for better understanding the negative impact of the planned and on-going urban development of adjacent settlement (Tzur Haddasa and Beitar Elite) within Wadi Fuqeen valley (Fig. 1), since the groundwater flow rates, spring discharge and the chemistry is altered.

The aim of this research is to improve understanding of the hydrologic processes controlling water quantity and quality of springs discharging small (<1E6 m3/yr) basin in the mountain aquifer. More specifically, the characterization of springs discharging an agricultural-dominated small basin (~8 km^2) - Wadi Fuqeen springs which represent local and regional water resources in the western slopes of the mountain aquifer.

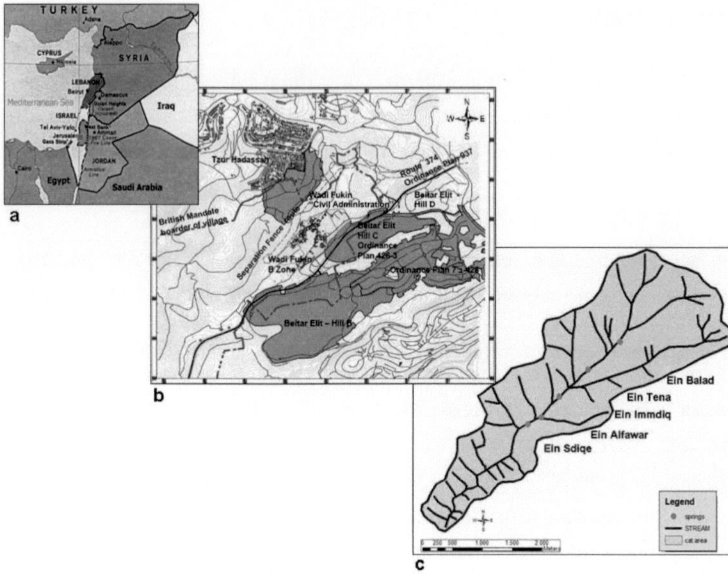

Figure 1 a: The study area in the West Bank, Palestine. **b:** A map showing the Wadi boundaries labelled with red line indicating the British mandate boarders for the year 1948. **c:** the catchment of Wadi Fuqeen springs group map showing the names and the springs location in the study area.

2 Study Area

Serar Al-Janobi catchment area covers an area of 25 km^2, located to the west of Bethlehem Governorate, which represents the southern part of the west Bank (Fig. 1). From the other hand the study area is part of the Auija Tamaseeh sub basin, which originally part of western Basin of the Mountain Aquifer (ARIJ, 1995).

The springs of the Serar Al-Janobi and its sub catchments especially the springs within Wadi Fuqeen belong to perched karstic aquifers. Wadi Fuqeen village is located on the top of the eastern trails of Hebron mountains, 16 km to the south eastern part of Jerusalem, and 12 km to the south east of Bethlehem. Crops in this area are watered by 9 springs using water channeled system and watering pools (traditional agriculture).

The stratigraphy of the West Bank includes several formations; each formation has different properties in rock type and aquifer potential. The lithology of the trails and slops along Wadi fuqeen and the surrounding area are Limestone, Dolomite, Chalk and Marl, comprising Yatta, Hebron and Bethlehem formation. These rocks determine the ability of formations to be aquifer or aquiqlude (Fig. 2). Yatta formation is considered as an aquiclude while Hebron and Bethlehem formation are considered as aquifers of the Turonian – Cenomanian age. The Alluvial formation is of Holocene (Quaternary) age, and it is considered as a good shallow aquifer. The deposits of this formation cover the flours of Wadi Fuqeen (Rabe M., 2007).

Formation	Lithology	Age	Character
Upper Bethlehem Fm		Upper Cenomanian	Aquifer
Hebron Fm		Lower Cenomanian	Aquifer
Yatta Fm		Lower Cenomanian	Aquiclude

Lithology Legend
Limestone
Dolomite
Marl

Figure 2: Generalized geological columnar section indicting the aquiferial characteristics of the various formations in the study area.

3 Results and Discussion

3.1 Groundwater chemistry

The discharge for each spring is variable depending on its emergence, elevation and its location at the upper or lower parts of the aquifer. This aquifer is considered as perched aquifer and its discharge fluctuates depending on the annual precipitation. Al-Quds spring emerges from Alluvial formation, and discharges to the west direction with an average value of about 0.26L/s, at an elevation of 690m.a.s.l. Al-Balad spring emerges from Hebron formation, and discharge to the west direction with an average value of about 0.38L/s, at an elevation of 670m.a.s.l. Al-Teenah spring emerges from Alluvial formation, and discharges to the west direction with an average value of about 0.1L/s, at an elevation of 625m.a.s.l. Immdeaq spring emerges from Hebron formation, and discharge to the west direction with an average value of abut 0.08L/s, at an elevation of 600m.a.s.l. Sdiq spring emerges from Alluvial formation, discharge to the west direction with an average value of about 0.06L/s, at an elevation of 560m.a.s.l, (fig. 3). The EC values ranges between (467-690 µS/cm) while the water hardness values ranges between (419 – 539 mg/l). The water of the springs is heavily polluted with Feacal Coliform > 1000 CFU / 100 ml.

Dorove diagram in (fig. 4a) show that the groundwater sample of the spring of Ein Immdeaq fail in field (2), while samples of the springs of Ein Albalad , Ein Teenah fail in field (5) indicating mixing with wastewater and accompanied with simple dissolution of carbonate minerals.

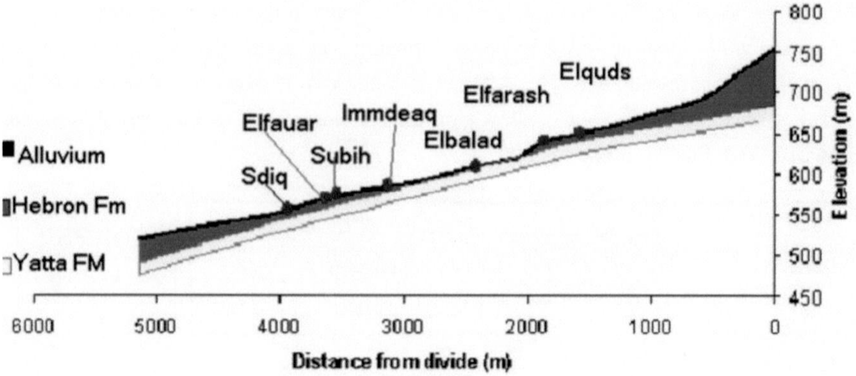

Figure 3: longitudinal cross section profile for the main stream in Wadi Fuqeen catchment presenting the flow direction from North to South and the perched aquifers.

The springs Ein Alfawar , Ein Sdiq plots in field (1) as presented in (fig. 4a) indicating ion exchange. On Piper diagram (fig. 4b) the ground water samples plots in (a, b and d) area of normal earth alkaline water with prevailing bicarbonate and with bicarbonate and sulfate or chloride. Relatively high concentration of chloride and sulfate is as a result due to washing the contaminants from animals and agricultural purposes as well as waste water from poorly designed cesspits.

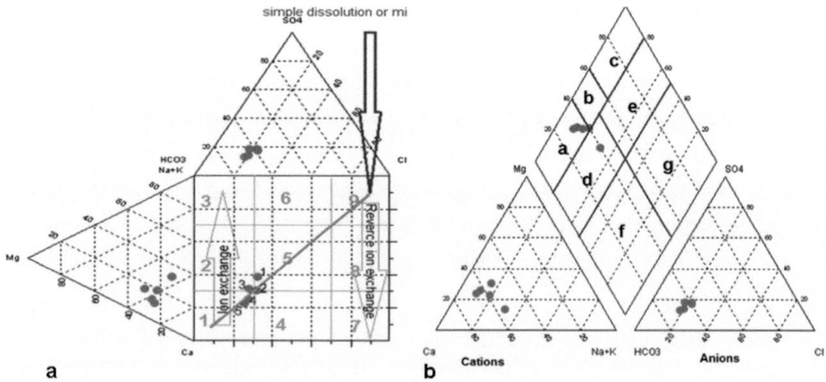

Figure 4: Hydrochemical data interpretation using aquachem, a: Durove diagram of ground water samples. b: Piper diagram for ground water samples.

3.2 Vulnerability of Wadi Fuqeen hydrological system

Wadi Fuqeen has a very small recharge area (8 km), with local and small underground catchment area that feeds the springs. Therefore it is considered highly vulnerable to land use modifications in the vicinity of the valley. The illegal intensive development and construction in the area over the last 20 years has damaged the Wadi due to the urban development within the recharge area of the springs, increasing surface runoff. As a result this urban development decreased the infiltration of rain into the underground catchment due to the increase of the sealed surfaces. The total planned area and the on going construction according the following plans, Plan 426/3/ hill C, Plan426/7/ hill B1, Plan 426/8/ hill B2, Plan -??unknown, Plan 520/Sansan hill B and Plan apartheid wall 937. All plans exceed 3 km^2 and constitute more than 70 % of the entire hill slope area within Wasdi Fuqeen recharge zone (fig. 5). These plans don't include the natural growth of the village itself:

The continues construction of Beitar Elit settlement to the east and the expansion of TzurHadassah settlement to the west (fig. 5) have harmed the valley's local aquifer and damaged the area's reinforcement slopes leading to a major decrease in the springs discharge and a complete destruction of three springs in the Wadi, not only but also it lead to Wadi soil erosion.

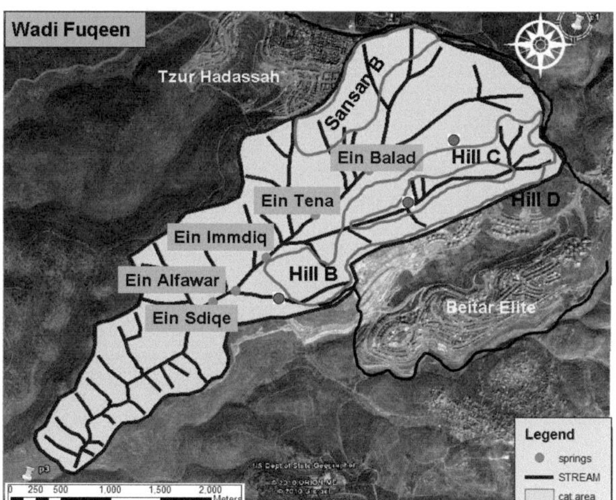

Figure 5: Planned and on going illegal urban development of the two settlements within Wadi Fuqeen hill slops.

4 Conclusion

The rain water is the only source of groundwater recharge. Any reduction in the discharge of the springs will decrease the total area which can be efficiently irrigated. Furthermore, the quality of the urban runoff as well as expected spills from the sewage system of Beitar Elite, mixing with the wastewater leaking from the poorly designed cesspits, infiltration leachat from washing the piles of animals dung, fertilizers by the rainfall in winter are responsible for introducing contaminants into the hydrological system. Currently, most of the recharge area is fully disturbed due to the unsustainable development of Betar Illit and Tzur Hadassah settlements.

In order to conserve and protect the fragile environment, we need to prohibit any farther illegal urban development along hill slopes (hill B, C and Sansan B). The planed apartheid wall within Wadi Fuqeen should be cancelled. Wastewater treatment and reuse, water harvesting and artificial recharge should be taken in consideration.

List of Literature

ARIJ- Applied Research Institute Jerusalem, Environmental profile for the west Bank District of Bethlehem, Jerusalem. 1995.

Y. Arkin, 1:50,000 Geological map of Jerusalem and its vicinity, The Geological Survey of Israel. 1976.

A. Ayalon, M. Bar-Matthews, and B. Schilman. Rainfall isotopic characterstics at various sites in Israel and the relationship with unsaturated zone water. GSI Report 16/04. 2004.

A. Bitan, and S. Rubin, Climatic atlas for physical and environmental planning in Israel, Ramot – Tel Aviv University. 1991.

COST 65. Hydrogeological aspects of groundwater protection in karstic areas, EUR 16547. European Commission, Directorate-General XII Science, Research and Development, Brussels, 446 pp.1995.

D. Drew and H. Hötzl. Karst Hydrogeology and Human Activities. Impacts, Consequences and Implications. International Contributions to Hydrogeology (IAH), 20. Balkema, Rotterdam, 322 pp. 1999.

I. Haviv and A. Lior. The Hydrological System of Wadi Fuqeen Springs. 2005. http://www.foeme.org/FOEME/SEKER/Hydrological_system_of_wadi-Fuqeen.htm

Hydrological Service of Israel. Development and Status of Israel's Water Resources as of Fall 2004. Report from Office of National Infrastructure, (in Hebrew). 2005.

M. Rabe and A. Marie. Hydrochemical and Hydrogeological Investigations of the Water Resources of Serar Al-Janobi Catchment Area. M.Sc. alqudes uni-Palestine. 2007.

Succes in IWRM in Germany and in the arid Arabic Countries

Is it possible to transfer German strategies and solutions to these countries? Practical insights.

Wolfgang Korthals

Consultant IWRM, 64297 Darmstadt, Deutschland,

E-mail: w.korthals@googlemail.com, Telefon: +49(0)6151/293407

Abstract

This paper develops a strategy how solutions of Integrated Water Resource Management (IWRM) can be transferred from a water-rich country like Germany to arid countries. These insights are drawn from many years of experience in the German water industry and in technical project management of large water supply installations, and from several months of training courses of Arab IWRM specialists. The paper identifies problems and mistakes that should not be repeated, and emphasizes the necessity to adapt and optimize. Public Relations and education play a central role, together with a broad view on sustainability, which includes culture, religion and virtual water.

Catchwords

Process oriented versus project oriented planning, virtual water, capacity building, data interchange.

1 Introduction

The German water management industry is proud of the level of their water supply and waste water facilities, both with regard to their construction and their management. This level has been reached within many decades of development and improvement. In many arid countries of the world, however, there are life threatening problems in water supply. Germany is a water rich country. Is it possible to transfer the successful strategies and solutions from Germany to the arid Middle East and North Africa (MENA) States to improve their water situation?

The challenges that we face are covered by section 4 of this congress. I argue however that many other aspects have to be taken into account if we strive for "integrated" water resource management: flood water, flood water protection, agricultural irrigation, ground water recharge, waste water, waste disposal, responsibilities of spatial planning, capacity building of the participants, the legal frameworks and their successful implementation, supervision and enforcement. In the last years new essential topics have came into focus: water reuse, flood water and wadi water harvesting for ground water recharge, and especially "virtual water" on a regional and global scale.

I argue that all the above mentioned fields need to be included into decision making. This insight developed during 30 years of work experience in the field of water supply in Germany. The last years I was the organizer and a lecturer in several long term courses of capacity building for Arabic junior managers from water and waste water companies. Theses courses were held in Germany on behalf of InWEnt (Internationale Weiterbildung und Entwicklung gGmbH, Deutschland) and lasted several months. During these courses it became evident that in the countries of the MENA region, it is not state of the art to take a comprehensive view on the necessary fields of IWRM.

2 Examples of projects

I will give two examples of severe problems of water management in Germany. I will illustrate how these problems were caused by mistakes in the past, and how solutions were developed over the years to remedy the situation.

1.) **The development of the river Rhine from a nearly biologically dead river to a river where salmon can breed again**.

In the years 1950 to 1970, industry developed quickly, and infrastructure was extended in Germany. In the course of this development, untreated industrial and communal waste water was lead into the river Rhine. This nearly caused the complete biological death of the river. It took two decades of rigorous efforts of the International Conference for the Protection of the Rhine (ICPR, German: IKRS) to improve the waste water treatment, not to forget the resistance against these projects. Since around 1995, the water quality of the river Rhine has recovered, such that today it is a habitat to salmon again [1].

2.) **Contamination of ground water and drinking water by excessive use of fertilizer and pesticides in agricultural practice and the return to non-contaminating usage**.

As in many countries of the world, too much nitrate and pesticides were, and still are, used in Germany for food production. As a result, groundwater was contaminated such that it could not be used for drinking water supply, or only with the aid of complex treatment. A law was passed (in 1989) that forced water companies to pay compensation to farmers, when they agreed to reduce their use of pesticides and fertilizer, turning the "polluter-pays principle" upside down. In this situation, some water companies took the initiative and hired agricultural engineers. It was their task to cooperate with the farmers and develop acceptable solutions for fertilization (organic and mineral) and pest management. Over time, a balance of trust and control developed, such that today it is possible everywhere in Germany to cultivate agricultural products in a groundwater neutral way. Still, the farmers and the Agro industry have to act responsibly! [2]

In the following, I will line out a number of strategic recommendations based on my experiences, how IWRM can be successfully planned and implemented.

3 Strive for optimized and adapted solutions with evaluation

Sustainable IWRM has to consider a large number of conflicting aspects mentioned above, and involves a large number of participants. Therefore, a primary precondition has to be understood and accepted: optimal solutions for a single aspect will be rare, if the overall problem is to be solved. Instead, all expectations and needs have to be balanced. Therefore, IWRM is about finding optimized solutions, rather than optimal solutions.

A second precondition is to recognize that solutions cannot be transferred straightforwardly from Germany to the MENA states, but have to be adapted to every single target country and situation.

This applies to the technical standard, to the requirements of operation and maintenance in different climatic conditions, and to the educational situation. The creation of local employment, also with low qualification, should be a priority.

All solutions have to be continuously and critically evaluated for their sustainability during development, during implementation, and afterwards.

It is common knowledge that facilities of European standard have been build in arid countries before, which never came into operation or quit operation shortly after their construction. Clearly, this cannot be the goal of cooperation and development.

4 Extend the concept of "sustainability": culture, religion, virtual water

Sustainable solutions do not only have to consider the well-known ecological, economical and social aspects, but also cultural and religious aspects. Culture and religion have affected water management for thousands of years with their experiences and traditions, but they have to consider the increasing world population, water stress, poverty, and insufficient qualification, and have to change. But I see chances and want to encourage.

It is obligatory nowadays to consider „virtual water" as a part of water management. Virtual water is the water that is used for the production of agricultural (and other) products [3]. As long as we restrict our view only to a water-rich country like Germany, the inclusion of virtual water is not very problematic. On a global scale, however, it becomes evident that water-rich countries are importing agricultural products from arid areas. This increases water stress in the exporting countries, particularly because there, irrigation water is competing with drinking water supply.

However it is still very difficult to initiate changes, as these countries have mainly a rural structure. One would have to intervene deeply into the social structures of the exporting countries, into global trade and customs regulations.

A quicker remedy is to improve irrigation systems in water poor countries. In the majority of cases 60% to 80% of irrigation water is wasted because this water has no root contact as a result of bad watering management [4]. With technical and financial aid, these problems are solvable, if they are connected with training and education of the farmers. These facts were often discussed in our IWRM courses.

5 Process oriented organization versus project oriented organization

To realize IRWM projects we need many skills of planning, implementation, operation, finances and skills of management. I argue, on top of this, there is the need for a skilled **decision management**. Decisions are much more effective if they result from mutual conviction rather than dogmatic or authoritarian behavior of a participant.

For many years, management consultants have told us to organize working fields in a **process orientated** way. The underlying assumption is that this will be effective, because there will always be an "expert" who has detailed knowledge of his field.

My view is that in order to work successfully in IWRM, we need to **start thinking project and task oriented again**, rather than process oriented. A process oriented organizational structure is not suitable to quickly finding optimized solutions: if a meeting between the building owner company and the planning consultant takes place involving 10 or 20 "experts" discussing at one table, decisions can hardly be found or made (fig.1).

In contrast, I propose a scenario with a superordinated project management with only two people sitting at a table (fig. 2). Naturally, the involved parties can send more experts to accompany the meetings, but there will be only one respective project leader who is responsible for decisions.

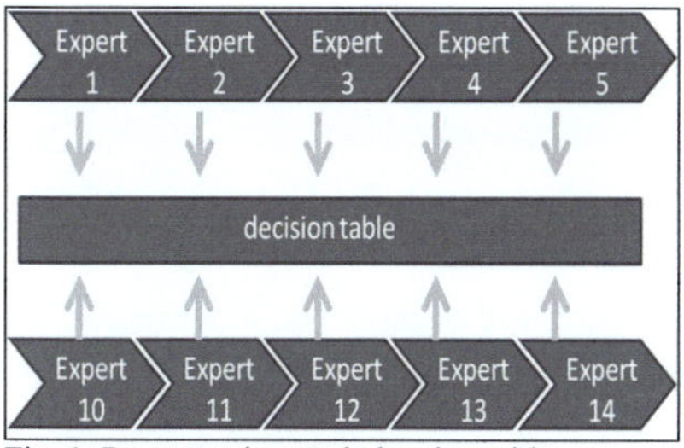

Fig. 1: Process orientated planning with complicated decision management

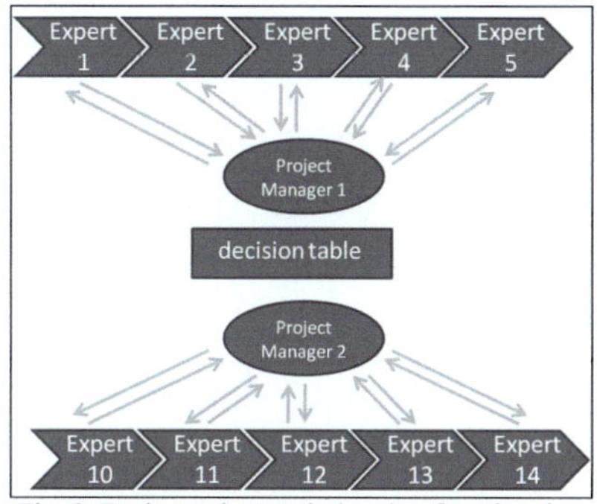

Fig. 2: Project orientated planning functional decision making

This means, though, that the project leaders and their staff have to have insights into more than one field of work, and have to be encouraged to acquire the necessary skills - to understand one another. In the language of management, staff will be "diversely positioned" and will develop these skills continuously. Today's practice shows, however, that this kind of behavior is often not desired or tolerated within the common hierarchical structures.

Let me illustrate this undesirable situation by means of an example from the practice. It could happen in MENA states but in Germany too: an operation manager of a waterworks would like to follow practical training in a laboratory in Germany. However, a different department is responsible in his country for laboratories and analyses. After a discussion with the employee's superior, he could not join this training program!

6 Responsibility and Leadership

In a small circle of responsible persons, as it was outlined above, issues can be discussed and decisions can be made efficiently. This involves good preparation and briefing by "experts". It also means delegating and accepting responsibility. For the staff, it means respecting decisions.

IWRM as a diverse field with many technological, operational, economical, cultural and religious requirements has a deep impact on landscape and society. In order to achieve the goals of IWRM, we need personalities with a clear vision and leadership, who have an overview of the processes, who keep up a dialogue with all participants, who can lead, can minimize conflicts and obstacles and who are able to find decisions.

7 Data management and unrestricted data interchange

Another aspect of IWRM is a qualified data management, which allows for comprehensive and unrestricted data interchange among all participants involved in project planning and implementation. Too often there are complaints that data are held back, in order to gain personal advantages or abuse them as a tool of power. Electronic databases are not yet used as much as they should. Particularly young employees also in MENA countries are used to using databases. A prerequisite is, however, that the data are collected reliably and safely and that there is the urgent needed money to collect data! A very good data management and interchange system for waterworks and authorities for the Frankfurt region can be found at www.grundwasser-online.de. The system is partially open to the public and serves thus as a tool of public relations at the same time. [5]

8 Capacity building

A prerequisite of successful IWRM is training and education on all levels.

Let me emphasize that it is not enough to educate engineers, because technically trained craftsmen and workmen will be needed for the operation of the facilities. There is a need to reinforce the training programs for this group of jobs, in Germany and abroad.

It is even better for a project if the technically trained craftsmen that are going to operate the facilities are part of the project team from the beginning. Experience has shown that this increases efficiency and saves money. Technical staff can also give valuable input during the planning phase, because they know best how installations should be designed in order to be usable in everyday operation, or in the event of a fault or emergency.

9 Public Relations (PR)

All steps of planning and implementation that are taken and that affect the public have to be accompanied responsibly by a clearly designed public relations campaign. People who are affected need to be carefully informed. Even in the case of difficulties in the project, it helps and creates trust to inform the public about these difficulties.

10 Financial Management

Naturally, financial management needs to be involved in all planning steps from the beginning. It is part of the strategic and coordinating duties of the project management to responsibly coordinate the financial department and all the technical departments. In my view, technical cost planning and cost control should remain within the responsibility of engineers – at last, engineers can make calculations, too.

11 Conclusion and perspectives

IWRM is a comprehensive and complex process which depends highly on the regional circumstances. Is it possible to transfer the successful strategies and solutions from Germany to the arid MENA States? I tried to illustrate how this is possible.

There are organisational and structural deficiencies in Germany as well as in the MENA States, but to a different degree. I explained that solutions from Germany need to be adapted, and that we always have to strive for optimized solutions. I identified a project oriented management of technical departments and finances as superior to a process oriented management. I also emphasized that the concept of sustainability needs to be extended to include culture, religion and virtual water. Furthermore, strong leadership and delegation of responsibility is required, in conjunction with unrestricted data management and good PR campaigning. Strong measures have to be taken to invest into education and training, not only of specialists, but also of the general public, in the MENA States especially women and children. The understanding of IWRM needs to be carried into society in order to be successful.

There are already positive developments which need to be reinforced, and conflicts need to be minimized. Change has to come also with aid of more money from the rich countries! Time is pressing, because on a global scale, water for water supply and food production is becoming sparser every day. We have been talking about the required actions in IWRM for more than a decade, it is time to act.

List of Literature

[1] Thomas P. Knepper, Otto Hutzinger, Editors. The Rhine, Springer 2006

[2] Wolfgang Korthals, Landwirtschaft und Trinkwasserschutz – vom Konflikt zur Kooperation. Bayerisches Landesamt für Wasserwirtscaftaft, Materialien Nr. 70, 1997

[3] Tony Allen, virtual water, www.kcl.ac.uk

[4] Peter-Jürgen Paschold, irrigation systems, www.forschungsanstalt-geisenheim.de

[5] Heiko Gerdes, Thomas Gutzke, Gerrit Seewald, www.grundwasser-online.de

Methods to evaluate and improve water allocation processes in a semi-arid watershed with given physical, environmental and social constraints and hydrological extreme conditions – Possibilities and Limitations

Authors:

- Nicole Kretschmer, Centro de Estudios Avanzados en Zona Áridas (CEAZA – ULS)), La Serena, Chile, Nicole.Kretschmer@ceaza.cl, Phone: +56- 51-334871, Fax: +56 – 51- 334741
- Ricardo Oyarzun , Centro de Estudios Avanzados en Zona Áridas (CEAZA – ULS)), La Serena, Chile, Ricardo.Oyarzun@ceaza.cl, Phone: +56- 51-204503, Fax: +56 – 51- 334741
- Manuel Eduardo Muñoz Zepeda, Administrador Junta de Vigilancia del Río Grande y Limarí y sus Afluentes, jvlimari@entelchile.net, Phone: +56-53-623833; Fax: +56-53-620098
- András Bárdossy, Institut für Wasserbau, Universität Stuttgart, Deutschland, andras.bardossy@iws.uni-stuttgart.de, Phone: -49 - 711-685 64679; Fax, +49-(0)711-685 64681

The management of water resources in semi-arid northern Chile has been driven on one hand by extreme climatic conditions, strong intra – and inter annual variations of water availability, and on the other hand by water rights and the rules of the water market. The watershed under study is located in the region of Coquimbo, the Limari watershed. The average annual rainfall does not exceed 120mm, whereas the evapotransperation exceeds 1.000mm. Agriculture is the prevailing activity in the watershed and only possible through a regulated hydrological and social system known as the "La Paloma System", which is in operation since 1972. It is composed of three reservoirs storing 1.000MCM and the associated channel network.
The system is administrated by nine private associations and the state due to an operation scheme established on the seventies. Since this time the system has been modernized in various ways but the operation hasn't been changed significantly; modernization due to more efficient irrigation technology, improvement of the channels as well as the extensions of plantations and the change of crops for exportation has been going on and with this development the demand structure has been altered. Furthermore the vulnerability of the system due to extreme climatic events (drought risk) has been increased.
Due to this complex system the demand of a transparent decision basis for the associations is increasing tremendously. The area suffered a severe drought from 1994 – 1997 which stimulated the water market enormously but nevertheless brought parts of the system to a collapse.
Although the communication among the associations and their decision makers is frequent, a common and transparent decision basis is missing and the effectiveness of new operational ideas and enlargements of the structures can't be tested so far.
The presented work tries to close this knowledge gap. The system has been extensively studied and through questionnaires the development of the area as well the actual indicators of decision making of the different levels of operation has been determined. A model of the system has been built up and simulated with the Water Rights Analysis Package (WRAP). Historical time series are used to evaluate the effectiveness and the limitations of the water allocation due to the water rights and the existent operation rules. A comparison with a recently completed study (conducted in part by the National Water Authority, DGA) which tried to evaluate the effectiveness of water use in the catchment with real historical flows has been done and analysed.
The model provides a range of optional capabilities to simulate a comprehensive range of complicated water rights and water management scenarios. Furthermore it operates with the data model WRAP Hydro which has been derived from the Arc Hydro model to develop input model parameters for the incorporated hydrological model out of a GIS environment.
Different operational schemes of water allocation as well the trade of water rights of the whole system has been simulated and analysed due to risk indicators (reliability, resilience and vulnerability).

THE STATE-OF-THE-ART IN THE COST-EFFICIENCY USE OF WATER HARVESTING TECHNIQUES IN AGRICULTURE

Luiz Rafael Palmier[1]
Rodolfo Nobrega[2]
[1]*Federal University of Minas Gerais, Department of Hydraulics and Water Resources Engineering, room 4509, Bloco 1, Escola de Engenharia, Avenida Antônio Carlos, 6627, Belo Horizonte, Brazil; +55.31. 3409-1003, palmier@ehr.ufmg.br*
[2]*Federal University of Campina Grande, Department of Civil Engineer, Campina Grande, PB; Brazil; +55.83. 3310-1155, rodolfolbn@gmail.com*

Abstract

Water resources management is of crucial importance for water assessments, water allocation, design and management of environmental systems. The overgrowing population, mainly in developing countries, and the prospective of climate change are calling for new approaches for water planning. Considering the persistently growing pressure on finite freshwater and soil resources, it is becoming increasingly clear that the challenge of feeding tomorrow's population is to a large extent about improving productivity of water within present land use, as new arable land is relatively limited. Water resources are very inefficiently used in both rainfed and irrigated agricultures. In fact, rainfed agriculture has generally been associated to low yield levels and high on-farm water losses. It is fundamental that crop output per unit of water input increases in both irrigated and rainfed systems, as per capita arable land area is declining even without considering the risks of soil degradation. New concepts of water management in scarcity regions are based on the use of water harvesting techniques. These techniques were already the basis of livelihood in arid and semi-arid areas many thousands of years ago, thus allowing even the establishment of cities in the desert. During recent decades the interest in water harvesting has increased and developing new or adapting old water harvesting techniques, as associated with the use of modern materials, has increasingly been apparently successful in several countries so that to increase the water availability, either by directly increasing the soil water content or by storing it for further application as supplemental irrigation in order to mitigate water stress periods occurring during the cropping season. Those cases tend to receive the most attention in the literature. However, the overall success is much less than expected in combining technical efficiency with low cost and acceptability to potential beneficiaries. Several national and international bodies have launched programmes to investigate the potential of water harvesting techniques but it is well recognised that much has to be done in order to clearly identify their real capabilities in several environments conditions. Notably, in several cases the costs of the water harvesting projects can be a major constraint for their widespread use. Therefore, a comprehensive computation of the costs to implement, operate and maintain a water harvesting project is crucial to support any economic analysis, which besides the direct benefits, should take into account the indirect ones provided to beneficiaries and society in general. There is a considerable variability associated with the costs of water harvesting projects, which means that the often required comparison between those and the costs of other supply technology should be undertaken with caution. In this paper an evolution of the state-of-the-art in the cost-efficiency use of water harvesting techniques – in order to restore the productivity of land which suffers from inadequate rainfall, to increase yields of rainfed farming, to minimize the risk in drought-prone areas and to combat desertification by tree cultivation – is presented based on the analysis of the later contributions available in the literature.

Catchwords

Water harvesting, supplemental irrigation, rainfed agriculture, economic analysis, agricultural production

1 Introduction

Irrigated lands, which account for almost twenty percent of world's cropland, consume around three-quarters of the annual renewable freshwater resources used by man and yield around forty percent of the world's food. For that reason, agricultural specialists are counting on irrigated land to produce most of the additional food that will be needed worldwide. However, in order to achieve this it is well recognized that irrigation efficiency must be greater and the low-cost irrigation developments must be available for poor farmers.

Meanwhile, it should be remembered that rainfed agriculture still plays, and will continue to do so, a critical role in food production as eighty percent of the agricultural land worldwide is under rainfed agriculture. Among the strategies to increase agricultural production in rainfed systems, the water harvesting practices for supplemental irrigation have increasingly been used in an effective manner in many countries. However, in numerous cases the water harvesting projects have not achieved their expected goal as the technologies and designs were not suitable for either the environment or the cultural habits of the beneficiaries. In addition, operation and maintenance of the schemes turned out to be either too costly and/or time-consuming.

For those reasons, especially in semi-arid environments, where agricultural investments are risky, farmers and decision-makers are reluctant to invest in new technologies. The process to convince them that water harvesting is yield-increasing practice depends on the availability of cost-benefit analysis for several crops and conditions. In this paper some current experiences are presented on increasing of agricultural yield by the use of water harvesting systems.

2 Economics of water harvesting for agricultural production

It is difficult to give exact figures on the world present total area under the various forms of water harvesting and its contribution to improve food production and food security at local and regional. Data on water harvesting for upgrading of rainfed agriculture in semi-arid regions are more common. As mentioned by Falkenmark et al. (2001), field studies suggest that the prospect of doubling yields, or even quadrupling, is realistic by producing more crop per drop of rain, but such large yield cannot be achieved with water management alone and at the establishment of water harvesting projects agronomic practices should also be improved by linking them with simultaneous soil fertility management – including the use of agricultural implements and addition of organic matter –, pest management, crop rotation, capacity building among farmers and extension services. In the former publication, several examples of yield upgrading in crop and fruit plantations were presented from different countries in the world.

Since them several researchers have been publishing the results of their studies.

Li & Gong (2002) conducted a field study in 1998 and 1999 with corn as an indicator crop to evaluate the use of ridge and furrow rainfall harvesting (RFRH) system with mulches to increase water availability in many areas of the Loess Plateau in northwest China and their results indicated that the ridge: furrow ratios had a significant effect on crop yield and yield components, with increased yield by up to 27.9%.

Fox & Rockström (2003) presented the results on the effects on Sorghum (*Sorghum Bicolor (L.)*) yields for three seasons under supplemental irrigation combined with soil nutrient application in an on-farm experiment in northern Burkina Faso, concluding that the average grain yield of 1403 kg ha^{-1} is higher than the farmer's normal practice by a factor of three. Total above ground biomass yields followed the same pattern as the grain yields for respective years.

Yuan et al. (2003) designed and tested a new planting pattern for potato production in semiarid areas of northwestern China. The purpose of the study was to examine the effect of the ridge with or without mulching on rainwater harvesting and the response in potato yield and to determine the optimal cross-sectional area of the ridge and furrow. The results showed that besides improving potato yield, the mulched ridge and furrow planting had the added advantage of collecting considerable rainwater for soil moisture.

Fooladmand & Sepaskhah (2004) conducted an economic analysis for temporal yield variations of four local grape cultivars grown in rain-fed microcatchments in the Bajgah area of Fars province, in Islamic Republic of Iran. Analysis of yield data with a multiple regression model indicated that a 9 m² (3×3 m) microcatchment area for each individual plant (vine) was the most appropriate area for vineyards in this region as yields (kg/ha) were 40% greater than those obtained for vines in the standard vineyard (without microcatchments) in this area.

Barron & Okwach (2005) presented the results of an on-farm study of the effects of supplemental irrigation (SI) on maize yield in semi-arid Kenya, where a surface run-off from a catchment of 2.7 ha was harvested in a hand-dug earth dam of 300 m2. Water harvesting of surface run-off added as SI resulted in improved maize yields in comparison with non-irrigated treatments as a result of dry spell mitigation, but only in combination with N fertilizer.

Wang et al. (2005) carried out an experiment with seven treatments to examine the yield promotion of potato crops cultivated with in plastic-covered ridges and supplied with water from a furrow rainfall harvesting (PRFRH) system, which increased temperature and availability of nutrients in the ridges, and maintained soil moisture in ridges. They concluded that the potato yield as well as the water use efficiency in the PRFRT system was significantly higher than in the other systems.

Fleskens et al. (2005) presented a method to measure increased water availability to olive (*Olea europeae*) trees grown on the terraced area of a traditional water harvesting technique site in southern Tunisia (jessr) and to translate these measurements into effects on yield. Although the technique was shown to greatly enhance possibilities for olive growing, yields remain dependent on water supply (rainfall+run-on) in spring. Based on detailed accounts for the costs and benefits of jessr construction and olive production, the authors concluded that on-site effects alone might not justify investing in the implementation of water harvesting techniques for that aim.

Xiao et al. (2007) conducted a field experiment at the Haiyuan Experimental Station, in a semiarid region of China, from 2000 to 2003 for rainfed spring wheat (*Triticum aestivum*) production by using rainwater harvesting with a sowing in the furrow between film-covered ridges (SFFCR) and with a sowing in the holes on film-covered ridges (SHFCR). The experimental result showed that the efficiency of water saving supplemental irrigation of field cultivation with SFFCR was 5.5–5.8%, and with SHFCR was 9.4–9.6%.

Wang et al. (2009) conducted field experiments in 2005 and 2006 at Gaolan, Gansu, China, to determine the influence of ridge and furrow rainfall harvesting system (RFRHS), surface mulching and supplementary irrigation (SI) in various combinations on rainwater harvesting, amount of moisture in soil, water use efficiency (WUE), biomass yield of sweet sorghum (*Sorghum bicolour L.*) and seed yield of maize (*Zea mays L.*) and their results suggest the integrated use of RFRHS, mulching and supplementary irrigation to improve rainwater availability for high sustainable crop yield. However, the same authors emphasized that the high additional costs of supplemental irrigation and construction of RFRHS for rainwater harvesting need to be considered before using these practices on a commercial scale.

3 Conclusions

Whether water harvesting measures are able to improve the water use efficiency of the entire catchment area will depend on the conditions of an individual case. Technical aspects of water harvesting projects are likely to be improved with further research on, for example, the development of methodologies to help the choice of a proper water harvesting techniques to be applied in a given region.

The analysis of the potential of upscaling of water harvesting from a field scale to a watershed scale is still a big challenge that must be overcome in order to convince decision-makers about the cost benefits of the water harvesting techniques. In assessing the potential contribution of water harvesting techniques to improving the water and food supply Rosegrant (1997) states that "… given the limited areas where such methods appear feasible and the small amounts of water that can be captured, water harvesting techniques are unlikely to have a significant impact on global food production and water scarcity." In a sense this is still verified in spite of the positive experiences carried out in the last decade on increasing crop yield associated with water harvesting adoption. However, it must be remembered that socioeconomic and environmental benefits of the extensive use of the water harvesting techniques are far more important than increasing agricultural water productivity (Oweis & Hachum, 2006).

List of Literature

[1] M. Falkenmark, P. Fox, G. Persson, J. Rockström, Water harvesting for upgrading of rainfed agriculture: problem analysis and research needs, SIWI report 11, Stockholm International Water Institute, Sweden. 2001.

[2] X. Y. Li, J. D. Gong. Effects of different ridge: furrow ratios and supplement irrigation on crop production in ridge and furrow rainfall harvesting system with mulches, Volume 54, Pages 243-254, Agricultural Water Management. 2002.

[3] P. Fox, J. Rockström. Supplemental irrigation for dry-spell mitigation of rainfed agriculture in the Sahel, Volume 61, Issue 1, Pages 29-50, Agricultural Water Management. 2003.

[4] T. Yuan, L. Fengmin, L. Puhai. Economic analysis of rainwater harvesting and irrigation methods, with an example from China. Volume 60, Issue 3, Pages 217-226, Agricultural Water Management. 2003.

[5] H. R. Fooladmand, A. R. Sepaskhah. Economic analysis for the production of four grape cultivars using microcatchment water harvesting systems in Iran. Volume 58, Issue 4, Pages 525-533, Journal of Arid Environments. 2004.

[6] J. Barron, G. Okwach. M. Rosegrant. Run-off water harvesting for dry spell mitigation in maize (Zea mays L.): results from on-farm research in semi-arid Kenya. Volume 74, Issue 1, Pages 1-21, Agricultural Water Management. 2005.

[7] X. Wang, F. Li, Y. Jia, W. Shi. Increasing potato yields with additional water and increased soil temperature. Volume 78, Issue 3, Pages 181-194, Agricultural Water Management. 2005.

[8] L. Fleskens, M. Stroosnijder, J. G. Ouessar. Evaluation of the on-site impact of water harvesting in southern Tunisia. Volume 62, Issue 4, Pages 613-630, Journal of Arid Environments. 2005.

[9] G. Xiao, Q. Zhang, Y. Xiong, M. Lin, J. Wang. Integrating rainwater harvesting with supplemental irrigation into rain-fed spring wheat farming. , Volume 93, Issue 2, Pages 429-437, Soil and Tillage Research. 2007.

[10] Y. Wang, Z. Xie, S. S. Malhi, C. L. Vera, Y. Zhang, J. Wang. Effects of rainfall harvesting and mulching technologies on water use efficiency and crop yield in the semi-arid Loess Plateau, China. Volume 96, Issue 3, Pages 374-382, Agricultural Water Management. 2009.

[11] M. Rosegrant. Water Resources in the Twenty-First Century: Challenges and implications for Action. - Food, Agriculture, and the Environment Discussion. Paper 20, International food Policy Research Institute, Washington D. C. 1997

[12] T. Oweis, A. Hachum. Water harvesting and supplemental irrigation for improved water productivity of dry farming systems in West Asia and North Africa, Water Management 80 (2006) 57–73. 2006.

Developing a soil map in a semi arid region using remote sensing-based approach for hydrological modelling-Wadi Kafrein / Jordan

William Alkhoury
Bianca Wagner and Martin Sauter

Department of Applied Geology, Georg-August-University of Göttingen, Goldschmidtstr. 3, 37077, Göttingen, Germany

Abstract

The purpose of this paper is to outline the development of a soil map of the Wadi Kafrein watershed in Jordan. No known soil map has previously been created for this watershed, yet a soil map is a necessary component for the construction of a comprehensive hydrological model of Wadi Kafrein. Soil types were studied in detail in some locations in Jordan by the "National Soil Map and Land Use Project of Jordan," but the catchment area of Wadi Kafrein was not included. Instead, the project surveyed a wide range of the adjacent catchment area, the Wadi Shueib. The soil of this catchment has also been studied in greater detail by several other research teams. The spatial distribution of soil units in the Wadi Kafrein catchment has now been determined, the map of which is presented in this paper (Figure 2). The approach used is based on remote sensing techniques, in which soil information from the adjacent catchment area, Wadi Shueib, has been extrapolated to Wadi Kafrein. Extrapolation has been done by studying and analyzing several satellite images of Cartosat-1 and ASTER data imagery. Field observations and data have been used to evaluate the remote sensing extrapolation techniques. Extrapolated soil units were compared with field measurements with attention focused on soil units' properties and their spatial distribution.
In this study, remote sensing techniques have been shown to be an effective and robust method for gathering spatial data on soils in arid and semi arid regions where information is generally rare. Due to the positive results from this study, future application of this approach is being planned to further develop and apply this approach and to research more expansive areas, which are in need of more hydrological studies but which lack such data on soils.

Catchwords

Arid and semi arid regions, hydrological modelling, remote sensing, soil map, Wadi Kafrein.

1 Introduction

Hydrological models are able to provide the user with solutions to many hydrological problems, and can assist in the management, protection, and ensured sustainability of surface water and ground water resources. The lack of detailed spatial and temporal data is the main obstacle in developing hydrological models and studies in arid and semi arid regions.
Several input components and data are needed to build up a detailed hydrological model, of which the soil map may be one of the most important as an interface to vegetation (transpiration) and atmosphere. Soil properties determine, to a large

extent, whether rain is either diverted as runoff or percolated and allowed to infiltrate into the ground.

Through the "National Soil Map and Land Use Project of Jordan," the soils of Jordan were widely studied. Three levels of study were involved; the resulting soil maps were published in the Jordan Soil Atlas (JSA). The soil of Jordan was classified according to the criteria and definitions of the USDA's Soil Taxonomy (1975) and the Keys to Soil Taxonomy (1990). In Level 2 of the project, five separate areas were selected for a semi detailed soil survey, starting from the North Western Area (Salt-Irbid-Mafraq) down to the Southern Highlands (Shoubak-Tafila) (Ministry of Agriculture, 1994). The semi detailed soil survey of Level 2 of the North Western Area did not cover the study area of Wadi Kafrein; but a wide part of the adjacent catchment, the Wadi Shueib, was included. Wadi Shueib exists under the same climatic conditions as Wadi Kafrein and both catchments have many physical characteristics in common. Also, the soil types and properties of Wadi Shueib have been studied in more detail by Kuntz (2003) for a wider study on the vulnerability of ground water aquifers in Wadi Shueib done by Werz (2006).

In order to present a spatial distribution of the soils, intensive soil mapping can be carried out in the catchment area, but this implies a tremendous effort and time in addition to cost. Therefore, a more typical approach is to make a few observations at designated representative sites within the catchment area, especially for the parameters which have the largest influence on the modelling results (Bathurst, 1986). In some cases where some areas are not accessible or are too remote to be able to be measured, remote sensing techniques may be applied in order to provide a spatial distribution of the catchment area (e.g. Rango, 1985, Schmuge, 1983).

As the soil map is a main input ingredient to the hydrological study of Wadi Kafrein, a remote sensing approach has been developed and the soil units from the well studied adjacent catchment of Wadi Shueib have subsequently been extrapolated to Wadi Kafrein. In this approach, satellite images were corrected and a supervised classification was performed after selecting appropriate training areas. Ground truthing and comparisons to the JSA were used to evaluate the results.

2 Description of used data

Within the course of the hydrological study of the Wadi Kafrein watershed, intensive field work and soil data acquisition has taken place. Satellite images of Cartosat-1 imagery data and ASTER have been acquired for this study as well. Cartosat-1 was used to develop a high precision Digital Elevation Model (DEM) with a final cell size of 5m x 5m (Figure 1) while ASTER was used to classify land use and to evaluate vegetation cover in detail (Alkhoury, 2010).

The inputs consist of a single stereo pair of Cartosat-1 acquired on Apr. 12, 2007 and multi-temporal ASTER satellite images from the dates Nov. 30, 2006; Jan. 17, 2007; Mar. 22, 2007; and May 25, 2007.

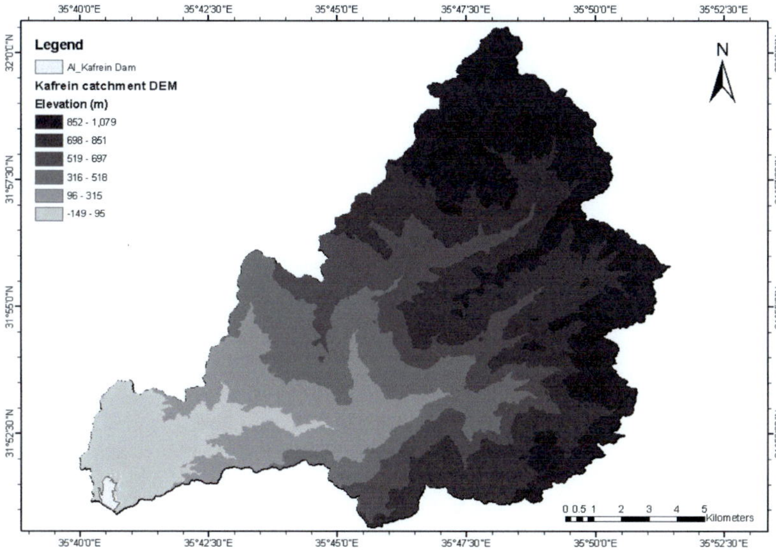

Fig. 1: High accuracy DEM extracted from Cartosat-1 imagery data with final cell size of 5m x 5m (Alkhoury, 2010).

The soil units of the "National Soil Map and Land Use Project of Jordan" and the results of the detailed soil map study of Wadi Shueib have been studied and analyzed as well. To further evaluate the preliminary results of the developing soil map, Google Earth images of Wadi Kafrein were acquired and for main channels or in high elevations, spatial distribution of identified soil units was studied. Within the "National Soil Map and Land Use Project of Jordan" and the Wadi Shueib soil map study by Kuntz (2003), soil properties and geomorphology are described and well documented.

3 Methodology of extrapolation

As a first step of extrapolation, all acquired satellite images of ASTER and Cartosat-1 were preprocessed, i.e. the images were converted to WGS 1984 coordinates and clipped and sorted to include those pertinent for the study area of Wadi Kafrein. Also, a total set of 47 ground control points (GCPs) were identified covering the whole catchment area of Wadi Kafrein, this being especially necessary to correct the Cartosat-1 data from which the high accuracy DEM was extracted. Several authors have recommended using at least 6 GCPs as a minimum in order to significantly increase the accuracy of the generated DEM (Kocaman et al., 2008; PCI Geomatics, 2006).
The multispectral ASTER images were subjected to radiometric and parametric corrections; later the urban areas and the water surfaces were masked and a master image was selected. Then all images and DEM were referenced to this master image.

Based on differing soil moisture regimes within Wadi Kafrein, the watershed was divided into two main areas: The Ustic-Aridic soil moisture regime and the Xeric soil moisture regime. The Ustic-Aridic regime is found in the lower part of the study area, extending from 35°43'30'' E westward to the Kafrein dam and further to the Lower Jordan Valley, which is covered by bare rocks. While to the east of 35°43'30'' E, the soil units are of a Xeric soil moisture regime.

In order to find the areal distribution of the different soil units, the DEM was spatially analyzed. Single raster layers of elevation, slope, aspect, and horizontal and vertical curvature were generated. Also, the satellite images were analyzed and band ratios were calculated to highlight iron oxides, carbonates, and clay minerals. All layers were later combined in one synthetic multi-band file with a spatial resolution of 15 m.

Finally, to extrapolate the various soil units from Wadi Shueib to Wadi Kafrein, training areas were selected based on distinctive soil units of the Wadi Shueib soil map of the scale 1:50,000. The supervised classification method was used, which is known for its accuracy for mapping classes compared to the unsupervised classification (Short, 2010), and was applied on both catchments so that the results of Wadi Shueib soil classification can be used for accuracy evaluation by comparing them to the previously available 1:50,000 soil map of Wadi Shueib.

4 Results

By following the approach described earlier, a soil map of Wadi Kafrein was developed with cell size of 15 m and high resolution (Figure 2). Seven soil units with Xeric soil moisture regime are shown. Five Ustic-Aridic soil units can also be

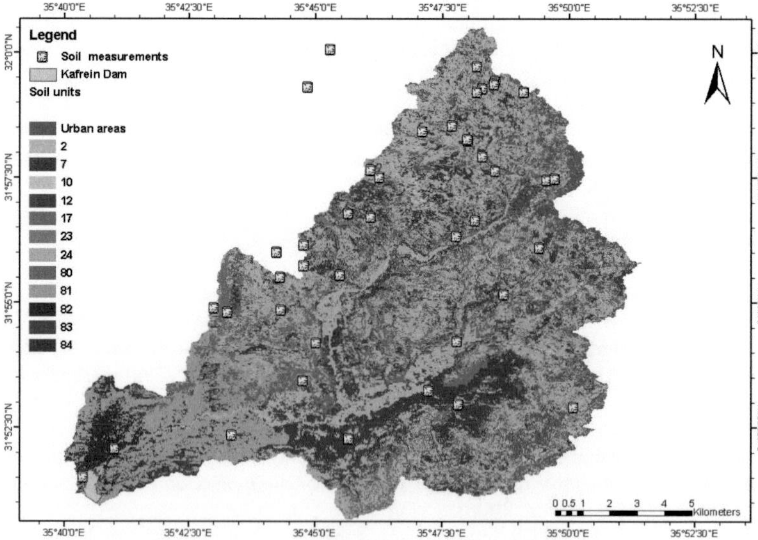

Fig. 2: Extrapolated soil map units of Wadi Kafrein using remote sensing approach

found in the lower part of the study area. Soil Unit 2 is distributed over low slopes and along the main Wadi channel in the valley floor. Soil Unit 2 is usually used for agriculture and olive trees and 70% are of the soil order Inceptisol. In total, 16% of the total catchment area of Wadi Kafrein is comprised of Soil Unit 2 and the mean annual precipitation over this unit ranges from 225-500 mm (Table 1). Soil Unit 17 is found over very steep slopes and covered mostly by grass. The mean annual precipitation over Soil Unit 17 exceeds 250 mm. Soil units 23 and 24 are distributed over shallow convex ridges and upper slopes with grass-like crops as dominant land covers with mean annual precipitation exceeding 300 mm. These two soil units cover 19% of the total catchment area.

Unit ID		M. annual ppt. (mm)	Elevation (m)	Area %	Km²	USDA Subgroup Classification*	WRB Classification**
2	Xeric	225-500	150-950	16	26	1-Calcixerollic Xerochrepts-50% 2-Typic Xerochrepts-20% 3-Lithic Haploxeroll-10%	1-Calcaric Cambisol-65% 2-Chromic Cambisol-35%
7		300-500	350-1030	15	24	1-Typic Xerochrepts-50% 2-Vertic Xerochrepts-15% 3-Lithic Xerochrept-10%	1-Chromic Cambisol-60% 2-Vertic Cambisol-20%
10		250-500	600-900	2	3	1-Typic Xerochrepts-60% 2-Vertic Xerochrepts-10% 3-Lithic Xerochrepts-10% 4- Lithic Xerorthents-10%	1-Chromic Cambisol-70% 2- Leptosols-10%
12		200-500	100-950	13	21	1- Typic Xerochrepts-45% 2-Lithic Xerochrepts-15% 3-Calcixerollic Xerochrepts-10%	1-Chromic Cambisol-50% 2-Calcaric Cambisol-30%
17		250-500	300-950	21	33	1-Typic Xerochrepts-30% 2- Lithic Haploxeroll-15% 3- Lithic Xerochrepts-15% 4- Lithic Xerorthents-10%	1-Mollic Leptosol-305 2-Dystric Leptosol-30% 3-Lithic leptosol-20%
23		300-450	500-900	2	3	1- Lithic Xerochrepts-30% 2- Typic Xerochrepts-20% 3- Lithic Xerorthents-20% 4- Lithic Haploxeroll-10%	1-Dystric Leptosol-35% 2-Rendzic Leptosol-35%
24		350-500	450-1000	17	27	1- Lithic Xerochrepts-40% 2- Typic Xerochrepts-30% 3- Lithic Xerorthents-10% 4- Lithic Haploxeroll-10%	1-Lithic Leptosols-50% 2-Mollic Leptosols-30%
80	Ustic-Aridic	200-250	150-300	1	1	1-UstochrepticCamborthids30% 2-Ustochreptic Calolorthide40%	1-Aridic Cambisols-40% 2-Lithic Leptosols-60%
81		150-250	-140-250	2	3	1-UstochrepticCamborthids30% 2-UstochrepticCalcoirothids30%	1-Aridic Cambisols-50% 2-Lithic Leptosols-50%
82		150-200	-140-50	1	1	--------	1-Lithic Leptosol-30% 2-Yermic Cambisol-40%
83		200-250	50-300	2	2	--------	1-Yermic Cambisol-60% 2-Lithic Leptosols-30%
84		200-250	50-150	1	2	--------	1-Takyric Calcisol-30% 2-Aridic Calcisol-40%

*** Ministry of Agriculture, 1994**
**** Kuntz, 2003; Werz, 2006 (World Reference Base)**

Table 1: Description of the soil map units of Wadi Kafrein.

The distribution of soils with Ustic-Aridic soil moisture regimes is limited to the lower part of the study area as shown in Fig. 2. These soils, with unit numbers from 80 to 84, belong to the Aridisols order with mainly Ustochreptic subgroups. The mean annual precipitation over these soil units' areas is less than 250 mm and generally no vegetation cover is present. Soil unit 80 is found on medium to low

slopes while soil unit 81 is found on very steep slopes, as can be seen in Figure 1 and Figure 2.
The Xeric or wet soil moisture regime soil units consist mainly of Inceptisols but also of Mollisols and Entisols. Together they cover 86% of Wadi Kafrein's total area. The Ustic-Aridic soil moisture regime soil units cover 7% of the study area and 7% are the share of the urban areas.

5 Conclusions and future perspectives

The application of remote sensing techniques is being more frequently employed in hydrological studies and is taking on a greater role due to easy acquisition of accurate data with high spatial and temporal resolution. Conventional mapping techniques cannot provide such accurate data within similar time frames or with such a relatively low budget. Preprocessing of the satellite images is vital in order to maximize results from data extrapolation. Type and resolution of data is also of utmost importance.
The extrapolated soil map of Wadi Kafrein (Figure 2) has been used in the detailed physically based hydrological model of Wadi Kafrein. After proper calibration, this model was validated and the results are in good agreement with observed field data, reflecting the quality of the developed soil units and their properties (Alkhoury, 2010).
Also, the geomorphology of the soil units and their spatial distribution is in good agreement with what is given in the references of the Jordan Soil Map Project and the detailed soil map study of Wadi Shueib. Another evaluation was considered here by the good agreement between the distributions of these soil units regarding the slopes with the slope map of Wadi Kafrein.
We hope by using this approach to apply this work and extend the study area to include wider catchments of the Lower Jordan Valley Basin.

List of Illustrations

List of Tables

List of Literature

[1] Alkhoury, W. (2010): Hydrological modelling in the meso scale semi arid region of Wadi Kafrein/ Jordan- Use of innovative techniques under data scarcity. PhD thesis (not published yet), Department of Applied Geology, University of Göttingen, Germany.

[2] Bathurst, J. (1986): Physically-based distributed modelling of an upland catchment using the Systeme Hydrologique Europeen. Journal of Hydrology, 87, pp. 79-102.

[3] Kocaman, S., Wolff, K., Gruen, A. And Baltsavias, E. (2008): Geometric validation of Cartosat-1 imagery. In Proc. 21th ISPRS Congress, 3-11 July, 2008 Beijing: International Archives of Photogrammetry, Remote Sensing and Spatial Information Sciences, Vol. 37. Part B1-3, pp. 1363-1368.

[4] Kuntz, D. (2003): Soils in the Wadi Shueib catchment area and their protective potential for the Groundwater- Salt area/ Hashemite Kingdom of Jordan. Master thesis (unpublished). Department of Applied Geology, University of Karlsruhe, Germany.

[5] Ministry Of Agriculture (1994): National Soil Map and Land Use Project- The Soils of Jordan. Level 2- Semi detailed studies, Hunting technical services Ltd., Volumes 2 and 3.

[6] PCI Geomatics (2006): Orthorectification and DEM extraction of Cartosat-1 imagery.

[7] Rango, A. (1985): Assessment of remote sensing input to hydrologic models. Am. Water Resources Association. Water Resour. Bull., 21, pp 423-432.

[8] Schmuge, T. (1983): Remote sensing of soil moisture with microwave radiometers. Trans. Am. Soc. Agr. Eng. 26, pp 748-753.

[9] Short, N. M. (2010): Remote Sensing Tutorial. Section 1: Image Processing and Interpretation-Morro Bay, California. The National Aeronautics and Space Administration (NASA). http://rst.gsfc.nasa.gov/

[10] USDA (1975): US Department of Agriculture, Soil Taxonomy. A Basic System of Soil Classification for Making and Interpreting Soil Surveys. Agric. Handbook No 436. Soil Survey Staff, USDA Washington DC.

[11] USDA (1990): US Department of Agriculture, Keys to Soil Taxonomy, SMSS. Technical Monographs Nos 18, 19. Blacksburg, Virginia.

[12] Werz, H. (2006): The use of remote sensing imagery for groundwater risk intensity mapping in the Wadi Shueib, Jordan. PhD thesis, Department of Applied Geology, University of Karlsruhe, Germany.

Evaluation and ranking of alternatives for the use of water of different qualities in semi-arid areas by means of a Muti-Criteria Decision Making framework.

Jacob Bensabat[1],
Joseph Guttman[2] & Tamar Milgrom[1]

[1] *Environmental & Water Resources Engineering Ltd.,*
PO Box 6770, Haifa 31067 Israel. Tel: 972-(0)4-8381515, jbensabat@ewre.com
[2] *Mekorot- Israel National Water Company*
Lincoln 9, Tel-Aviv 61201, Israel.

ABSTRACT

In many semi-arid and arid areas, water availability, particularly freshwater, is the key limiting factor of their economic development. Freshwater resources are so stressed that they cannot be further exploited in a sustainable way and stakeholders in these areas are often left with rather bad choices: 1) over-exploitation of fresh water resources thus jeopardizing the sustainability of their utilization or 2) to combine the utilization of the freshwater with other marginal sources like brackish water and reuse of sewage and/or treated effluent for agricultural purposes.

The current study focuses on the area between Jerusalem-Bet Lehem in the west and the Dead Sea and the Lower Jordan Valley in the east (Marsaba- Feshkha basin). This study is part of the SMART project activities, funded by the German Ministry of Science and Education (BMBF). Freshwater is mainly allocated to the large population centers in the western part, and to a much smaller extent for agriculture in the Lower Jordan Valley. Using brackish water, which is relatively abundant in the Lower Jordan Valley and treated effluent, which is always produced in volumes that proportional to the domestic and industrial consumption for agriculture, release additional freshwater for domestic use.

This paper summarizes the groundwater resources in the basin and the alternatives to explore and utilize the various water resources for all purposes, the methodology aimed at evaluating and ranking such kind of alternatives by means of a multi-criteria decision making procedure based on the Analytical Hierarchical Process (AHP, Saaty) and the challenges faced by the stakeholders in defining the various alternatives. The paper describes the process work-flow of the AHP and the resulting recommendations. The selected alternatives resulting from the AHP is reviewed and recommendations are made for its application to other basins of the project area and in similar semi-arid areas in the world that have similar conditions.

Catchwords: IWRM, Lower Jordan Valley, Water Supply, SMART, AHP, DSS.

1. Introduction

In many semi-arid and arid areas, freshwater availability is the key limiting factor to economic development. Population growth and the improvement of life standards are key factors of stress on freshwater sources, in terms of both quantity and quality ([3], [7], [8] and [9]). These include peak demands for agriculture (during the irrigation periods), drinking water supply for domestic and industrial needs, energy production, environmental protection, tourism, rural and urban shifts, and population migration ([1]). All these inevitably result in the over-exploitation of freshwater resources. However, over-exploitation is not a long-term solution as it is often associated to technical problems (pumping) and degradation of quality (salinization). One way to alleviate the problem is to consider using water sources of different qualities, such as brackish water and treated effluent, as these are available in relatively large amounts. The purpose of this paper is to demonstrate ways to combine the use of waters of different qualities in order to reduce the water stress. Specific choices are determined by means of a multi criteria decision-making approach (MCDM). In this work, we use the Analytical Hierarchy Process (AHP, [10] to [13]). Water management decisions usually involve a number of alternatives and multiple, often conflicting, evaluation criteria, which may not be of equal importance to the decision makers.

2. The study area

The study area is the Marsaba-Feshkha basin, which is located in the eastern slopes of the Judea anticlinorium (Figure 1) and in the western margin of the Lower Jordan Valley and in the western shore of the Dead Sea. The area is arid and suffers from an acute deficit of freshwater. It is isolated from the national water supply system and thus its development must rely (at least in the near future) on its local water resources. Most of the population is concentrated in major cities located in the western part of the basin (Ramallah, Jerusalem, Bet Lehem, Male Edumim), and the remainder in the Lower Jordan Valley (LJRV). Agriculture in the LJRV is the main economic activity. While drinking water is the major commodity and is first in priority, the economic development critically depends on the availability of water for irrigation and to a lesser extent for tourism. The water sources that are considered include freshwater, brackish water and effluent. Their proper utilization could greatly increase the overall water yield of the basin. The replenishment and the outcrops of the water bearing layers of the Judea Group (Lower Cenomanian-Turonian) are located in the western part of the basin. There, the average rainfall is approximately 600-700 mm/year. To the east (the LJRV and the Dead Sea), the aquifer becomes confined and rainfall reduces sharply with precipitations of 50 to 150 mm/year and potential evaporation of up to 2,600 mm/year.

Figure 1: The Marsaba-Feshkha basin, the underground water flow direction, and existing and proposed wells.

The most important hydro-geological feature in this basin is the Marsaba, anticline. It is assumed that this anticline and its associated flexure act as the eastern boundary of a large fresh water body, which dictates the groundwater flow and the outlet, the Feshkha Springs ([6]). There are two regional aquifers: the Upper and the Lower one. Both are fresh in the west, while in the east (from the Marsaba anticline) the lower aquifer is highly saline. The total discharge of the Feshkha springs is 60-65 MCM/year with salinity between 1500-6000 mg/l Cl-. Sewage generated in the eastern neighborhoods of Jerusalem is stored in the Og reservoir (Volume of 1.3 MCM), located north-west to the Dead Sea. The secondary treated wastewater mixed with brackish water is used for agriculture (palm trees).

3. Research Methods

Making decisions in any complex environment with multiple criteria requires first to consult the relevant stakeholders in order to understand the water supply needs, problems and constrains. The next step is to define the water supply alternatives and the set of criteria according to which they are rated. The final decision involves many intangibles that need to be traded-off. Through traded-offs it is possible to clarify the advantages and disadvantages of policy options under conditions of risks and uncertainty ([13]). The methodology chosen for breaking down the multiple problems of the Marsaba-Feshkha study area and aggregating the solutions into a conclusion is the Analytic Hierarchy Process (AHP). According to this method, the prioritization process is carried out through pair-wise comparison of alternatives and is measured by rating. An AHP model typically consists of an overall goal, a set of criteria to specify the overall goal decomposed to sub-criteria, and finally, at the lowest level of the hierarchy, the decision alternatives to be evaluated ([2], [10] and [11], [12]). The hierarchy can be illustrated as the diagram in Figure 2.

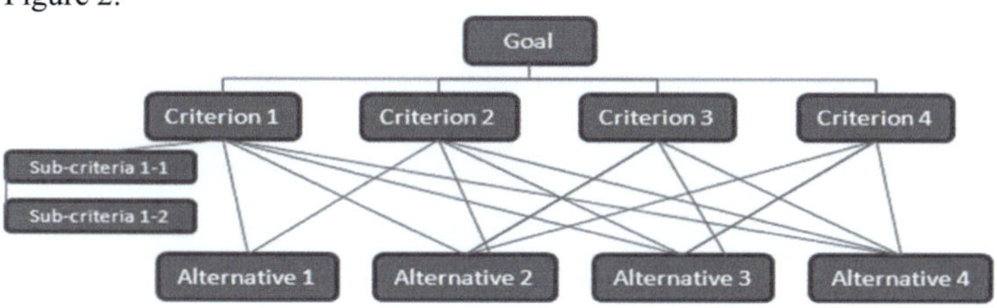

Figure 2: Illustration of the Analytic Hierarchy process Based on: Saaty, 2008

The pair-wise comparisons made by the decision maker express the preferences between two elements on a ratio scale from equally important (i.e., equivalent to a numeric value of one) to absolute preference (i.e., equivalent to a numeric value of nine) of one element over another ([11]). The ratings are organized as numerical numbers in a comparison (judgment) matrix. Based on this, relative weights for all elements of the hierarchy are calculated with the Eigenvalue method, indicating the priority level for each element in the hierarchy ([10]).

4. Problem formulation, results and discussion

In order to construct an comprehensive water supply plan, it is important to identify the water resources, their potential and the locations for utilization, the customers and their needs (drinking, agriculture), the agricultural planning (types of crops, irrigation periods etc.). Previous studies ([4], [5]) show that by drilling new wells (red triangle in Figure 1) additional freshwater of approximately 25 MCM/year can be produced. This additional amount adds to the current water abstraction from Israeli and Palestinian wells of approximately 25 MCM/year. The water from the wells in the western and central parts can be supplied to Ramallah, Jerusalem, Bet Lehem area and Male Edumim, while the water in the eastern part can be supplied to the Lower Jordan Valley.

After discussion with the stakeholders, we defined four criteria (C1 to C4):

Criteria 1 (C1): hydrogeological, engineering and technical uncertainties.

- **Hydrogeological y:** It relates to the depth of each well and on its final construction (length of casing and perforation), the limited availability of data on the expected static water level and on the hydraulic properties of the target water bearing layer.
- **Engineering and technical uncertainties**: difficulty of access and lack of infrastructures that will have to be constructed in parallel to the drilling.

Criteria 2 (C2): Costs from increasing and improving the utilization of additional water sources.

Criteria 3 (C3): Protecting the natural environment.

Criteria 4 (C4): Aquifer sustainability. Protecting the aquifer (water levels and quality).

In the prioritization process and the pair-wise comparison between the criterions, relative weights for all elements of the criteria hierarchy were given. According to the criteria rating results, the criteria hierarchy is: C1>C2>C4>C3. These results indicate that hydro-geology, engineering and technical uncertainties are viewed as more important than the other criteria.

The uses of different water resources that are accessible in this region and the water mixing between them are defined as alternatives (A1-A5):

Alternative 1 (A1): Fresh water - supplying all the additional 25 MCM/year from wells for drinking purposes and for agriculture.
Alternative 2 (A2): Desalinated water - supplying all the additional 25 MCM/year from desalination of the brackish water from the springs.
Alternative 3 (A3): Fresh water and effluent dilution - supplying 20 MCM of fresh water from wells mixed with 5 MCM of effluent from the sewage treatment plant.
Alternative 4 (A4): Desalinated water and effluent dilution - supplying 20 MCM of desalinated water mixed with 5 MCM of effluent.
Alternative 5 A5): Freshwater, desalinated water and effluent dilution – supplying 20 MCM of fresh water from wells diluted with 2.5 MCM of desalinated water and 2.5 MCM of effluent.

Prioritization process was held through comparison of all five alternatives for each criterion. The results are presented in the following table. The results (table 1) show the overall alternative hierarchy: A2>A4>A5>A3>A1. These indicate that any alternative that has higher proportion of desalinated water is more favorable than the fresh water alternative. The reason is that in the freshwater alternative the weight of criteria C1 is very significant. While in the desalination alternative the C1 is less significant because the brackish water is "visible" in the springs.

	Weight	A1	A2	A3	A4	A5
C1	**0.572**	0.015	0.246	0.022	0.129	0.159
C2	**0.234**	0.036	0.048	0.076	0.022	0.051
C3	**0.047**	0.002	0.017	0.004	0.021	0.002
C4	**0.148**	0.005	0.056	0.008	0.066	0.013
Summary		**0.058**	**0.367**	**0.110**	**0.238**	**0.225**

Table 1: The outcome comparison of water resources alternatives for all policy criteria

5. Summary and Conclusion

Previous studies indicate that in the Marsaba- Feshkha basin, there is an opportunity to increase the utilization of freshwater by drilling additional wells. On the other hand, desalination of brackish water from the springs can be an alternative resource to the drilling. A Multi Criteria Decision Making (MCDM) process was implemented for qualifying and ranking various alternatives of water utilization of approximately 25 MCM/year, according to four criteria, based on hydrogeology, engineering and technical uncertainties; economic costs; natural environment protection; and aquifer sustainability. These four criteria were compared between themselves and were given relative weights for all elements of the criteria hierarchy. The research results indicate that hydro-geology, engineering and technical uncertainties are the most important criteria. The alternatives (freshwater or desalination water solely and/or mixed with effluent or brackish water) were mutually compared giving an overall priority for each alternative. This process was repeated for each one of the five alternatives for the four criteria. The results show that the alternatives A2 and A4 that are based on desalination of 20 MCM/year of brackish water in Feshkha springs are prioritized. The alternatives A1 and A3 that are based on fresh water exploitation by drilling new wells are less valued. This prioritization is an outcome due to the hydrogeology, engineering and technical uncertainties from drilling new wells. In the existing knowledge, the alternative of desalination water in the Feshkha Spring is prioritized because the water source is known. Reducing one of the uncertainties from drilling new wells may raise the ranking of fresh water exploitation alternatives (by wells in the Judea Desert) and it may become as the favorable solution.

6. Acknowledgement

This study is part of SMART (Sustainable Management of Available Resources with Innovative Technologies) project, founded by the German Ministry of Education and Research, BMBF IWRM R&D Program, Funding No.: FKZ 02WM0800

7. References

1. Al Radif, A. (1999). Integrated Water Resources Management (IWRM): An Approach to Face the Challenges of the Next Century and to Avert Future Crises. Desalination, 124, 145-153.
2. Barzilai, J. and Golany, B. (1994). AHP rank reversal, normalization and aggregation rules. Infor, 32(2):57-61.
3. Bouwer, H. (2000). Integrated water management: emerging issues and challenges. Agricultural Water Management, 45, 217-228.
4. Flexer, A., Hotzl, H., Ali, W.,Bensabat, J., Guttman, J., Yellin-Dror, A. (2001). A 3_D Hydrological model of the Marsaba Feshcha Region. Annual report 2001, German- Israel Bi-lateral research.
5. Flexer, A., Hotzl, H., Ali, W.,Bensabat, J., Guttman, J., Yellin-Dror, A. (2002). A 3_D Hydrological model of the Marsaba Feshcha Region. Annual report 2002, German- Israel Bi-lateral research.
6. Guttman J., Rosenthal A. (1991): Mizpe-Jericho Area - Study of Salinization Mechanisms and Water Potential, TAHAL Consulting Engineers Ltd. 01/91/46 (in Hebrew).
7. Jønch-Clausen T. & Fugl J. (2001). Firming up the Conceptual Basis of Integrated Water Resources Management. International Journal of Water Resources Development. 17(4), 501-510
8. Jorgensen, S. E. & Rast, W. (2007). The use of Models for Synthesizing Knowledge for Integrated Lake Basin Management, and Facilitating Implementation of the World Lake Vision. Lakes & Reservoirs: Research and Management, 12: 3–13.
9. Keskinen, M. (2006). The Lake with Floating Villages: Socio-economic Analysis of the Tonle Sap Lake. Water Resources Development, 22(3), 463-480.
10. Saaty, T.L. (2001). Decision making with dependence and feedback: the analytic network process. RWS Publications: Pittsburg, USA.
11. Saaty, T.L. (2005). Theory and applications of the analytic network process. RWS Publications: Pittsburg, USA.
12. Saaty, T.L. (2006). Rank from comparisons and from ratings in the analytic hierarchy/network process. European journal of operational research, 168: 557-570.
13. Saaty, T.L. (2008). Decision making with the analytic hierarchy process. International journal of services sciences, 1(1): 83-98.

List of Illustrations

List of Tables

Handling of uncertain information in modelling complex water management systems *)

Prof. Dr. Michael Birkle
Fraunhofer Institute of Optronics, System Technologies and Image Exploitation IOSB, Fraunhoferstrasse 1, 76131 Karlsruhe, Deutschland,
E-Mail: michael.birkle@iosb.fraunhofer.de, Telefon: +49 (0) 721 6091-380

Abstract

The water management in complex water supply systems are usually based on models of water resources. Using these models we can simulate the change from a given status of water resources into a future situation as a result of man made or natural activities. It is obvious, that the quality of the models and, of greater importance their parameterisation, determine the quality of simulation results. Thus, it is of great importance to get a true transformation of a more or less known reality into model parameters and boundary conditions.

Catchwords

Water resource modelling, model parameters, water flow balances, flow balance equations, specially distributed parameters

1 Introduction

The water management in complex water supply systems are usually based on models of water resources. Using these models we can simulate the change from a given status of water resources into a future situation as a result of man made or natural activities. It is obvious, that the quality of the models and, of greater importance their parameterisation, determine the quality of simulation results.

In most cases the water supply systems of large urban areas are complex multi resource systems. For water resource modelling the available information is usually more or less incomplete or uncertain and often the data are inconsistent or even contradictory, in particular for the groundwater resources of the large mega cities. Additional difficulties result from the fact that normally we have large modelling areas. In the case of Beijing Municipality area (Fig. 1) for example we have a total area of about 16,800 km². In this area 6,300 km² are part of the North Chinese Loess Plain with irrigated intensive agriculture and a coherent high yield but still overexploited groundwater aquifer. The

*) The project was funded by the German Federal Ministry of Education and Research Reference Number 02WA 0849

remaining 10,500 km² are dry and barren mountainous regions with rock and karst aquifer of minor importance. That means that we have to deal with large areas of several thousands of km². Therefore, it was a great challenge to generate a realistic and consistent information basis for water resources modelling and model parameter determination for Beijing.

2 Methods and Solutions

A strong and efficient method to get a consistent and complete information basis for water resources modelling is water flow balancing on different levels [Chen 2003].

Figure 2 shows the system of water flows in Beijing as a comprehensive water budget, representing a system of inter-dependent balance equations. These balance equations start with an input/output balancing of groundwater surface water and waste water on one side and consumption of households, industry and agriculture on the other side. Due to irrigation, agriculture is the largest consumer group. And balancing goes in detail in land use dependent balance equations on soil surface between water supply from precipitation, irrigation and waste water reuse and the use of water for surface water runoff, infiltration, evapotranspiration, and vertical groundwater recharge. Using these balance equations implausible or incorrect data could be identified and information gaps could be closed.

The water flows in Figure 2 are calculated on the basis of yearly mean values valid for the period 1995 – 2000. In general such a water budget has to be balanced and therefore it is consistent and in first approximation complete. But it is more or less uncertain, because only a minor part of the different flux quantities can be gained from actual measured data. Some of the most important flux quantities, for example ground water recharge, cannot be measured directly and have to be estimated from balancing procedures and/or plausibility considerations. Therefore, we have substantial differences in quality of the numerical values in Figure 2 affecting the quality of water management activities. In Figure 3 the numerical values of the different water fluxes are arranged according to the quality of their numerical values and to the origin of information used for their determination.

In category A are numerical vales derived from measured data, for example precipitation. The yearly precipitation is easy to monitor as well as its areal distribution, using an adequate number of monitoring equipment. Usually, geographical maps and land use maps are also available in good quality and relevance to present status. Also the exploitation and consumption for drinking water and industrial water is measurable in most cases. But the exploitation of water resources for agricultural irrigation can only be estimated, because this kind of exploitation is related to a large number of surface water pumping facilities and groundwater wells (about 50,000 in the Beijing area) distributed according to the agricultural demand. Agricultural irrigation is the largest consumer group and, therefore, the total resource exploitations are affected with their uncertain estimations and therefore arranged in Category B. In the water budget of Figure 2, the agricultural water consumption is deduced from values for evapotranspiration of wheat and maize cultivation in the North Chinese Plain, assuming two harvests/year. [Foster 2004]

An example for uncertain estimations in Category C is groundwater recharge consisting of vertical recharge and horizontal recharge. Both recharge contributions must be deduced indirectly from input/output considerations including the published ground water level decrease. [Foster 2004]

A water budget as shown in Figure 2 is the first step for a modelling data base. Because most of the fluxes in Figure 2 are regional distributed, the spatial mean values have to be regionalised in a second step. Surface water fluxes as well as wastewater fluxes are restricted on different water bodies or waterways. They can be described via input/output data of the different water bodies. The spatial mean values of the surface or wastewater fluxes in the water budget (Fig. 2) therefore have to be distributed on the different water bodies or waterways (rivers, channels, pipelines, lakes etc.) of the model area, representing them in a good approximation with concentrated input/output parameters.

The groundwater flow on the other side is a water flux continuously distributed over the groundwater model area. As a groundwater model therefore a Finite Elements model is used with spatially distributed model parameters, representing the parameter values for each of the model elements. In the groundwater model of Beijing we have about 150 000 elements. Therefore it seemed reasonable to present the spatial distribution of the parameters as functional maps. Functional maps were generated for

- hydro geological parameters
- boundary conditions for groundwater model
- vertical groundwater recharge
- groundwater exploitation.

As a first example of a functional map for hydro geological parameters, Figure 4 shows the distribution of the hydraulic conductivity of the loose stratum layers in the plain area in the depth of 30 – 120 m. The hydro geological conductivity maps were determined on the basis of borehole data, geological and hydrological maps and profiles, water abundance map and lithology surface soil map with its ancient river zones. From the hydraulic conductivity maps we deduced the specific yield maps using European experiences [Hölting 2005], because only a rough estimation of about 0,05 – 0,20 for the specific yield in the North Chinese Plain was available. [Chen 2003]

The boundary conditions of the groundwater model are the horizontal groundwater inflow and outflow passing the geographical boundaries of the model area. In the Beijing ground water model area we have a resulting ground water flow from WNW to ESE, from the western and northern mountainous regions to the Bohai Sea. The water budget needs round about 0.7 km^3/y inflow to get a groundwater overexploitation of 0,8 m^3/y groundwater level decrease, mentioned in various publications for the period 1990 – 2000.[Foster 2005] [Chen 2003] [Zhang 2002] In detail this inflow is assumed to be split in two components. The first component is the slope runoff in the transition zone between bedrock and loose stratum. The amount is dependent on precipitation. The second component comes from the fractured bedrock or karst aquifer in greater depth and is assumed to be independent from yearly precipitation in first approximation. This kind of inflow will fluctuate only over decades of years. Due to the different dynamic behaviour, the quantity of the components could be estimated roughly.

The groundwater outflow was assumed to be near zero (0.03 km^3 /y) because of the fact that a severe continuous decrease of the groundwater table during the last decades is reported for Beijing area in publications. Inflow and outflow of groundwater was regionalized according to hydraulic conductivity maps resulting in boundary condition functional maps.

Vertical ground water recharge results from infiltration of water into the soil surface respective into the unsaturated zone. The contributions of this infiltration are

- infiltrated precipitation
- infiltrated irrigation water
- water percolation from surface water areas (perennial and ephemeral)
- water percolation of leakages from water supply and waste water systems.

Infiltration is dependent on the land use, therefore, vertical ground water recharge depends on land use. [Lerner 2002] [Bouwer 2002] [Chen 2003] [Liu 2002] [Sophocleous 2002] [Scanlon 2002] The land use, documented in the available land use maps, were classified in the example of Beijing municipality into 5 land use classes

- urban regions (1,980.88 km^2)

- irrigated farm lands and paddy fields (3,718.47 km^2)
- irrigated orchards and open woodland (309.63 km^2)
- meadows, shrubbery and wood (54.98 km^2)
- water areas (237.39 km^2).

For each of these classes, the water supply (precipitation, irrigation, waste water reuse/diffuse losses) and water application (infiltration, surface water run-off) were balanced on the soil surface. This kind of balancing was necessary, because the vertical groundwater recharge in total cannot be measured directly [Scanlon 2002] [Lerner 2002].. Thus, it has to be estimated from its contributions mentioned above. In Figure 5 the land use dependent balance equations on soil surface to generate the functional groundwater recharge map are shown for the first two land use classes as examples.

In urban areas and in residential zones of rural areas a reduced infiltration of rainwater and an increased rainwater run-off for surface water recharge is assumed, due to building density and sealed surfaces. Therefore, we have only a small contribution of rainwater to the vertical groundwater recharge in these areas. A contribution of fresh water irrigation to vertical groundwater recharge is neglected in urban regions. But losses of water from fresh water and waste water sub-surface pipes and distribution facilities (including some reuse of treated waste water for irrigation) is assumed as an additional source of infiltration in urban regions, described as "waste water losses".

Infiltration from precipitation or irrigation into the soil surface of open, irrigated farmland is the source of evapotranspiration, on one side, and of vertical groundwater recharge, on the other side. It is assumed that round about 80 % of the infiltrated water are consumed by evapotranspiration and 20 % are groundwater recharge. These values are estimated annual mean values used in temperate climate zones. [Foster 2005] [Liu 2002] They also seem reasonable for infiltrations into the Loess soils of the North Chinese Plain, from rainwater as well as from usual irrigation techniques like temporal flooding of agricultural fields. For open land (rural area) a uniform ratio of rainwater infiltration to rainwater runoff into surface water is assumed. The runoff of irrigation water into the surface water is neglected in the plain area. Assumed are two harvests a year so we have to postulate a minimum evapotranspiration of about 870 mm/year and a groundwater recharge of 218 mm/year.

Using the land use map and the balance equations on soil surface, the groundwater recharge functional map shown in Figure 6 was generated.

3 Conclusions

These functional maps are the input data of the groundwater model. Fulfilling the water budget and the soil surface water balances these data are complete and consistent, and they are true within the range of uncertainty of the fluxes of the water budget and its basic assumptions. Therefore a future improvement of modelling results is not primarily a question of the mathematical features of the models, it has to start with an improvement of the water budgets and their regionalisation for the years under consideration.

The validity of the mentioned assumptions and estimations is proven through the plausibility of the modelling results. We reached a good correspondence of calculated and measured groundwater table (Figure 7).

4 List of Literature

[Foster 2004] Foster, S; Garduno, H.; Evans, R.; Olson, D.; Tian, Y.; Zhang, W.; Han, Z., Quaternary Aquifer of the North China Plain – assessing and achieving groundwater resource sustainability, Hydrogeology Journal (2004) 12 P. 81-93

[Höltling 2005] Höltling, B.; Goldewey, W. G., Hydrogeology 6. Auflage 2005, Elsevier GmbH München

[Chen 2003] Chen, J. Y.; Tang, C.Y.; Shen, Y. J.; Sakura, Y.; Kondoh, A.; Shimada, J., Use of water balance calculation and tritium to examine the drop down of groundwater table in the piedmont of the North China Plain. Environmental Geology (2003) 44 P. 564-571

[Zhang 2003] Zhang, X.; Pei, D.; Hu, C., Conserving groundwater for irrigation in the North China Plain, Irrig. Sci. (2003) 21 P. 159-166

[Lerner 2002] Lerner, D.N., Identifying and quantifying urban recharge. A review Hydrology Journal (2002) 10 P. 143-152

[Bouwer 2002] Bouwer, H., Artificial recharge of groundwater: Hydrogeology and engineering. Hydrogeology Journal (2002) 10 P 121-142

[Liu 2002] Liu, C.; Zhang, X.; Zhang, Y., Determination of daily evaporation and evapotranspiration of winter wheat and maize by large scale weighing lysimeter and miro-lysimeter, Agricultural and Forest Meteorology (2002) 111

[Sophocleous 2002]Sophocleous, M., Interaction between groundwater and surfacewater: The state of the science, Hydrology Journal (2002) 10 P. 52-67

[Scanlon 2002] Scanlon, B. R.; Healy, R. W., Choosing appropriate techniques for quantifying ground water recharge, Hydrology Journal (2002) 10 P. 18-39

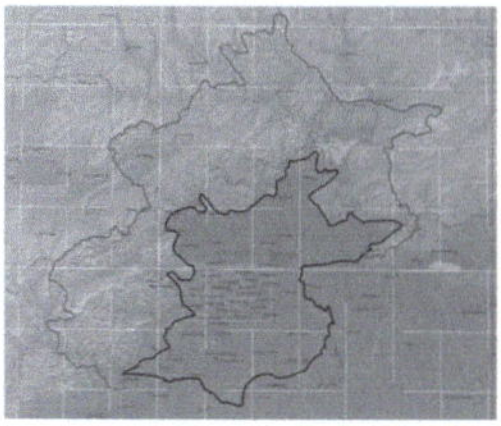

Fig. 1 Beijing Municipality Area

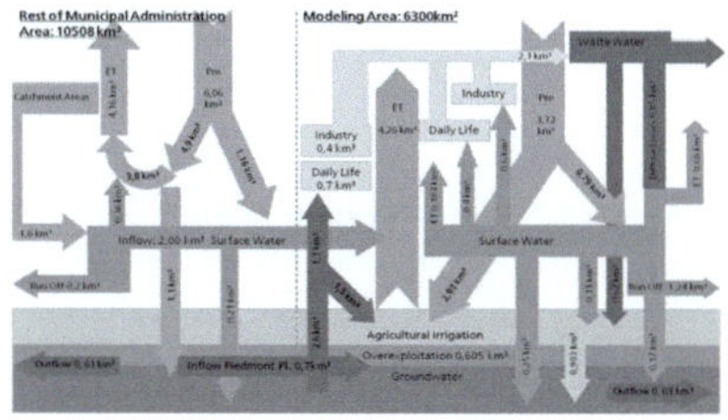

Fig. 2 Yearly Water Budget Beijing – Mean Values for the period 1996-2000

A Deduced from actual measured values (good quality)	B Deduced from publications, research efforts in other countries, extrapolations, etc. (reasonable and plausible estimations)	C Deduced from A and B type values using balancing considerations (uncertain estimations)
• Precipitation • Industrial water consumption • Daily life water consumption • Surface water inflow into model area • Geographical maps • Land use maps (data) • Geological and hydrogeological maps	• Agricultural water consumption • Diffuse waste water losses • Evapotranspiration data • Groundwater exploitation • Surface water exploitation • Waste water discharge • Infiltration and the ratio evapotranspiration/ground-water recharge • Hydrogeological data (hydraulic conductivity, specific yield, etc.)	• Groundwater inflow • Groundwater outflow • Groundwater recharge • Surface water recharge • Agricultural waste water reuse • Groundwater overexploitation • Surface run-off of precipitation • Infiltration data

Fig. 3 Deduction and Quality of Numerical Values in the Water Budget

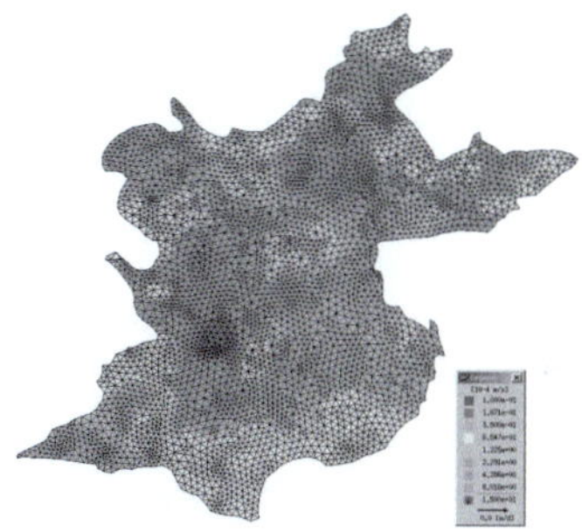

Fig. 4 Hydraulic Conductivity Map (30 – 60/120 m in depth)

Water Balance for urban regions (1981 km²)

Precipitation	+ Waste Water	= ET_{Pr}	+ ET_{WW}	+ GWR_{Pr}	+ GWR_{WW}	+ SWR	
590	+ 429	= 418	+ 343	+ 22	+ 86	+ 150	mm/y
1.169	+ 0.85	= 0.828	+ 0.680	+ 0.044	+ 0.170	+ 0.297	km³/y

Water Balance of irrigated farm land & paddy fields (3718 km²)

Precipitation	+ Irrigation	= ETPr	+ ETIrr	+ GWRPr	+ GWRIrr	+ SWR	
590	+ 583	= 404	+ 466	+ 101	+ 117	+ 85	mm/y
2,194	+ 2.168	= 1.502	+ 1.733	+ 0.376	+ 0.435	+ 0.316	km³/y

Fig. 5 Water Balancing on Soil Surface (Examples)

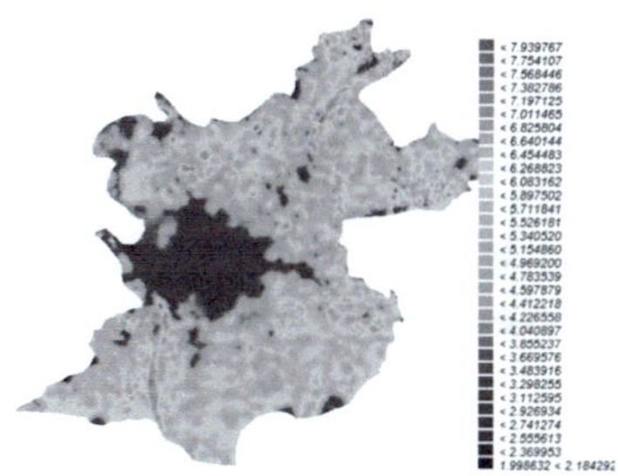

Fig. 6 Groundwater Recharge Map

measured simulated

Fig. 7 Groundwater Table Measured and Simulated for Year 2000

Integrated Water Resources Management in South Africa

- Maximizing Water Efficiency by Interlinking Hydrological and Economic Modelling -

Dipl.-Ing. Markus Bombeck, Dipl.-Vw. Michael Harbach,
Prof. Dr.-Ing. Dr. rer. pol. Karl-Ulrich Rudolph
*Institut für Umwelttechnik und Management an der Universität Witten/Herdecke gGmbH, 58455 Witten, Deutschland, E-Mail: mail@uni-wh-utm.de,
Telefon: +49(0)2302/91401-0*

Abstract:

With funding from the German Federal Ministry for Education and Research (BMBF), the Institute of Environmental Engineering and Management (IEEM) and the Center for Development Research (ZEF) jointly develop methods and tools for an IWRM concept, which - if implemented properly - will contribute to the improvement of today's water supply and which evaluates - in economic terms - the optimal allocation of the water resources in the project region, the Middle Olifants sub-catchment in South Africa (Grant Nos.: FKZ 0330734 A & D).

Catchwords:

IWRM, South Africa, Middle Olifants, Modelling, Water Use Efficiency, Technology Transfer, Water Franchise, Optimal Allocation

1 IWRM

Integrated Water Resources Management (IWRM) is a fundamental idea in water management, basing on the principle of overall cost-efficiency optimization in water catchment areas [1]. There are several definitions of IWRM which promote the efficient water use (e. g. [2], [3], [4], [5]), supposing that economic aspects play a vital role in IWRM. One of the major goals of IWRM is consequently to maximize the net value of water within a catchment basin. This is described by the following formula (simplified):

$$\sum\left[TV-\left(P_Q+P_Z\right)\right] \overset{!}{=} max\ \pi$$

The benefit or net value (π) of the water use is equal to the total value of the water use (TV) minus the costs arising from that use, where P_Q are water-relevant costs (e. g. pumping) and P_Z are non-water relevant costs (e. g. labour). The most beneficial water use equals the maximized net value when the water resources are optimally allocated. Regarding the calculation of TV, P_Q and P_Z, the economic value and the marginal costs of water for all water users must be accounted for.

Catchments often deal with competitive water use, which means that the demand for water is higher than the sustainable yield. To correspond to the complex system of water resources and different types of water users, it is important to verify the valuation methods and the parameters used, which makes the participation of all affected parties very important.

2 Middle Olifants, South Africa

South Africa is both a developed and a developing country: On one hand, there are rapidly growing industries (mining, power plants, steel manufacturing ...) and a sound institutional framework. On the other hand, especially the rural areas are characterized by a very low level of economic and infrastructural development. South Africa, due to its development, is a mind setter for the whole African continent. An IWRM concept developed and verified in South Africa will be the best option for a replication in other developing and transformation countries.

Figure 1: Target area: The Middle Olifants sub-catchment

The new South African National Water Act (NWA), which became effective in 1998 after the end of Apartheid and start of democracy in South Africa, claims that all water users should *"use water productively and in a sustainable manner for social and economic activities; in a manner that promotes growth, development and prosperity of all people to achieve social justice and equity"* [6]. An IWRM concept is necessary to approach the problems in the water sector.

South Africa early saw future problems in the water sector, so the project team could be supplied by the South African project partners with a number of water-related models and a reliable database, which was updated during the project where necessary.

The Middle Olifants is a sub-catchment in the Olifants catchment north-east of Johannesburg/Pretoria. The Olifants River flows through the Krüger National Park and Mozambique into the Indian Ocean. The Middle Olifants is characterized by highly developed irrigation, a dynamic growth in the mining sector (especially platinum in the Bushveld-complex), and over 1.5 m inhabitants, mostly rural, often settled remote from surface water resources in the former Homeland-areas.

Ecologic aspects play a vital role due to the fact that the Olifants River flows through the Krüger National Park. Quantity and Quality problems arise due to over-allocation, mining (active, abandoned and defunct mines), agricultural practice and wastewater treatment plants out of operation. All this promotes the Middle Olifants as an attractive research area for an IWRM project of applied research for practical implementation and replication all over Africa and other developing regions, later on.

3 Research Objectives and Innovative Approach

The structure of this project reflects this fundamental aim of IWRM, consisting of

(1) WRM (Water Resources Module, simulating the change in quality and quantity induced by whatever influence),

(2) WAM (Water Allocation Module, maximizing the water utilization value, depending on the available resources), and

(3) WIM (Water Intervention Module, program scenario of measures, undertaken to enhance e. g. water supply, sanitation).

The IWRM-approach interlinks the three modules reflecting the mutual influences. The first phase of the research project, therefore, focuses on setting up a hydrological-economic simulation of the river catchment area. The hydrological simulation analyses also the different types of water demand and the yield for different types of resources, and afterwards balances the water on quaternary catchment level in the project region. The economic part of the simulation contains in-depth information on the productivity of each of the demanding water users.

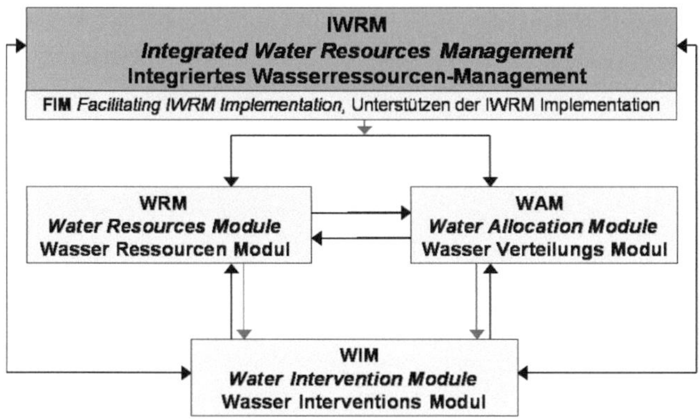

Figure 2: Project Structure

By combining these modules, the simulation visualizes on how to best (in terms of economic efficiency) allocate the available water among different water users – taking into account current South African legislation. Further, the simulation helps to single out areas for targeted water intervention measures (e. g. extension of a wastewater treatment plant).

4 Project Participants

The project is funded by the Bundesministerium für Bildung und Forschung (BMBF), Germany, through Projektträger Jülich, and on the South African side by the Water Research Commission (WRC). Complementary funding came from the "Development Marketplace" under the World Bank and IFC. The academic partners in Germany are the Institute of Environmental Engineering and Management at the University of Witten/Herdecke and the Center for Development Research of the University of Bonn. The partners in South Africa are the Department of Water Affairs (DWA) and the School of Agricultural and Environmental Sciences at the University of Limpopo. Industrial partners are HUBER SE and REMONDIS Aqua International GmbH. Stakeholders in South Africa are municipalities, industries and mines, irrigation boards and farmers.

5 Findings and Results

Three major water user groups in the Middle Olifants were identified, which are the domestic, mining and irrigation sector with a total demand of approx. 400 m m³/a. The biggest share is contributed to agricultural irrigation with more than 86 %. For public water supply, a quantity of about 232 m m³/a was found to be available in an acceptable water quality, after deduction of the reserve for the human and ecological minimum quantity, according to the priorities of the NWA. Because of the Status Quo with water users abstracting water to their demand, the water resources are over-exploited, water for the human and ecological reserves is not provided, and all users have to live with frequent supply shortages, leading to production downfall, especially in dry periods (affecting the domestic and agricultural sector, leading to tremendous losses of economic output). To gain social equity and ecologic sustainability, complying with the NWA, the overall demand will have to be cut by more than 38 %, through enhanced water efficiency (e. g. reduction of water losses [7]).

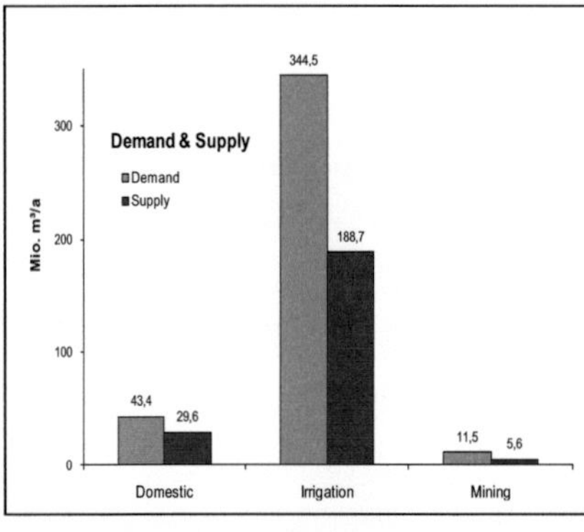

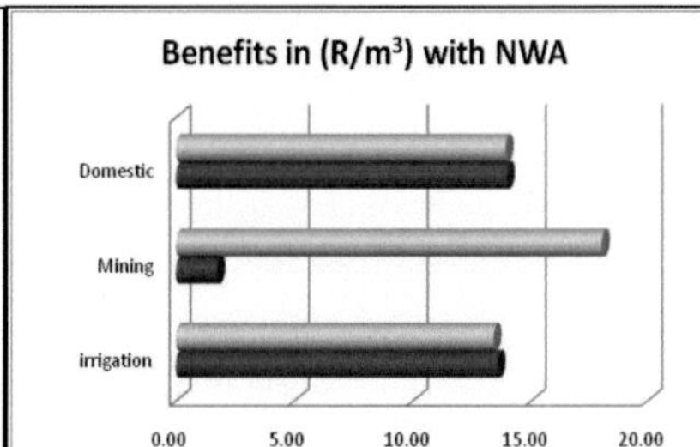

	irrigation	Mining	Domestic
Interregional	13.39	18.03	13.90
Intraregional	13.56	1.75	13.95

Figure 3 & 4: Demand & Supply Status Quo, Benefits for the Status Quo at Optimal Allocation

The results of WAM are based on a pure economic assessment. Based on the proportionality of the allocation for the Status Quo, a reallocation of water from irrigation to the mining sector would be necessary to increase the economic benefit. The amount of water that would have to be re-allocated from irrigation to mining was found to be lower than expected, due to the high level of irrigation technology (a lot of farmers use drip-irrigation, which tremendously increases the water use efficiency). The water quantity allocated to the domestic sector would need no change.

Besides the analysis of different scenarios (e. g. for the development of groundwater resources or for climate change based on the expectations of the IPCC [8]), different options for executive and institutional measures were investigated to remedy the water crisis in the Middle Olifants. Evidenced by the results of WRM and WAM and confirmed through an international survey among senior water experts [9], in most cases the water crisis is caused by the shortage of management rather than the shortage of water resources. Especially management and ownership are two non-technical issues that strongly influence the water supply apart from physical availability. If resources were managed and allocated more efficiently, water availability would improve. Especially, the know-how-factor influences the service quality the most likely and the most strongly. Therefore, and as a response to the observed conditions, a new management model for the water sector (Water Franchise) was developed that emphasizes on bridging the existing know-how gap by an experienced partner empowering a local company to deliver water services.

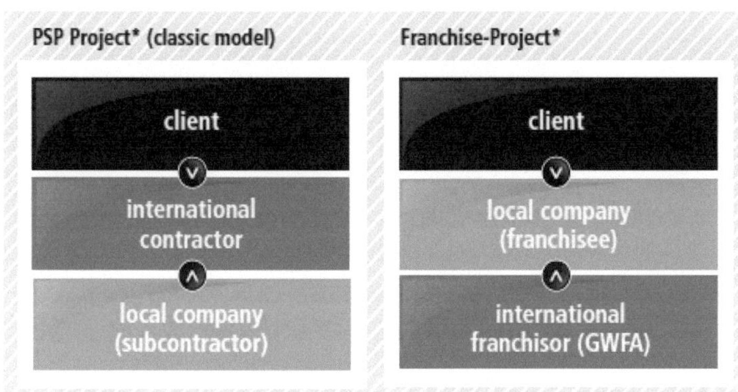

Figure 5: Water Franchise concept (source: www.waterfranchise.com)

The advantages of the water franchise are that the local partner (franchisee) is self-employed and, hence, has a strong incentive to deploy its resources at maximum efficiency. Furthermore, the experienced partner (franchisor) is strongly incentivized to transfer its know-how as its remuneration depends on the use of his knowledge by the franchisee (remuneration is largely performance based). A pilot implementation was described for a small community in the larger project region, objectives were so-called „near-to-customer-services“, i. e. meter reading, billing & collection, small repairs etc. [10]

6 Conclusion

It was one of the fundamental findings of the researchers so far, that insufficient management of the water resources is often the reason for deficits. As part of this - especially in developing and transition countries - it is not a lack of technology (which may be imported), but a lack of technical quality during installation and maintenance, which causes problems on the water supply and water treatment side.

A more sustainable operation and maintenance of the existing water facilities could make a major contribution to improve the current situation (preventing wastewaters to spoil downstream water resources, reusing wastewater to increase downstream water resources etc.), and to mitigate the water crisis, effectively.

All these features are analysed and assessed through IWRM, so that the available (although quite scarce) water resources of the Middle Olifants can be better protected and more efficiently allocated (in terms of economic efficiency). In this regard, the identification with the project, needing and leading to capacity development and ownership as part of the participation process, are important aspects. With a technical and economic approach of IWRM, the water resources can be used productively in a sustainable manner for social and economic activities promoting growth, development and prosperity for all people to achieve social justice and equity, as claimed by the NWA of South Africa.

7 Acknowledgements

The IWRM Project Team from Germany and its Project Partners in South Africa are grateful that the Bundesministerium für Bildung und Forschung supported this IWRM project through two research grants (0330734A & D), and that further support came from Water Research Commission, the South African Department of Water Affairs, and local municipalities in South Africa. Additional thanks are for the World Bank/IFC support and industrial efforts from Biwater, REMONDIS Aqua and HUBER.

List of Illustrations

List of Literature

[1] K.-U. Rudolph. Optimizing Wastewater Costs in River Catchment Areas - about the principle of water efficiency by Rincke (1968) and the new EU water directive. Wasser Abwasser Praxis, October 1999.

[2] Global Water Partnership, Integrated Water Resources Management, TAC Background Paper Nr.4, SE 105 25, Stockholm. 2000.

[3] GWC - Global Water Council (2004): www.worldwatercouncil.org

[4] Directive 2000/60/EC of the European Parliament and of the Council of 23 October 2000 establishing a framework for Community action in the field of water policy, Official Journal of the European Communities, L327/1. 22.12.2000.

[5] International Conference on Water and the Environment: The Dublin Statement and report of the conference. Geneva: World Meteorological Organization. 1992.

[6] Department of Water Affairs and Forestry, National Water Act, RSA 1998. www.dwaf.gov.za/documents

[7] K.-U. Rudolph, R. Ardakanian, J. L. Martin-Bordes. Economic Aspects of Water Loss Reduction, Proceedings of International Workshop on Drinking Water Loss Reduction: Developing Capacity for Apllying Solutions. Proceedings No. 1, UNW-DPC Publication Series. UN Campus, Bonn. 3-5 September 2008.

[8] Intergovernmental Panel on Climate Change: "Climate Change 2007: Impacts, Adaption and Vulnerability", Fourth Assessment Report, Summary for Policymakers, Working Group II of the IPCC, Brussels. April 2007.

[9] M. Harbach, K.-U. Rudolph. International Water Utility Management – Survey 2009/2010, Institut für Umwelttechnik und Management an der Universität Witten/Herdecke. 2010.

[10] K.-U. Rudolph, M. Harbach. Water Franchise – The Local Solution for Global Problems, Challenging Water Utilities and Water Industry World-Wide, bbr 03/2009, Extra International Issue, pp. 8 - 11. 2009.

A systematic approach for model-based integrated water resources management

Jens Tränckner
Frank Blumensaat, Björn Helm, Tatjana Terekhanova and Peter Krebs
Institut für Siedlungswasserwirtschaft, TU Dresden, 01062 Dresden, Deutschland,
E-Mail: jens.traenckner@tu-dresden.de, Telefon: +49(0)351/463 360 24

Abstract

To identify efficient IWRM options models of different complexity are applied. This paper proposes to combine the advantages of MFA and detailed process models using a stepwise differentiation of pressures and affected subsystems. This way IWRM focuses efficiently on relevant pressures and potentially suitable measures. Only for these detailed process analysis using process models is carried out. The method is exemplarily applied for a catchment in Flanders. A multi-objective evaluation framework is applied for cross-comparison on different management options.

Catchwords

IWRM, integrated modelling, mass balancing, water quality impacts

1 Background and Objective

Integrated Water Resource Management (IWRM) is a powerful framework for the multi-objective evaluation of human water use and its social, economic and environmental consequences. To capture all relevant entities of mass and energy fluxes IWRM it should be applied to river basins. However, the design of concrete measures requires focusing on subsystems and relevant pressures. To understand and address the complex interaction of processes IWRM is in most cases supported by models of different systems scope, spatial and temporal resolution and different ability to assess cause-effect relationships. Mass balance models operate on lumped aggregation units, e.g. on natural subcatchment or raster format basis; they describe annual or monthly input-output relations with more or less integration of physical boundary information. They require relatively few input data and give a consistent evaluation for the entire basin and serve thus for system analysis and identification of "hot regions" rather than hot spots. Due to their drastic simplifications, they show however deficits in plausibility, accuracy and the assessment of measures [1]. Process models, on the other hand, describe individual aspects using physically-based approaches. They are well suited to develop and assess measures, however limited to their respective systems boundary and scope. Regarding the various number of aspects in river basins, a single model approach necessarily fails to equally describe the diversity of processes and to represent them with the adequate accuracy. The planning of consistent sets of measures on a river basin scale must therefore combine the advantages of the different model approaches. In this paper an appropriate approach is proposed and exemplarily applied.

2 Method

The approach is strongly related to the DPSIR framework. The key idea is to reduce complexity by a stepwise differentiation of relevant pressures and affected subsystems to allow an efficient analysis.

Step1: Preliminary systems analysis, screening: It sets the systems boundary, compiles obvious deficits. Main drivers and the potentially affected subsystems are put in relation. The stakeholders affecting and using the river system are identified. Based on the compilation and the discussion with the stakeholders the project objective is defined.

Step 2: Identification of relevant pressures: Inside the defined boundaries all potential drivers are collected. Available data which can be used to describe arising pressures are investigated. As umbrella model a mass flow analysis (MFA) is applied to relate the different pressures with regard to the river system. Selection of an appropriate MFA tool should be based on the interested pressures/parameters and the temporal and spatial resolution (input/output data). Since MFA tools are generally not intended to describe highly dynamic processes. related pressures (like short term toxicity, hydraulic stress) cannot be described at this stage. Calculated emissions are validated with ambient water quality which may require additional field measurements. The contribution of single emissions to the physical-chemical and hydrological status of the receiving water is estimated. As result, a differentiated matrix of subsystems and potentially relevant pressures is obtained.

Step3: Assessment of the ecological impact: The so far emission based pressures are compared with the ecological state indicators (flora, benthic invertebrates, fish fauna) of the ambient water at the spatially differentiated subsystems. Based on this comparison probable cause-effect relationships are identified and a combined prioritisation of most relevant pressures is achieved.

Step 4: Detailed process analysis: For the identification of suitable response measures, remaining unclear pressures have to be quantified and cause-effect relationships between relevant pressures and impacts are investigated in detail. This assessment can be efficiently supported by detailed process models capable to address the relevant processes and parameters and regarded response measures . To calibrate the model an additional measurement campaign is mostly indispensable.

Step 5: Multi-criteria evaluation and design of response measures: Outcome of process models are in most cases time series of physical-chemical parameters in a predefined spatial and temporal resolution but not the ecological status itself. The assessment of the ecological status is obtained by transforming these data based on available guidelines [2-4]. into characteristic evaluation criteria. [5; 6]. The different evaluation criteria are compiled in a consistent multi-objective benefit analysis [6-8]. All stakeholders, namely those who have to implement or who are affected by the identified measures, are incorporated in the final definition of response measures.

3 Integrated Case Study

Study Site description – scope statement

The examined catchment area is situated in the Southern part of Flanders in Belgium, North-East of the city of Leuven in a flat to hilly region. The topography holds for low spatial rainfall variability in winter but rainfall intensity may vary for storms occurring in summer. Natural catchment is drained by 3 main creeks. These are *strongly modified water bodies* with mostly a strict trapezoidal shape including a reinforced river bed invert and a *rather low base flow*. Analysis of routine sampling for typical water quality parameters shows a rather *poor physico-chemical status* (high nutrient concentrations, high organic pollution, low DO) at most rivers sections. *Black coloured organic sediment layers* were observed during the field trip. According to Flemish water quality standards the *status in all investigated rivers is insufficient.*

Urban System: The area is characterised by dense to scattered housing – 90% of the households are connected to the central combined sewer network but untreated wastewater from ~2600 PE is *directly discharged* to adjacent rivers. The sewer network shows typical properties of a Flemish sewer system: small slopes, considerable influence due to backwater effects, throttle pipes instead of conventional throttle devices, numerous backwards working overflows. *Increased CSO activity* could be observed at several CSO structures. The catchment's sewer network drains into the Leuven wastewater treatment plant which furthermore receives wastewater from surrounding municipalities, representing a non-negligible interaction with an adjacent wastewater system.

Step 1: Figure 1 shows a flow scheme that illustrates the spatial differentiation of potential pressures on adjacent rivers based on 'qualitative' information from the local wastewater operator and the environmental authority. Black and solid squares symbolise remaining wastewater outlets and black, dashed arrows represent existing CSO structures. The contribution of diffuse pollution and further pressures is not clear

Step 2: The overall substance flow analysis using a modified MONERIS allows the estimation of pollutant loads within the catchment. According to this pollutant flows for total phosphorous (TP) / total nitrogen (TN) are as follows: CSO discharges – 14/12%, untreated WW discharges – 65/55%, contributions via groundwater – 6/11%, erosion – 15/22%. Results underline that pollutant emissions via soil erosion are rather low compared to point source emissions. Therefore diffuse pollution issues are here excluded from further investigations. To further assess the relevance of spatially distributed point source emissions average wastewater discharge and annual base flow minimum are put into relation (Figure 2). Specific pollutants are not (yet) considered; as most relevant system states organic pollution, nutrients (TN, TP) and NH_3-N are defined. Potential pressures introduced via CSO could, at this stage, only be identified based on qualitative observations from the local operator with increased activity at CSO#6 and CSO#8. The analysis of physico-chemical indicators (routine sampling) shows similarly poor water quality at most river sections. The morphological status at all river sections according to [9] is very poor (class 4-5). This underlines an area-wide susceptibility at all sections regarding potential pressures.

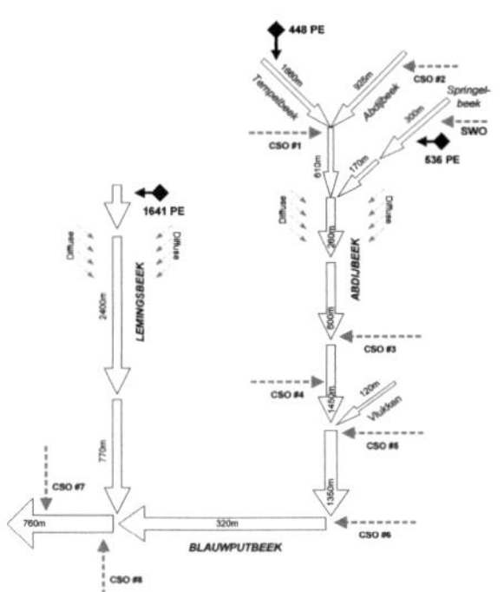

Figure 1: river system scheme with potential pressures.

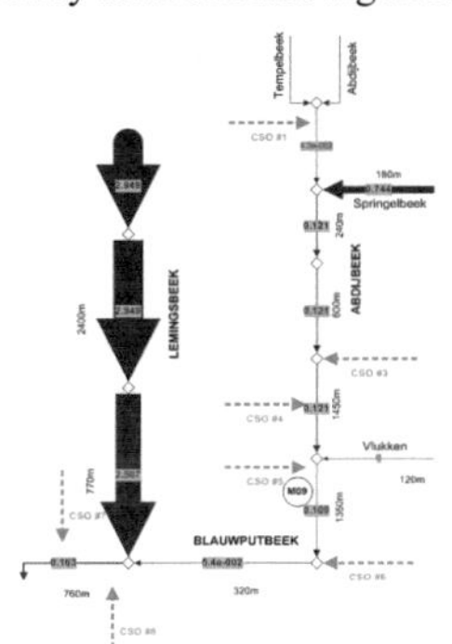

Figure 2: ratio between continuous wastewater discharge and minimum natural base flow.

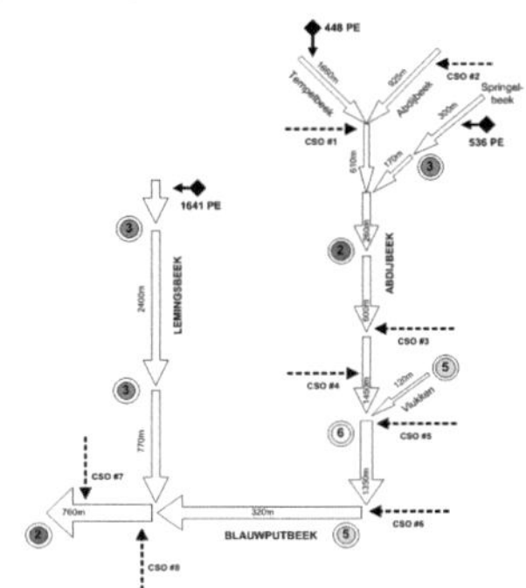

Figure 3: Comparison with values of ecological status.

Step 3: Biological water quality is assessed at 8 locations by means of the Belgian Biotic Index BBI, ranging from 1 ('very poor') to 10 ('very good'). Figure 3 shows the reported BBI values including potential pressures. The combined analysis leads to the qualitative assessment matrix in Table 1. The direct discharge of untreated wastewater at the upper part of the catchment clearly leads to a poor ecological water quality. Even there is no significant contribution of additional wastewater along the river courses, the streams obviously do not recover from the high wastewater loading in the upper part. The effect of the intermittent discharge from CSO#6, CSO#8 is not fully clear, even though the worsening of the biological water quality indicates some influence.

Pressure \ Spatial Unit	Upper Lemingsbeek	Lower Lemingsbeek	Upper Abdijbeek	Lower Abdijbeek	Blauw-putbeek
Diffuse Pollution	Low	Low	Low		
Wastewater	High		High		
Stormwater			unclear (?)		
CSO			unclear (?)	unclear (?)	unclear (?)
Morphology constraints	High	High	High	High	High

Table 1: Assessment matrix of spatially differentiated pressures.

Step 4: Further analysis focuses on the investigation of short-term dynamics and the spatially differentiated superimposition of all relevant pressures to ecological effects. For this an integrated model application is developed including only relevant processes and parameters. The integrated model consists of i) a hydrodynamic pollution transport model that interacts with adjacent sewer networks and river systems, and ii) a hydrodynamic water quality model describing rainfall-runoff processes, flow/pollution transport and conversion processes in the main river network. Parameter estimation and model validation is carried out separately for each compartment model and as lumped model application based on data obtained by comprehensive measurement campaign (4 months in total, online monitoring incl. frequent grab sampling). Its application allows an accurate quantification of the spilling activity and a representative assessment of the combined impact of CSO's and continuous wastewater discharges on the receiving water ecosystem.

Step 5 – Multicriteria Evaluation: Different management options in different 'affected areas' are tested to various degree of intensity, separated as single applications and in a reasonable combination. Options include i) centralised or decentralised treatment of the waste water discharges, ii) on-site infiltration of roof runoff, decentralised treatment and retention of road runoff, and iii) CSO reduction by RTC strategies. To address continuous/intermittent pollution and hydraulic impacts water quality based evaluation criteria are chosen at relevant locations. These include i) the relative exceedance duration of the Continuous Impact Level (CIL) and the Maximum Concentration Limit (MCL) for (DO and NH_3-N, and ii) the discharge maximum in the river. Emissions from CSO discharges are separately evaluated through the total spill volume and the maximum spill rate. For each management option a 'water quality index', between 0 (status quo) and one (desirable status), is calculated to provide a harmonised measure for criteria-specific performance assessment of management options ('cross scenario comparison'). The desirable status is defined differently depending on the type of criterion: i) for water quality criteria (DO and NH_3-N): full compliance with given standards [2], ii) for the maximum discharge (Q_{max}) near natural system status (from hydrologic simulations), iii) for CSO emissions: total elimination of spills. Total spill volume and maximum spill rate are assessed relative to the difference between status quo (x0) and 0. Here, negative values *reflect a worsening of the system state* (i.e. a relative increase compared to the status quo)

Results are illustrated as: i) equally configured 'SpiderWeb' charts (Figure 4) allowing a spatially differentiated and criteria specific assessment and ii) mean of the criteria-specific indices (without any weight allocation) reflecting a global performance measure of the regarded management option, whereas only water quality criteria are taken into account (Figure 6). Emission based aspects (here: annual overflow volume) are considered separately. The grey, right-wing striped scenario 'x0' represents the present situation failing in compliance of water quality standards at most considered locations and is assessed with 0.39. The second, green shaded bar represents the 'near-natural' status ('z0') that reflects an 'undisturbed' situation in terms of water quality without any CSO spills (= 1).

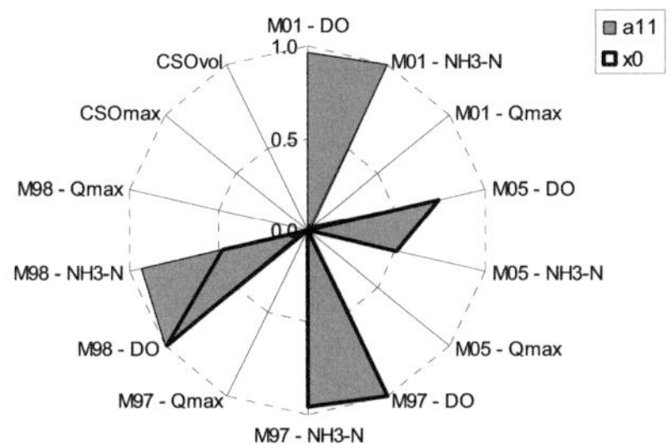

Figure 4: 'SpiderWeb' illustrating a criteria-specific assessment of one scenario; the bold line indicates the status quo.

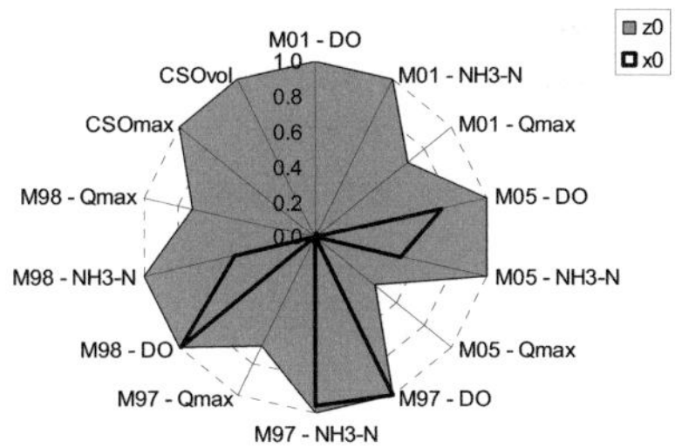

Figure 5: results illustrating the criteria-specific assessment for the 'near natural' status ('z0'); the bold line indicates the status quo.

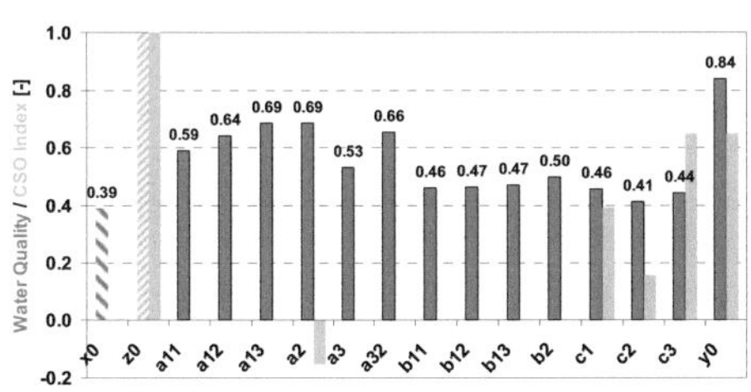

Figure 6: final scenario assessment regarding 'mean' water quality and CSO spill volume.

All tested scenarios lead to an improvement compared to the water quality for the status quo (x0). Single measures being applied separately (group a, b, c) lead to a considerable improvement. Best overall performance is achieved for the combination of the scenarios 'a13', 'b2' and 'c3' (-> 'y0').

Sewer management measures (group c) lead to spill volume reduction but do not significantly improve water quality. It is also found that water quality and emission based performance cannot be considered isolated as the improvement of the water quality does *not* automatically mean the emission-based performance improves. Considering scenario 'a2' a worsening of the sewer system performance (-0.15) is observed. That is the simulated spill volume increases by 15% compared to the status quo, while overall water quality improves! This again underlines the complexity of an integrated assessment as cause-effect relationships do not only affect several compartments (sewer, river, etc.) but also vary concerning the type of effect.

4 CONCLUSIONS

The proposed stepwise approach combines MFA and process models to efficiently identify relevant pressures appropriate IWRM measures. This way the effort for detailed process analysis can be optimised. Although the approach looks very model-oriented, application at the case study showed that it demands and facilitates a continuous and targeted discussion with stakeholders to find efficient solutions. A current application at the much bigger and more complex catchment of Western Bug (Ukranian) proves its generic applicability.

List of Literature

[1] ATV-DVWK, ATV-DVWK-Themen Signifikante Nährstoffeinträge aus der Fläche, ATV-DVWK-Arbeitsgruppe GB-5.6 „Diffuse Stoffeinträge", ISBN:3-937758-27-5, 2004.

[2] FWR, Urban Pollution Management Manual - 2nd edition. Report No. FR/CL 009, 1998.

[3] BWK, BWK-Merkblatt 3. Ableitung von immissionsorientierten Anforderungen an Misch- und Niederschlagswassereinleitungen unter Berücksichtigung örtlicher Verhältnisse, Bund der Ingenieure für Wasserwirtschaft, Abfallwirtschaft und Kulturbau (BWK) e.V., Düsseldorf, 2001.

[4] BWK, BWK-Merkblatt 7. Detaillierte Nachweisführung immissionsorientierter Anforderungen an Misch- und Niederschlagswassereinleitungen gemäß BWK-Merkblatt 3, Bund der Ingenieure für Wasserwirtschaft, Abfallwirtschaft und Kulturbau (BWK) e.V., Kassel, 2007.

[5] N. Schindler, J. Tranckner, and P. Krebs, Extreme value statistics for river quality simulations. Water Science and Technology 61 (2010) 397-406.

[6] F. Blumensaat, J. Tranckner, S. Hoeft, N. Jardin, and P. Krebs, Quantifying effects of interacting optimisation measures in urban drainage systems. Urban Water Journal 6 (2009) 93-105.

[7] B. Helm, J. Tränckner, H. Sieker, and P. Krebs, Flexibilität als Bewertungskriterium bei der Planung von Systemen der Regenentwässerung. Korrespondenz Abwasser, Abfall 56 (2009) 372-381.

[8] R. Merz, and W. Buck, Verfahren zur Entscheidungsfindung bei Mehrfachzielsetzung sowie Zielsysteme und Zielkriterien für die Planung und Bewertung wasserwirtschaftlicher Maßnahmen, Institute für Hydrologie, Wasserwirtschaft und Kulturtechnik Universität Karlsruhe, Karlsruhe, 1997.

[9] HMULV, Leitfaden für das Erkennen ökologisch kritischer Gewässerbelastungen durch Abwassereinleitungen. in: D. Borchardt, and C. Günther, (Eds.), Wiesbaden, 2004.

Overall-effective measures for integrated water resources management in the coastal area of Shandong Province, China

Project Overview and Monitoring Achievements

Monninkhoff L.
Kaden S., Bossel K.
DHI-WASY GmbH, Waltersdorfer Strasse 105, 12526 Berlin, Germany, E-mail: bmo@dhi-wasy.de, Telephone: +49(0)30/6799980

Abstract

The Huangshui River Basin, located in the Northeast of Shandong Province, China, and spanning 64 km of coastline, is an outstanding example for water conflicts arising from piecemeal action as well as rapid growth of population, industry, and agriculture, that can only be solved by an integrated water resources management (IWRM) approach. To tackle these conflicts, a four year joint BMBF / MOST founded Chinese-German project was launched in June 2008. In this ongoing project, an IWRM approach has been developed, including a modelling system, in which all single relevant physical and socio-economic processes are treated as a part of a complete and interactive system. The interaction of runoff, recharge, surface and groundwater, both from the quality and the quantity point of view, were the main processes to focus on. To take into account the interests of different stakeholders as well as future social and economic developments, a GIS supported DSS which is closely linked to the hydrological models has been developed. The model results will be the main input to the multi-criteria analyses used to predict the impact of possible measures and scenarios. Furthermore, the existing monitoring system has been analyzed and extended.

Keywords

IWRM, China, saltwater intrusion, DSS, groundwater, monitoring

1 Introduction

The Shandong Province, especially the Huangshui River Basin and Longkou County (Figure 1), is an out-standing example for water conflicts arising from piece meal action as well as fast growing population, industry and agriculture. In the coastal catchments of the Shandong province the water scarcity is even increased due to salt water intrusion, reducing the usability of available water resources. Furthermore, social status and income of farmers in the Shandong province is significantly below a level that would allow them to keep up with the technology development in irrigation and farming. The socio-economic problems can only be tackled by truly integrated water management approaches (IWRM). In the 90's already many measures against the saltwater intrusion have been implemented. In 1995 for example, an underground dam was finished in the downstream part of the Huangshui River (Liu et al., 2003). Even after this project was finished, the area affected by salt water intrusion continued to increase. It is part of the project to investigate this area and to analyze whether its extent can be reduced.

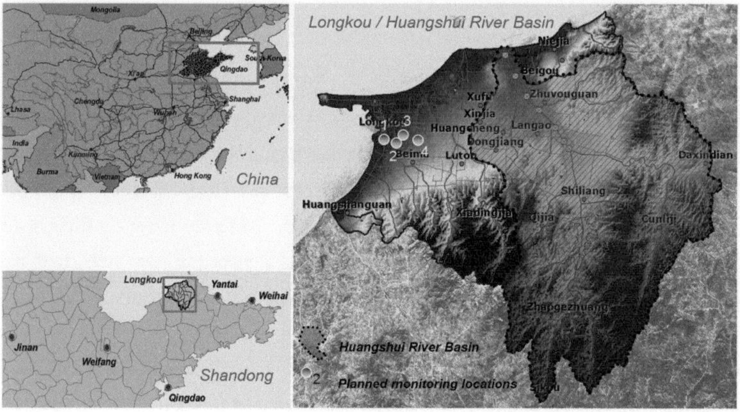

Fig. 1: Project area

The total water consumption within the project area (ca. 1560 km²) during the years 2000 until 2007 was about 205 million m³/a. It is composed by: agriculture (irrigation) 73 %, domestic 10 %, industry 16 % and environment 1 %. With a usable runoff of about 190 million m^3, the water demand exceeds the water resources, in average by about 8 %.This problem is even more severe considering the monthly and annual distribution of water resources and water demand (Kutzner et al., 2006).

2 IWRM method

The steps and components applied in this project basically can be separated in four parts (see Fig. 2). At first, a so-called measures catalogue has been set up. From the analyses of the present situation, all potential measures for the project area, especially in the field of water saving, groundwater recharge, water recycling, measures against salt water intrusion and institutional measures, are collected, sorted and combined in a logical and functional measures catalogue.

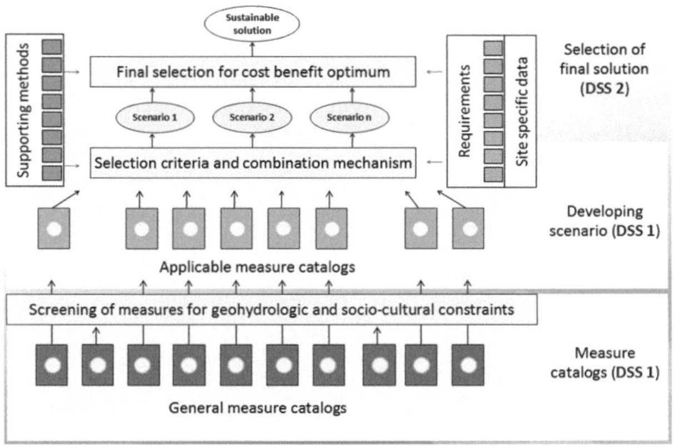

Fig. 2: IWRM Concept to find a sustainable solution (after Geiger, 2006)

Secondly, a Decision Support System (DSS1) has been developed to find the most promising combinations (measures scenarios) of these potential measures, only partially taking into account their geographical location. This DSS1 is based on a water balance for the complete region. The water balance has been integrated in a convenient Excel-based application in order to see the effect of each measure. For this, all measures had to be analyzed in detail and the minimum and maximum effects to each of the defined sources and destinations (a source can be rainfall, a destination can be groundwater) had to be estimated. Besides this water balance, the DSS in this stage is supported by other criteria, like costs and social acceptance. In this way, a complex dataset is decomposed into smaller constituent elements between which pair-wise comparison is elicited, enabling well-founded assessments of performance for each measure.

Thirdly, a GIS-based information system is under development, which incorporates:

- Information to all selected measures scenarios from DSS1 and the possibility to specify each measure according its location, intensity or size.
- A number of hydrological models, which together describe all relevant hydrological processes and which are coupled on the fly or by automatic data exchange. These processes include runoff, evaporation, recharge, surface water flow and groundwater flow. This integrated model represents the fourth part of the overall IWRM concept.
- Tools which automatically implement the measures within the geographical setup of each hydraulic model (for example weirs or reservoirs).

- Criteria and multi-criteria evaluation routines to analyze the simulated measures scenario (DSS2). The basic criteria are similar to the ones of DSS1, but they will be separated in a number of components, which each can be evaluated by automatic routines using the results of the models together with relation tables representing the socio-economic situation in the area.
- Tools which visualize (maps and time series) the results of these evaluations and enable the user to compare them with previous analyzed measure scenarios.
- Documentation routines which summarize the measures which were analyzed and give a clear overview of the effectiveness of these measures.

In the next figure the basic concept of the GIS-based and model-supported system (DSS2) is displayed.

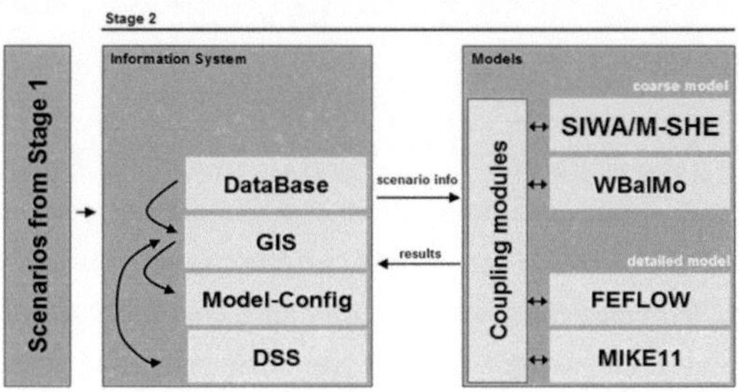

Fig. 3: Concept of the GIS-based and model supported DSS2 system

3 Models

The integrated hydraulic model concept includes two different model approaches; a relatively coarse and a detailed model. The coarse model is used to verify the general feasibility of the water usage proposed in the selected measures scenario. This model consists of the simulation packages SIWA and WBalMo (both DHI-WASY). Furthermore, a MIKE SHE model (DHI) has been set up to verify the results of SIWA and to support mass transport modelling also for the unsaturated zone. SIWA is a 1-D empirical soil water model, based on land-use, soil, slope and groundwater depth data. The model provides a fast and robust way to calculate spatially distributed monthly groundwater recharge and runoff rates. WBalMo (Water Balance Model) is a simulation system for river basin management. Recording relevant system characteristics, probability estimates can be provided for water deficits. If the resulting deficits for all water users of a coupled WBalMo and SIWA simulation are within the limits set in the

GIS environment, the specified measures can be accepted and relevant results can automatically be transferred to the detailed model. This detailed model also consists of two modules; FEFLOW (DHI-WASY) and MIKE11 (DHI). FEFLOW provides an advanced 3D environment for performing complex groundwater flow and contaminant transport modelling, including density dependent flow processes. MIKE11 simulates unsteady flow in regular as well as looped river networks. MIKE11 and FEFLOW can be coupled using the module IfmMIKE11 (Monninkhoff and Li, 2009). The detailed model gives information about the impact of the proposed measures on the groundwater levels and salinity values. All models have been set up and their results could already be used to support the development of the water balance (Chapter 2).

4 Monitoring system

Considering that effective water management and the usability of any modelling rely crucially on the available data input, project efforts aimed at creating an advanced monitoring system for the study area. Weak spots were found above all in the coverage of discharge volumes, water quality of surface and groundwater.

One of the main data gaps concerned discharge volumes, particularly for the largest tributary to the Huangshui River, the Huangchengji. A customized measurement system has been designed to meet the requirements of strongly varying discharges, high sediment contents, and an irregular riverbed. The Huangchengji River Bridge provides the best location to measure discharges along this river. The bridge consists out of 13 separate gates. To provide a comparatively cheap measurement method, a measuring device will only be installed at one specific gate, and from this, the total discharge of the river will be derived. Measurements will be done by an OTT ADC (Portable Acoustic Digital Current Meter). In order to derive a sound relationship between discharge at one gate to the total river discharge, a 3D model was developed with the hydraulic three-dimensional software MIKE3. An overview of the model and the bridge is shown in Figure 4.

To identify spatial and temporal development of changes in water quality a monitoring concept was developed that enables depth-dependant measurements. The concept was technically conveyed by designing multi level measuring points that, as a combination of stationary measuring and mobile sampling technology, continually record water levels and quality parameters in different aquifer depths. Two well holes were drilled and filtered in different aquifer depths (Points 1 and 2 in Figure 1).

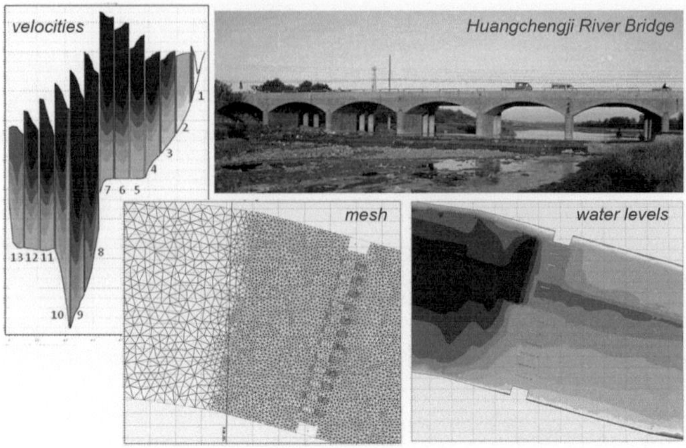

Fig. 4: Overview of the MIKE3 model

The first measurement point was equipped in the bottom aquifer with a solar-powered probe that continually transmits five water quality parameters to a website (Figure 5).

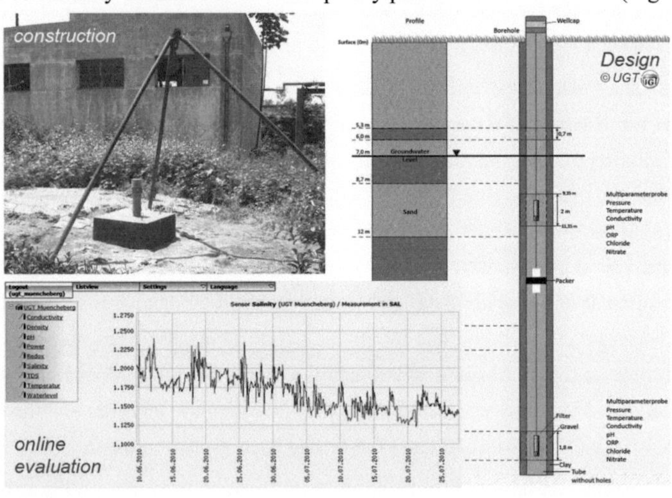

Fig. 5: Construction, design and online evaluation of a groundwater measurement point

On-site work also included testing a newly developed sampling device called the Mobile Monitoring Shuttle (MMS, developed by our project partner UGT). Water samples were taken convection free and pressure neutral. The MMS is intended to supplement the automatic monitoring whenever needed. In case any of the automatically measured parameters shows conspicuous behaviour, water samples can flexibly be taken and chemically analyzed. To help ensure steady data flow in the future, a Chinese staff member was introduced to handle and maintain the technical equipment.

5 Outlook

A prototype of the GIS-based DSS is planned to be finished until the end of 2010. The concept guarantees that different measure scenarios and criteria can easily be integrated, that additional tasks can be implemented or tasks can be exchanged and that hydrological models can be altered or exchanged.

Acknowledgements

BMBF Germany and MOST China are acknowledged for their support. Thanks also go to the Province of Shandong and Government of Longkou City, which takes part in the financing of the project as well.

The overall project is done in co-operation with the Ruhr-University Bochum, IB Schlegel Munich, IÖW Institute for Ecological Economy Berlin and UGT Müncheberg. For project details see www.wasy.eu/shandong.html. The project is support by BMBF under the project number: 02WM0924 (DHI-WASY).

List of Literature

[1] Geiger W.F. et al., 2006, Evaluation and indicator system for sustainable water management. – In: 1st UNESCO/UNEP Training Course on Sustainability in Water Management for urban and rural development, June 23rd -27th, Shanghai, China

[2] Kutzner, R., Zhang, B., Kaden, S., and Geiger, W.F., 2006, Sustainable and integrated water resources management for the coastal areas of Shandong Province, China, Water science and technology, ISSN0273-1223, Nr 11-12(54): 57-64.

[3] Liu, Z., Meng, F., and Zhang, S., 2003, Coastal underground reservoir system engineering in Longkou city, Site Investigation Science and Technology, Nr 6: 47-52.

[4] Monninkhoff, B. L. and Zhijia Li, 2009, Coupling FEFLOW and MIKE11 to optimise the flooding system of the Lower Havel polders in Germany, International Journal of Water, Volume 5(2), pp. 163-180(18), Inderscience Publishers

Drinking Water Supply in a Rural Area of Indonesia:

Lessons learned from the optimization of ground water and river water treatment plants

Stauder, Stefan and Eggers, Jutta

DVGW Technologiezentrum Wasser, Karlsruher Str. 84, 76139 Karlsruhe, Deutschland, E-mail: stefan.stauder@tzw.de, Telefon: +49(0)721/9678-122

Abstract

In a project, funded by BMBF within the framework of "IWRM Gunung Kidul" and a group of German water works the drinking water treatment under tropical monsoon conditions in the special region of Yogyakarta, Indonesia have been investigated. For this purpose, three water works of the local water supply company Bantul were selected. One objective is the short term improvement of the existing water supply, but the activities are also pilot projects in character. The goal is to develop suitable treatment technologies for the three main drinking water resources in the tropics: river water, groundwater and bank filtrate. This paper summarizes the findings of a survey on ground- and river water quality and the efficiency of the belonging treatment processes. Moreover strategies for optimized drinking water treatment are presented, including the results of their full scale implementation.

Catchwords

Drinking water treatment, water supply, groundwater, river water, manganese removal, flocculation

1 Introduction

The rural region of Bantul is located in central Java, south of Yogyakarta, and with an area of approx. 500 km² home to approx. 800 000 people. The local water supplier PDAM Bantul operates 12 water treatment plants (WTP) as well as the belonging networks with a total of approx. 11 000 house connections, supplying approx. 70 000 people. This corresponds to a connection rate of only 9 %. The majority of people in Bantul, as in Indonesia in general, obtain the water for private usage from shallow wells, i.e. "dug wells" with a depth of a few metres, located in or near the kitchen and thus also near the septic tanks, where the waste water is discharged.

Roughly 5 mio m³ water are delivered by the PDAM per year. With approx. 40 %, the water losses are very high and the delivered water quality does not meet the WHO drinking water standards. Both drawbacks are common in the public water supply of Indonesia. The drinking waters often have increased turbidity and are bacteriological contaminated. Eight of 12 WTP in Bantul use groundwater, two bank filtrate, one spring water and one river water. With capacities of 5-25 L/s the plants are rather small.

2 Groundwater Treatment

Groundwater chemistry

In February 2009 a survey on the water chemistry of two shallow wells (depth: 4 and 10 m) and three deep wells (depth: 80-100 m) was performed in Bantul region. The results demonstrate that the five ground waters have a similar chemism. They show an average mineralization (sum of Ca^{2+} and Mg^{2+}: 1.5-2.0 mmol/L, TDS 350-600 mg/L). The pH is rather low (6.5-6.9) and the concentration of natural organic matter is within a common range (TOC: 0.5-1.8 mg/L). Neither oxygen nor nitrate are dissolved and the waters exhibit increased concentrations of ammonia, Fe(II) and Mn(II). Noticeable are the water temperature of approx. 28° C and slightly increased concentrations of potassium and silicium. Toxic substances, e.g. fluoride, arsenic and heavy metals are not present. As expected, the examined water from shallow wells show a clear microbial contamination (> 20 E. coli/100 mL), whereas in the deep well waters neither E. coli nor other Coliforms were detected.

Existing ground water treatment

The treatment technology in the eight ground water plants in Bantul is nearly identical consisting of aeration, sand filtration and chlorination. Customer complaints on high turbidity in tap water are common in the belonging supply areas. More over serious technical problems arise e.g. from clogging of pipes by black precipitations (manganese dioxide). This is illustrated in the left photograph of figure 1, showing a pipe removed from the distribution system of WTP Sewon after approx. 10 years of operation. The right photograph shows the cleaning process of the filter nozzles which have to be done 2-3 times every year.

Manganese dioxide precipitations also clog house hold water meters, hindering correct recording of water consumption and thus causing financial penalties for PDAM. As a result successful ground water treatment is considered as not feasible by local experts. Moreover it is assumed that all ground waters are highly contaminated by faecal bacteria. As a consequence, the PDAM, as well as other water suppliers in the Yogyakarta region, plan to abandon deep groundwater treatment plants and replace them e.g. by river WTPs.

Fig.1: Pipe from Sewon supply area (left), cleaning of filter nozzles (right)

Optimization of the treatment process

In order to investigate and implement optimization measures for ground water treatment in general, WTP Sewon was chosen as a pilot site. Water analysis at this plant had shown that the raw water content of 3 mg/L manganese is very high and the efficiency for manganese removal of the three sand filters in Sewon is rather poor (20-70 %).

According to German experience in ground water treatment [1, 2] a two way strategy was pursued in order to implement appropriate conditions for biological manganese removal:

1. Increase of pH to a value > 7.0 during filtration by installing a spray aeration device into the existing aeration tower (target: > 50 % CO_2-removal)
2. Increase of EBCT in the filter unit by filling in additional sand

These simple measures were completed in 06.02.2010. After seven weeks of optimized operation water analysis were carried out. According to the results in table 1 the removal of manganese improved drastically (> 99 %) due to the mentioned measures.

Since March 2010, the WTP Sewon is delivering drinking water according to Indonesian and WHO standards [3, 4]. This improvement was positively recognized in an instant by the customers who noticed the striking absence of the brownish colour of their tap water.

		Raw water	Before optimization		After optimization	
			Aerated	Filtered	Aerated	Filtered
Oxygen	mg/L	<0.3	4.2	3.5	7.5	4.8
Diss. CO_2	mg/L	57	55	-	28	-
pH	-	6.8	6.8	6.8	7.2	7.2
Ammonium	mg/L	0.6	-	0.4	-	<0.05
Iron	mg/L	0.44	-	0.04	-	0.02
Manganese	mg/L	3.1	-	**2.1**	-	**0.02**

Tab. 1: Results of water analyses in WTP Sewon before (1.2.2010) and seven weeks after optimization (22.03.2010)

For sustainable improvement of the treatment process a new blower and a new pump for appropriate filter backwash will be installed in autumn 2010. In the near future, the capacity of WTP Sewon shall be doubled from 10 to 20 L/s. The new spray aeration device is already designed for this and the sand filters could handle an increased throughput. The latter was demonstrated by identifying the dependency of manganese removal from EBCT in WTP Sewon (e.g. EBCT 12 min for 95 % manganese removal).

3 River Water Treatment

River water chemistry

In Table 2 selected analytical results of the Progo and Opak river waters during dry and rainy season are compiled. The two river waters show a similar chemism, with a temperature of approx. 28°C and a rather low mineralization (sum of Ca^{2+} and Mg^{2+}: 1.0-1.2 mmol/L, TDS: 160-260 mg/L). Noticeable is the fast fluctuating turbidity (suspended solids) from 30-50 FNU (Formazine Nephelometric Units) up to > 1000 FNU (see figure 1). The concentration of natural organic matter (TOC 1.2-3.2 mg/L) is on a comparatively low level. Heavy metals and organic micropollutants such as pesticides and pharmaceutical residues were not detected in relevant concentrations (129 different organic substances measured by HPLC/DAD, HPLC/MS-MS, GC/MS, GC/ECD, detection limits: DL = 1-100 ng/L). Also the values for adsorbable sulfur and halogenic compounds (AOS and AOX) as sum indicators for anthropogenic pollution are low. Solely the complexing agent EDTA in the Opak river water is slightly increased. It has to be noted that in both river waters the indicator bacteria E.coli and coliforms are present in high concentrations.

		Opak River (Trimulyo)		Progo River (Sedayu)	
		Dry season	Rainy season	Dry season	Rainy season
Oxygen	mg/L	6.7	7.8	7.1	5.8
pH	-	7.6	7.8	8.2	7.6
TDS	mg/L	255	228	201	161
Hydrogencarbonate	mg/L	-	161	149	94
Calcium	mg/L	26.8	28.4	26.1	20.1
Ammonium	mg/L	-	0.13	0.08	0.25
TOC	mg/L	1.2	2.2	1.3	3.2
EDTA	µg/L	9	3	-	< DL
AOS	µg/L	13	13	-	39
AOX	µg/L	9	< 5	-	< 5
Pesticides	µg/L	< DL	-	-	< DL
Pharmaceutical	µg/L	< DL	-	-	< DL
E. coli (MPN)	1/100mL	> 2420	41000	> 2420	18600
Coliforms (MPN)	1/100mL	> 2420	198600	> 2420	1414000

Tab. 2: Selected results of the river water analyses

Existing river water treatment

For optimization of river water treatment WTP Sedayu at the Progo river was chosen as a pilot site. This treatment plant comprises the two units "south" (nominal capacity: 20 l/s) and "north" (nominal capacity: 10 l/s). The treatment steps in both units include flocculation, sedimentation, sand filtration and chlorine disinfection. The technology applied is typical for river water treatment in rural areas of Indonesia and rather basic compared to German standards [5-7]. However, the main problem of WTP Sedayu is a deficient design of the sedimentation tank in the unit "south". As a consequence no sedimentation of suspended solids/flocs takes place and the subsequent filters clog rapidly. Because of that the unit "south" was put out of operation and the throughput in the unit "north" was increased by PDAM Bantul as a compensatory measure.

The unit "north" is now running with twice of its nominal capacity leading to a surface load in the tube settler of 9 m/h and in the sand filters 35 m/h, respectively. Due to this overload the removal of suspended solids is hindered. The clear water turbidity is therefore rather high, usually in the range of 2-4 FNU, and short term break trough of suspended matter happens frequently (> 5 FNU clear water turbidity).

Optimization of the treatment process

As an immediate measure to improve clear water quality in WTP Sedayu, the dosage of aluminium sulfate-solution (produced onsite by dissolving granular aluminium sulfate) was replaced by dosing liquid poly-aluminium-chloride. Moreover a procedure for adjusting the flocculent dosage depending on the raw water turbidity was introduced in March 2010 (5 -15 mg/L Al).

After these measures the turbidity of raw and clear water in the unit "north" was recorded over a period of 3 months (solitax/ultraturb, Hach Lange GmbH). According to the results, the turbidity removal improved significantly. Figure 2 shows exemplarily the turbidity data from 13-25.05.2010. For the correct interpretation of this figure the different scale of the raw water and the clear water turbidity has to be considered (factor 100!). As can be seen, and mentioned before, the turbidity in the raw water often increases from below 100 FNU to values above 1000 FNU within a few hours. Moreover the plant is regularly shut down for some hours. Despite these unfavourable conditions and the above mentioned overload the internal goal of clear water turbidity below 1 FNU could be achieved most of the time. Elevated values on 15 and 16.05.2010 were caused by delayed filter backwash.

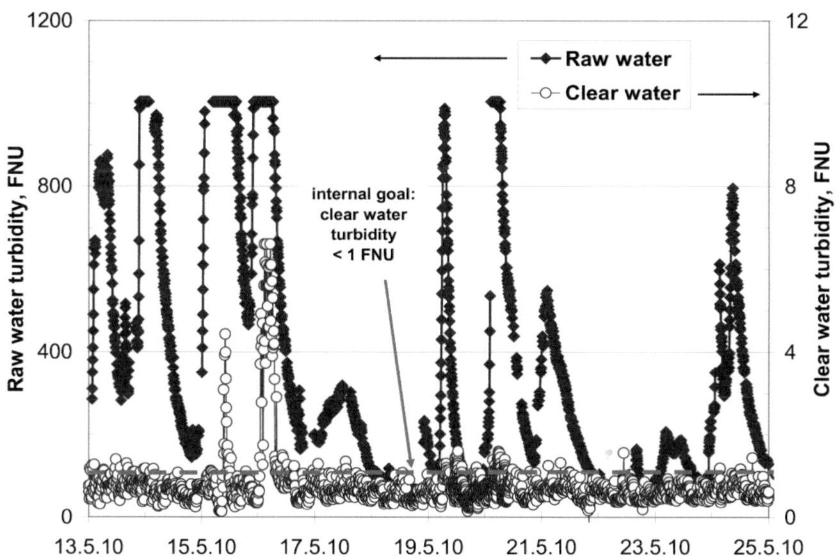

Fig.2: WTP Sedayu: Turbidity in raw and clear water (13-25 May 2010)

The effluent turbidity of the tube settler was measured manually twice a day (Nephla, Hach Lange GmbH). The results in the range of 2-5 FNU, demonstrate a high efficiency of the sedimentation unit. This is presumably due to the high water temperature (28°C) which enhances the flocculation/sedimentation process [8].

Further upgrading of WTP Sedayu will be done in autumn 2010 by installing an appropriate tube settler device into the sedimentation tank of the unit "south". Moreover a retrofit of the filters for double layer filtration and the installation of a chlorine dosing pump is planned (replacement of the existing gravity dosing system). After implementation of these measures unit "south" can be put into operation and optimized river water treatment in Indonesia will be investigated. One main goal is the testing of alternative flocculent and the use of sludge-refeed in order to minimize the running costs.

4 Conclusions

For geogenic reasons, the deep groundwaters in the province Yogyakarta contain elevated concentrations of CO_2 and manganese(II). In the majority of the existing water works, no sufficient manganese removal is achieved. This causes severe technical problems such as clogging of transmission pipes and consumer complaints regarding „dirty" water.

In order to demonstrate adequate groundwater treatment, an existing plant (WTP Sewon) was upgraded by low-cost measures. The goal was to accomplish favourable conditions for biological manganese removal. This is a simple, reliable and cost effective process. The approach was successful and WTP Sewon is now able to eliminate 3 mg/L Mn(II), thus producing drinking water according to Indonesian and WHO standards. Moreover, the technical upgrade of the plant enables to double its capacity from 10 to 20 L/s.

The two examined river waters were found to have a rather low content in dissolved minerals and organic matter and exhibit sufficient concentrations of hydrogen carbonate (precondition for a stable flocculation process). No relevant amounts of anthropogenic micropollutants (e. g. heavy metals, pesticides and pharmaceutical residues) were detected. However, the river waters exhibit a high and fast fluctuating content of suspended solids (max turbidity > 1000 FNU) as well as a severe contamination with faecal bacteria. Considering this background, an insufficient removal of turbidity, as found in the examined river water treatment plant Sedayu, is problematic. As increased turbidity can impair the effect of final disinfection, it is imperative to minimize the clear water turbidity in river water treatment plants.

Despite several technical and operational drawbacks surprisingly good treatment efficiency could be achieved by a simple optimization of the flocculation process. Further upgrading of WTP Sedayu will include implementation of double layer filtration and a dosing pump for hypochlorite-solution. These measures shall guarantee the achievement of the internal goal for clear water turbidity (< 1 FNU) and a safe disinfection.

List of Illustrations

List of Tables,

List of Literature

[1] DVGW-Arbeitsblatt W 223 (Teile 1 und 2) – Enteisenung und Entmanganung (2005)

[2] Böhler, E. (1999). Optimierung der Enteisenung und Entmanganung in Filteranlagen, in Veröffentlichungen aus dem Technologiezentrum Wasser Karlsruhe, Band 8

[3] World Health Organization (2006). Guidelines for drinking-water quality: Vol. 1, Recommendations. – 3rd ed.

[4] Menteri Kesehatan Republic Indonesia (2002). Keputusan Menteri Kesehatan RI NOMOR 907/MenkesENKES/SK/VII/2002 Tanggal 29 Juli 2002 Tentang Syarat- Syarat Dan Pengawasan Kualitas Air Minum, [Indonesian Drinking Water Ordinance]

[5] DVGW-Arbeitsblatt W 217 - Flockung in der Wasseraufbereitung Teil 1: Grundlagen (1987)

[6] Jeckel, M. (2004). Flockung, Sedimentation und Flotation in Lehr- und Handbuch Wasserversorgung Bd. 6 Wasseraufbereitung – Grundlagen und Verfahren, Oldenbourg Industrieverlag München

[7] Werner, G. (1982). Kompaktflockungsanlage mit Schlammabscheider und Eindicker im Wasserwerk Langenau. Schlußbericht BMFT-Vorhaben 02 WT 720 I, 02 WT 901I. Zweckverband Landeswasserversorgung Stuttgart

[8] Kühn et al (2006). Leitfaden. Exportorientierte F&E auf dem Gebiet der Wasserver- und -entsorgung. Teil I: Trinkwasser, Band 2. Technologiezentrum Wasser Karlsruhe

Concept of appropriate water and waste water treatment in the karst region Gunung Kidul, Southern Java, Indonesia

Kerstin Matthies[1]
Susanne Fach[2], Stephan Fuchs[2] and Ursula Obst[1]
Karlsruher Institut für Technologie,
[1] Institut für Funktionale Grenzflächen, 76344 Eggenstein-Leopoldshafen,
[2] Institut für Wasser und Gewässerentwicklung 76131 Karlsruhe, Deutschland,
E-mail: kerstin.matthies@ifg.fzk.de, Telefon: +49(0)7247/826847

Abstract

The district of Gunung Kidul in Southern Java is located above a big karst formation, the Gunung Sewu. Due to the underground drainage system of the Gunung Sewu, rainfall rapidly seeps into the ground. This leads to acute water shortages, especially during dry season [5]. The deficient waste water treatment results in a contamination of the groundwater that is impaired by the poor filtration capacity of the karstic underground.

During several sampling campaigns, the reservoirs and pump stations along the Bribin distribution network as well as the Gua Bribin and a second cave (Gua Seropan) have been examined. The focus has been primarily on microbial contamination, especially fecal bacteria.

To overcome the circle of contamination a concept for water and waste water treatment has been developed. Considering the economical and technical circumstances of the region, low cost solutions with simple technical requirements were chosen for water and waste water treatment.

Catchwords

aerob, anaerob, appropriate technology, coliform bacteria, waste water treatment, water treatment, microbiological monitoring, Southern Java

1 Introduction

In the province of Yogyakarta in Java, the district of Gunung Kidul is considered one of the poorest areas in Indonesia. It is located above a big karst formation, the Gunung Sewu. Due to the underground drainage system of the Gunung Sewu, rainfall rapidly seeps into the ground. This leads to acute water shortages that greatly affect the population, especially during dry season, which lasts from May to September [5].

Part of the IWRM[1]-Indonesia project that is funded by BMBF[2] was the accumulation of water in a cave, the so-called Gua Bribin and feeding of this water into the existing water distribution network. Within the project, water supply will be increased significantly This increase in water supply results in increasing waste water flow which affects the contamination of water sources. Furthermore, waste water is either discharged directly into the ground or into unsealed septic tanks. For instance, more than 80% of the septic tanks in Wonosari (Gunung Kidul, Indonesia) have an unsealed bottom [4], which causes a contamination of groundwater with nutrients like nitrogen, phosphorus and also bacteria, especially during the rainy season.

A lack of sufficient water treatment and the partially dilapidated water distribution system make the situation even worse. Not only is the water highly contaminated by fecal microorganisms, but it also contains a lot of pathogens. This is why consumers are forced to use barely sustainable methods like boiling, to disinfect the water and avoid illnesses (mostly diarrhoea). Within the scope of the research project IWRM-Indonesia, appropriate and sustainable treatment concepts should be developed, to supply water in sufficient quantity and quality. Therefore, a thorough water treatment system is essential.

2 Materials and methods

Monitoring of water sources

During several sampling campaigns, the reservoirs and pump stations along the Bribin distribution network as well as the Gua Bribin and a second cave (Gua Seropan) have been examined. Although several physiochemical parameters were measured, the focus has primarily been on microbial contamination, especially fecal bacteria. Apart from the sampling campaigns for the distribution network, the two caves have been sampled once a month for one year.

The microbial analysis was mainly done by using the ColiLert-System from IDEXX Laboratories following the manufacturer's instructions. The quantitative detection of *E.coli* and total coliforms is based on an enzymatic cleavage of synthetic substrates [6].

---

[1] Integrated Water Resources Management

[2] Bundesministerium für Bildung und Forschung = Federal Ministry of Education and Research

Therefore 100 ml samples were taken from each sampling point, transported in a cooling box for protection against the sun and analysed within 6 hours.

Waste water Treatment system

Considering the economical and technical circumstances of the region, low cost solutions with simple technical requirements are essential for water as well as waste water treatment. To determine an appropriate solution for waste water treatment systems, available technologies were examined and a survey questionnaire was conducted. The questionnaire inquired about the technical, social, and economic aspects of the people, such as their water demand, sanitary habits, and income. Furthermore, two different laboratory experiments were carried out, anaerobic and aerobic treatment of waste water.

Waste water of septic tank or 3-chamber pit latrines in Karlsruhe was used in laboratory experiments in Karlsruhe. This kind of waste water is comparable to the waste water of Indonesian septic tanks, as it consists only of human waste, (black- and greywater), excluding any rainwater or industrial sewage which could inhibit the activity of biomass in the reactors. The entire waste water system comprises of two treatment units, one for anaerobic treatment of septic sludge and one for aerobic treatment of septic tank overflow.

The first experiments were performed to determine the speed of the hydrolysis reaction. The hydrolysis experiments were conducted at mesophilic temperatures (35 °C) with a retention time of 1 day.

With the aid of biogas counter, the progress of hydrolysis was traced continuously. Analysis of the total solids (TS), total volatile solids (TVS), and organic acids indicates the degree of sludge stabilization. Further parameters such as ammonium, COD[3], N_{total}, and pH-value were analysed to understand the processes within the reactors. All parameters were determined according to DEV [3].

The second stage consists of a vertical flow sand filter, because of their simple maintenance and high quality effluents. Experiments involved a lab-scale vertical flow sand filter. It comprised of a perspex column, 10 cm in diameter and 100 cm in height, that was filled to 15 cm with 2/8 mm round gravel for the bottom layer. The main layer contained 50 cm of 0/2 mm lava sand, a type of sand with a high specific surface area. Suspended solids and organic matter are removed by both biological degradation and physical processes such as, adsorption, filtration, and trapping. While adsorption occurs throughout the entire media bed, the biodegradation of the organic matter occurs mainly

[3] Chemical oxygen demand

within the first 20 cm of the filter surface. This surface layer of the sand filter is considered the biologically active layer, containing much of the bacterial mass and other microorganisms. So, Experiments involved a lab-scale vertical flow sand filter.

3 Results and Discussion

Monitoring of water sources

The ColiLert analysis of the sampling campaigns in October 2009 and July 2010 clearly showed a development of the coliform contamination within the course of the Bribin distribution system (Figure 1). While every sample was contaminated by coliforms, the bacterial count rose within the distribution system and reached a maximum at the end of every branch.

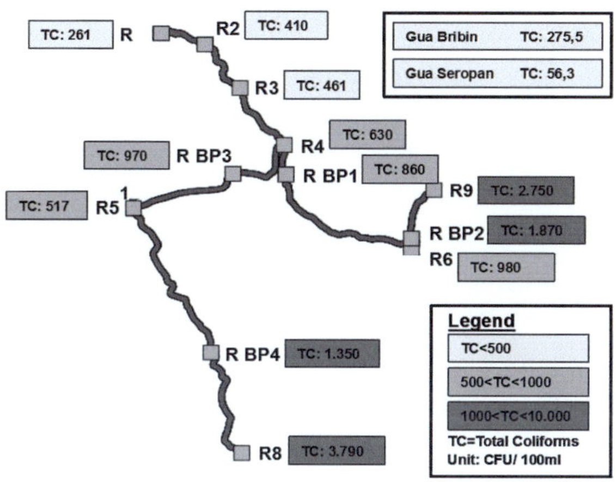

Figure 1: Distribution of total coliform data in July 2010

R1-R9 = reservoirs, R BP1-4 = pump stations

In the first reservoir (R1) only 261 CFU[4]/100 ml could be detected, but the total coliform count reached 2,750 CFU/100 ml at the reservoir R9 and even 3,790 CFU/100 ml at the reservoir R8. This circumstance probably results from the dilapidated pipelines and the fact that they run mostly above-ground and are heated up by the sun. The relatively high temperatures in tropical Java then contribute to a better growth of bacteria.

[4] Colony forming units

The monthly analysis of the two caves (Figure 2) showed that during dry season the contamination by coliforms was unsteady, though relatively low in comparison to the values for the beginning of wet season, where the coliform count suddenly increased dramatically.

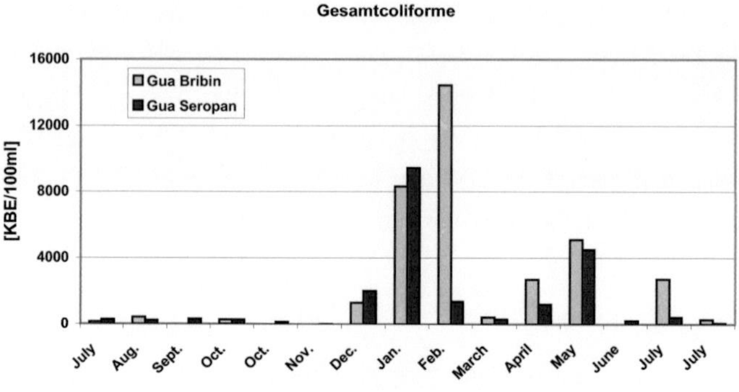

Figure 2: Total coliform data from July 2009 to July 2010

The average value for total coliforms in dry season was about 250 CFU/100 ml. In December wet season began and coliform count began to rise. For the Gua Seropan total coliform values reached a maximum of 9,500 CFU/100 ml in January 2010. The maximum of 14,500 CFU/ 100ml total coliforms in the Gua Bribin could be detected in February 2010. The reason for this increase can be found in the high rainfall and resulting inflow of bacteria in addition to the poor filtration capacity of the karst underground.

Waste water Treatment system

Results show that the anaerobic treatment of septic sludge older than 3 months is possible but not very effective when compared with literature [2]. This study achieved gas production between 30 and 60 ml/g TVS_{inflow} in the hydrolysis stage by co-digestion of primary sludge and food waste. The maximum gas production of the hydrolysis experiments in Karlsruhe was 15 ml/g TVS for the co-fermentation of sludge and food waste and 4 ml/g TVS for the fermentation of just sludge.

Therefore, the anaerobic treatment of septic sludge older than 3 months should be utilized only if high stabilization is needed. But high biogas production should not be expected.

If biogas production is desired, it is necessary to use "fresh" sludge for anaerobic treatment. Adding food waste to sludge increases biogas production.

Intermittent sand filtration is an appropriate solution for septic tank overflow. High removal efficiencies were achieved, including, at maximum, 99 % for ammonium and 86 % for COD at a hydraulic loading rate of 75 l/(m²*day) and 140 g COD/(m²*d). Despite the high COD loading, no clogging occurred.

4 Conclusion and Forecast

The monitoring results proved that water quality is deficient and degrades within the course of the Bribin distribution system (Figure 1). The detected contamination with fecal bacteria is caused by the poor waste water treatment and the inflow of resulting contaminations into the groundwater through the karstic underground. This fact is underlined by the dramatic increase of contamination in the beginning of wet season (Figure 2).

Based on the monitoring results the water treatment concept was devided into three parts. Right after the first reservoir, a central sand filtration will be implemented to eliminate turbidity and bacteria. As the contamination increases within the distribution network, a second treatment step shall be established central, but close to the consumer. In order to find the most appropriate treatment for this hygienisation, several disinfection techniques will be compared in a field laboratory, using different water sources. The third treatment step will be a point-of-use ceramic filtration.

To protect the water sources from contamination through waste water, a thorough waste water treatment will be established. Instead of building new treatment systems, the old septic systems will be improved, sealed and implemented into the chosen adapted treatment system.Regarding the community's toilet water usage, a treatment system using anaerobic processes was found to be most suitable. Though experiments show small biogas production, anaerobic treatment of waste water is necessary for stabilization, especially if reused in agriculture. In addition, as mentioned before, the tested storage tanks developed anaerobic conditions after long retention times, stabilizing the waste water successfully without any mechanical stirring or heating. Thus the anaerobic treatment process comprises of an anaerobic septic tank, instead of an actual energy powered digester. This also eases the community's operational costs and requirements. Once the solids settle, it is estimated that two times per year (especially after the rainy season) it can be reused in agriculture. The lab experiments reveal that, this is possible even considering the hygienic requirements, due to reducing E.coli concentration in the sludge and also in the supernatant down to 10 CFU/ml. The anaerobic tank is expected to remove almost all settable solids and to reduce pathogens (<10 CFU/ml) and organic compounds significantly.

The second stage is to treat the septic tank overflow, the supernatant, with aerobic sand filtering. According to design parameters, the supernatant is intermittently discharged, (twice a month), to the aerobic sand filtration unit for a further reduction of solids, pathogens, and COD. The sand filter effluent then can be diluted and used as fertilizer.

Involving the local people is crucial to the success of installing a sustainable water and waste water treatment system. The first step is to integrate the staff, pupils and residents of the surrounding areas in the planning and implementation process. Increased awareness and participation of the people ensures an understanding and acceptance of water and waste water treatment applications.

List of Illustrations

List of Literature

[1] M. Bahgat, A. Dewedar, A. Zayed, Sand-filters used for wastewater treatment: buildup and distribution of microorganisms, Water Research. 1999.

[2] A. Blank, Durchführung halbtechnischer Untersuchungen zur zweistufigen Vergärung auf der GKA Baden Baden – Sinzheim. Institut für Wasser und Gewässerentwicklung, Universität Karlsruhe (TH) Zwischenbericht 2008

[3] DEV, German Standard Methods for the Examination of Water, Wastewater and Sludge (Deutsche Einheitsverfahren zur Wasser-, Abwasser- und Schlammuntersuchung). Wasserchemische Gesellschaft in der GDCh, DIN (Editor) 2008

[4] K. Müller, Strukturerhebung der Wasserver- und Abwasserentsorgung der indonesischen Stadt Wonosari und ihrer ruralen Umgebung, Diplomarbeit, 2009

[5] F. Nestmann, P. Oberle, M. Ikhwan, T. Lux, U. Scholz, Bewirtschaftung unterirdischer Fließgewässer in Karstgebieten – Pilotstudie auf Java, Indonesien, WasserWirtschaft 2009

[6] M. Wiegand-Rosinus, U. Obst, Practical Experience with Rapid Methods for Controlling the Microbiological Quality of Drinking Water and Disinfection Efficiency, Proceedings of the Specialised IWSA Conference „Rapid Microbiological Monitoring Methods", 1999

Problems with cyanobacteria in a water supply river: case study of River Velhas, Brazil

Eduardo von Sperling
Marcos von Sperling
Department of Sanitary and Environmental Engineering,
Federal university of Minas Gerais, 30170-111 BeloHorizonte, Brasilien,
E-mail: eduardo@desa.ufmg.br, tel. 55 31 34091019

Abstract

The paper describes the development of algal blooms in River Velhas (State of Minas Gerais, Brazil) and discusses the possible causes of these pollution episodes. Since this river is a relevant water supply source for the metropolitan region of the capital city, Belo Horizonte (3 million inhabitants), the presence of cyanobacteria could strongly impair the water use for human consumption. The results of the monitoring program carried out in River Velhas show that this water course is strongly subject to eutrophication problems coupled with special climatic conditions. The prevalence of long dry periods clearly favour the onset of cyanobacteria blooms. The spatial distribution of cyanobacteria populations shows a concentration at the lower reaches of the river, i.e., downstream the discharge of treated and untreated sewage.

Catchwords

Water supply, cyanobacteria, water treatment

1 Introduction

The discharge of untreated sewage in lakes and in running waters is still one of the main issues regarding anthropic environmental impacts on the water quality. This situation is particularly severe in the case of tropical regions, where all metabolic processes, including here the decomposition of organic matter, are performed in accelerated rates. Moreover in these regions there is a chronic lack of financial resources in order to prevent and to restore the water quality of polluted water bodies. One of the main problems in relationship with sewage discharges in rivers and lakes is the possibility of cyanobacteria growth as a consequence of the presence of nutrients (nitrogen and phosphorus). The effects of the eventual release of cyanotoxins are potentialized in tropical climates, since warm water conditions clearly may accelerate the dynamics of the eutrophication process. A deep concern is dedicated to cyanobacterial blooms in Brazil, since it was the first country in the world to register human deaths in a dialysis unit caused by the presence of cyanoprocaryota toxins

[1]. Blue-green algae or cyanobacteria are primitive microalgae with plant chlorophyll. These ancient and remarkable organisms may inhabit quite diverse environments. They have long been recognized as a water quality problem in lakes and reservoirs due to their potential toxicity and to their capacity to impact off-flavours to drinking water. Consequently many water utilities are concerned about controlling cyanobacteria input to the treatment plant. Cyanobacteria present a range of characteristics that give them a clear competitive growth advantage over planktonic algae in certain environmental conditions. They are not favoured by high light intensity and require little energy to maintain cell structure and function. Moreover some species present a buoyancy regulation capacity due to the possession of gas vacuoles within their cells. This is important in avoiding light damage in high-light environments, such as tropic lakes, or in gaining access to light in turbid or low-clarity water. Cyanobacteria are also able to store phosphorus (luxury uptake), what is useful to allow continued growth under conditions of fluctuating nutrient concentrations. They are also poorly grazed by the zooplankton, since they are not a preferred food for this aquatic community.
Most cyanobacteria have maximum growth rates above 25^0 C and are therefore favoured by higher temperatures. Several papers in the literature present evidences of the marked influence of the local climatology over the onset of cyanobacteria blooms. There are also evidences that cyanobacterial blooms may be minimized by limiting the light availability for these organisms. The use of shading of surface waters as a lake restoration technique is discussed in [2]. Moreover some experiments have shown that light limitation may effectively act for restriction of cyanobacteria growth [3].or even for establishing competition between toxic and nontoxic strains [4].

2 Methods

River Velhas is located in the state of Minas Gerais, Brazil, in the southeastern part of the country. Its length is 801 km, with a mean slope of 0.08 % (1100 m at the sources and 464 m in its mouth). The river drains a total area of 27868 km^2 with around 4.5 million inhabitants, which are distributed over 51 municipalities. Around 70 % of this population is concentrated in the metropolitan area of the city of Belo Horizonte. The mean flow of the river is around 300 m^3/s. Local climate presents two well defined periods: rainy season (October to March) and dry season (April to September), with an average yearly precipitation of 1500 mm. The drainage basin of River das Velhas is mostly occupied by agricultural use and cattle breeding. Urban areas, which extend at around just 2% of the total surface, are however a relevant concern regarding water quality. The water course, which is partially navigable, is scarcely used for recreational purposes. Several municipalities, including Belo Horizonte, abstract water from River das Velhas.

A regular monitoring programme is being carried out since 13 years at several sampling points in River das Velhas under the supervision of IGAM (Water Management Institute of the State of Minas Gerais) (www.igam.gov.br) . Water samples are collected 4 times in the year (January, April, July and October) and the main physico-chemical, bacteriological and hydrobiological parameters are analysed according to *Standard Methods for Water and Wastewater Analysis* (APHA/WEF, 2005).

3 Results and discussion

Figure 1 shows the distribution of cyanobacteria populations (sampling points follow downstream direction). Higher values are found downstream the discharge of treated and untreated sewage (after station 083), stressing the role of nutrients (phosphorus and nitrogen) as one of the key factors for the onset of cyanobacteria blooms.

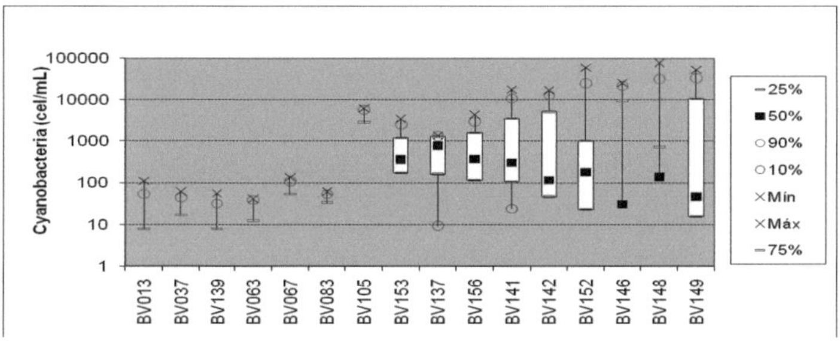

Figure 1: Cyanobacteria populations in River Velhas

Another striking influence is due to the local climatology, which acts according the following scheme:

1) Higher air temperatures ⇨ Higher water temperatures → higher phytoplankton growth
2) Low precipitations ⇨ Low water velocity→ less shoreline erosion → higher water transparency → higher phytoplankton growth
⇨ *Lower dilution capacity* → higher nutrients concentrations → higher phytoplankton growth
⇨ *Higher water residence time* → higher phytoplankton growth
⇨ Lower air humidity → higher phytoplankton growth
3) Stronger solar radiation ⇨ higher phytoplankton growth → favour cyanobacteria

The stochastic nature of these climatic phenomena difficults the configuration of reliable predictions regarding phytoplanktonic blooms. The striking role played by climatology on the dynamics of algal growth has been also confirmed in temperate water bodies [5,6].

4 Conclusions

The results of the monitoring program carried out in River Velhas show that this water course is strongly subject to eutrophication problems coupled with special climatic conditions. The prevalence of long dry periods clearly favour the onset of cyanobacteria blooms. Together with the presence of nutrients, the conjunction of dry periods and low water velocity are the basic causes of eutrophication processes in warm waters. The spatial distribution of cyanobacteria populations shows a concentration at the lower reaches of the river, i.e., downstream the discharge of treated and untreated sewage.

List of Literature

[1] Azevedo, S.M.F.O., Carmichael, W.W., Jochimsen, E.M., Rinehart, K.L, Lau, S., Shaw, G., Eaglesham, G. Human intoxication by microcistins during renal dialysis treatment in Caruaru-Brazil, Toxicology, 181, 441-446.2002

[2] Cooke, G., Welch, E., Peterson, S., Newroth, P. Restoration and Management of Lakes and Reservoirs, Lewis Publishers, Ann Arbor. 2005

[3] Mur. L. Light as a selective factor in the distribution of phytoplankton species. Water Science and Technology, 32, 25-34.1995

[4] Whitton, B., Potts, M. (eds.) The ecology of cyanobacteria: their diversity in time and space. Kluwer Academic Publishers, Dordrecht. 2000

[5] Phlips, E., Hendrickson, J., Cichra, M. Meteorological influences on algal bloom potential in a nutrient-rich blackwater river. Freshwater Biology, 52, 2141-2155. 2007

[6] Sobafer, S. Effect of climate on the trophic structure of temperate forested streams: a comparison of Mediterranean and Atlantic streams. Science of the Total Environment, 390, 475-484. 2008

Water Quality and Aquifer Recharge: Studies on Emerging Pollutants and Viruses in the Jordan Valley

A. Tiehm[1]
N. Schmidt[1], C. Zawadsky[1], P. Lipp[1], N. Seder[2], M. Ghanem[3], L. Wolf[4], H. Hoetzl[5]

1 Water Technology Centre (TZW), Department Environmental Biotechnology, Karlsruher Str. 84, 76139 Karlsruhe, Germany (E-mail: tiehm@tzw.de)

2 Jordan Valley Authority, P.O. Box 2412, 11183 Amman, Jordan

3 Palestinian Hydrology Group, P.O. Box 323, Ramallah, West Bank/Palestine

4 CSIRO, Queensland Biosciences Precinct, 306 Carmody Rd St Lucia Queensland, Australia 4067

5 Department of Applied Geology (AGK), University of Karlsruhe, Kaiserstraße 12, 76128 Karlsruhe, Germany

Abstract

In arid regions the use of non-traditional water resources such as waste water is an important issue. However pollutants such as pathogens and pharmaceutical residues can become potential health risks if adequate treatment is missing. As part of the SMART Jordan Valley project [1] it is the objective of this study to assess the occurrence of emerging pollutants and selected pathogens and to examine the elimination of key compounds in waste water treatment and aquifer recharge. Among the most frequently detected compounds in the area were pharmaceutical residues such as lipid regulators (e.g. gemfibrozil, bezafibrate), antiepileptics (e.g. carbamazepine) and antiphlogistics (e.g. naproxen, ibuprofen, diclofenac) that previously also have been reported in European and US surface waters. In batch and soil column studies, e.g. ibuprofen and bezafibrate proved to be biodegradable. Treating waste water in a membrane bioreactor (MBR) in combination with powdered activated carbon (PAC) reached considerable removal rates also for the more persistent compounds. Also viruses were shown to be present in most of the Jordan Valley surface water samples. MBR treatment and soil infiltration resulted in a decrease of pathogenic microorganisms. Future studies will focus on biodegradation mechanisms, and evaluation of the removal efficiency of pilot and demonstration plants in the area.

Keywords

Groundwater recharge, MBR, pharmaceutical residues, viruses, waste water reuse

1 Introduction

The Jordan Valley is one of the driest areas in the world and groundwater is being exploited at about twice its recharge rate to meet the increasing water demand [2]. Waste water reuse can lead to the accumulation of persistent organic pollutants (POP) and their release into the aquatic environment has become an issue of increasing concern over recent years. More than 80 pharmaceuticals and some of their metabolites have been detected in influents and effluents of European and US waste water treatment plants (WWTPs), in the range of ng/L up to µg/L. Another aspect of artificial groundwater recharge is the unintentional introduction of pathogenic microorganisms contained in waste water. The objectives of our studies are (i) to assess the occurrence of emerging pollutants and selected pathogens in water samples of the Jordan valley, (ii) to obtain more insight into biodegradation processes, and (iii) to monitor and improve the removal of key compounds in treatment processes such as membrane bioreactors (MBR) and soil-aquifer-treatment (SAT).

2 Material and Methods

2.1 Sampling Campaign: Chemical Analysis and Detection of Microorganisms

Grab samples of wastewater, surface water and groundwater have been collected from different sites at the Lower Jordan River Valley. Analysis of pharmaceutical residues was done with a HPLC-ESI-MS-MS following a previously described procedure [3]. Analysis for MS2 bacteriophages, adenoviruses, rotaviruses and noroviruses was done via the Polymerase Chain Reaction method (PCR) after enrichment of viruses by the cation coated filter method [4].

2.2 MBR treatment plant

Huber SE Germany provided a MBR pilot plant consisting of a process tank (800L) in which two membrane modules (2 m^2) were submerged (fig.1). The filtrate flux was adjusted to 15 $L/m^2/h$ and filtration time was set to 2 min. Sludge retention time was 25 d, hydraulic retention time HRT was 24 h and total suspended solids content (TSS) within the process tank was adjusted to 8-12 g/L by weekly removal of access sludge. The MBR was continuously operated since November 2007 with the effluent of a full-scale WWTP after primary treatment. To improve the elimination of persistent pollutants, the MBR filtrate was further treated with 10 mg/L powdered activated carbon (PAC) [5].

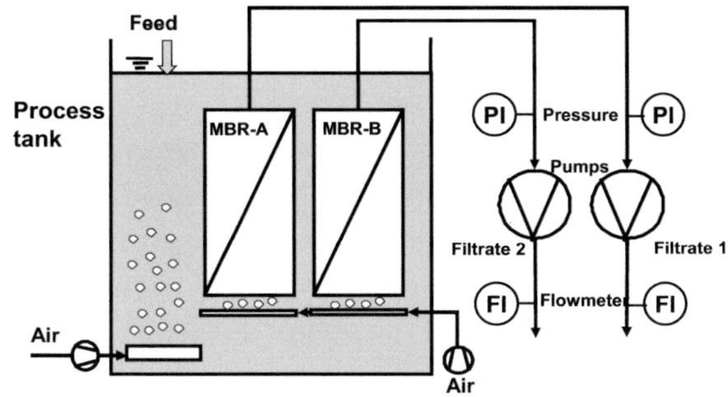

Fig. 1: Schematic illustration of the MBR pilot plant

2.3 Batch Experiments into Biodegradation of POPs

Using filtered raw waste water and treated waste water, batch tests were set up to examine biodegradation in the presence of two different background concentrations of organic load. A mixture of pharmaceutical residues (5 µg/L of each substance) was added. The batches were inoculated with activated sludge from a full-scale WWTP and incubated at room temperature under aerobic conditions. Controls were autoclaved and incubated cooled at 4°C.

2.4 Elimination of Emerging Pollutants during Soil Passage

Unsaturated soil columns were irrigated with treated waste water from a full-scale WWTP (fig. 2). HRT in the columns was 6-7 days. By comparing leachate concentrations of the biologically inhibited (2°C) and the bioactive column (20°C), biodegradation of trace organics (5µg/L) during soil passage was assessed.

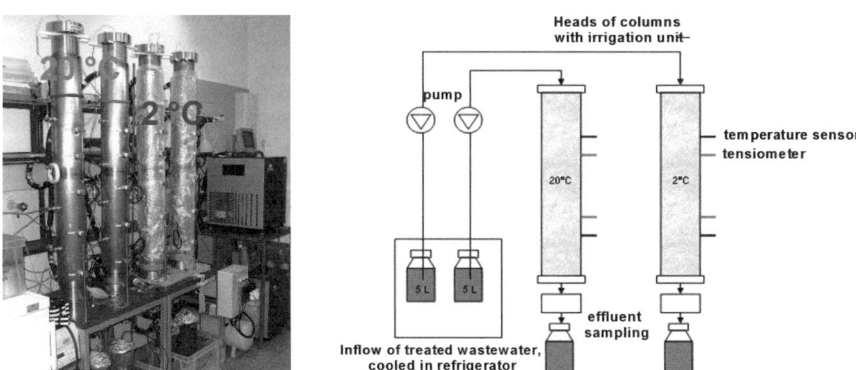

Fig. 2: Laboratory setup and scheme of the column experiment

3 Results and Discussion

3.1 Screening of viruses and emerging pollutants

In total 23 Jordan valley water samples were examined for viruses and MS2 bacteriophages (fig. 3). In 69% of the samples at least one group of viruses was detected. Rotavirus group A was dominating (72%). The results are consistent with previous reports since rotaviruses seem to be ubiquitous: 50% to 60% of cases of acute gastroenteritis of hospitalized children throughout the world are caused by human rotaviruses [6].

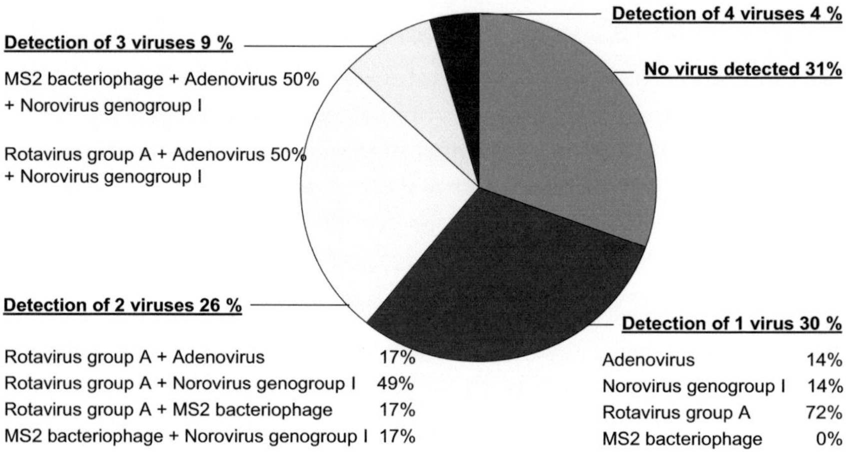

Fig. 3 Detection of viruses and the MS2 bacteriophage in Jordan valley water samples

The concentrations of some of the most frequently detected substances among pharmaceuticals in Jordan are presented in table 1. Comparison with literature data shows that the detected values for pharmaceuticals in Jordan are in the same range as in Europe and the USA. Carbamazepine is a quite persistent compound that also was found in groundwater (GW) with up to 180 ng/L, which is in accordance with findings in German groundwater samples [3]. On the other hand ibuprofen, naproxen and bezafibrate were not detected in groundwater samples but still with up to 1400 ng/l, 550 ng/L and 390 ng/L, respectively, in surface water samples.

Tab. 1 Minimum, maximum and median concentrations of selected pharmaceuticals in the Jordan Valley compared with data from Europe and the USA (GW = groundwater, SW = surface water, WW = waste water), adapted from Tiehm et al. 2010 [7]

Group	Substance	Min/Max concentration (Median) GW	SW	WW	Literature*
		[ng/L]	[ng/L]	[ng/L]	[ng/L]
		LOD 20 ng/L	LOD 20 ng/L	LOD 50 ng/L	
Lipid regulators	Gemfibrozil	<LOD/70 (<LOD)	<LOD/2100 (150)	<LOD/4800 (1400)	SW: 840 – 4760, (b)
	Bezafibrate	<LOD	<LOD/390 (23)	<LOD/430 (270)	SW: 130, (c)
Antiepileptic	Carbamazepine	<LOD/180 (<LOD)	<LOD/1600 (240)	<LOD/3600 (1800)	GW: max. 900, (a) SW: 870 – 6300, (b)
Antiphlogistics	Naproxen	<LOD	<LOD/550 (<LOD)	<LOD/240 (<LOD)	SW: max. 32, (d)
	Ibuprofen	<LOD	<LOD/1400 (<LOD)	<LOD/750 (140)	SW: 1000, (e) WW: 230 – 6400, (f)
	Diclofenac	<LOD	<LOD/160 (<LOD)	<LOD/390 (<LOD)	SW: 680 – 5450, (b)

* Literature data represent mean values, if not marked otherwise (max. = maximum values); a: Sacher et al. 2001, b: Andreozzi et al. 2003, c: Möder et al. 2007, d: Benotti et al. 2009, e: Kolpin et al. 2002, f: Ternes 1998

3.2 Elimination of emerging pollutants during MBR treatment, soil passage and in batch studies

Analyses of samples from the influent and effluent of the MBR showed that some substances (ibuprofen and naproxen) are removed, whereas others (carbamazepine and diclofenac) passed the MBR [8]. However, addition of PAC to the MBR filtrate improved elimination of the most persistent pollutants, i.e. carbamazepine (86% removal).

In batch assays, bezafibrate, diclofenac and ibuprofen proved to be biodegradable (fig. 4).

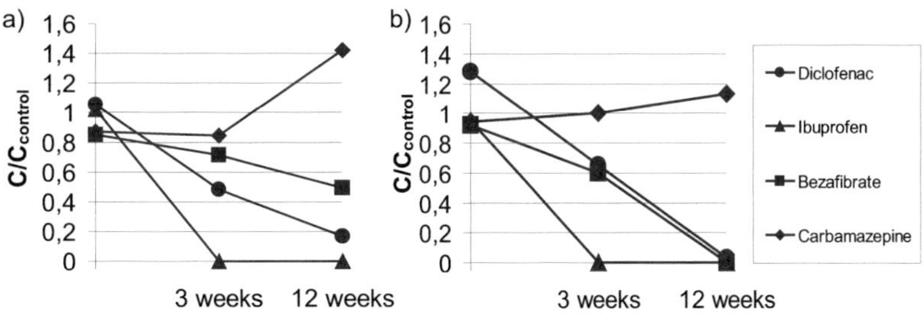

Fig. 4 Biodegradation of 4 pharmaceuticals during batch experiments with filtered raw waste water (a) and effluent from a waste water treatment plant (b)

Increase of carbamazepine during the batch experiment is probably due to the conversion of glucuronide conjugates to the parent compound as it has been reported previously [9].

In the soil column study, complete elimination of diclofenac and bezafibrate was observed at 20°C whereas significantly less pronounced removal occurred at 2°C (fig. 5). The temperature controlled difference in elimination rates clearly indicated biological degradation of these compounds at room temperature. Carbamazepine showed moderate to no removal at 2 and 20°C, sometimes exceeding influent concentrations (depicted as negative elimination rates) due to sorption/desorption events. Ibuprofen was completely eliminated even at 2°C.

Also a removal of viruses has been demonstrated in the soil columns and MBR (data not shown).

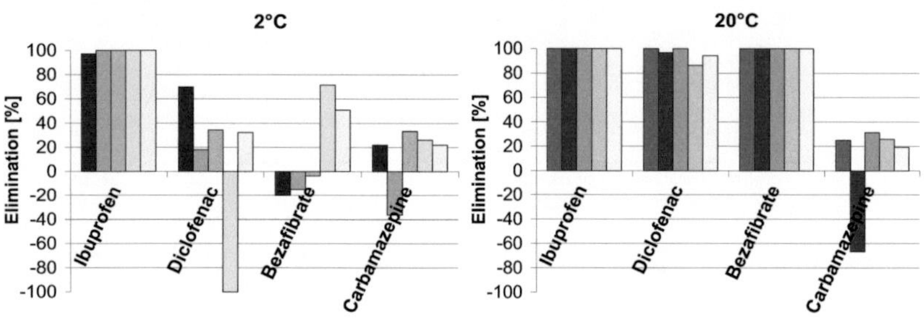

Fig.5 Elimination of selected pharmaceutical residues after trickling through soil columns operated at different temperatures

4 Conclusion and Outlook

Emerging organic pollutants such as pharmaceutical residues and pathogenic viruses have been assessed in the Lower Jordan Valley. The concentrations detected were in the same order of magnitude as reported for Europe and the USA. However, intensive water reuse in arid regions as well as water recycling strategies under moderate climatic conditions can result in an accumulation of pollutants. The experimental results demonstrate biodegradation of emerging pollutants in waste water and during soil infiltration under aerobic conditions. With the process combination of MBR and PAC the concentrations of persistent organic pollutants can be further diminished. Future studies will include biodegradation of pharmaceutical residues in the presence of potential auxiliary substrates to stimulate co-metabolic transformation, continue monitoring campaigns in Palestine and Jordan and examine pollutant removal in pilot-scale plants operated in the Jordan Valley.

Acknowledgements

The authors would like to thank all involved partners of the SMART ("Sustainable Management of Available Water Resources with Innovative Technologies") project for their support. Funding by the German Federal Ministry of Education and Research (project no. 02WM1082) is gratefully acknowledged.

List of Literature

[1] SMART– Sustainable Management of Available Water Resources with Innovative Technologies. http://www.iwrm-smart.org/. 2007

[2] MWI – Ministry of Water and Irrigation, Jordan. Water for Life – Jordan's Water Strategy 2008 - 2022. http://www.idrc.ca/uploads/user-S/12431464431JO_Water-Strategy09.pdf. 2009.

[3] F. Sacher, F. Th. Lange, H.-J. Brauch and I. Blankenhorn. Pharmaceuticals in Groundwaters – Analytical Methods and Results of a Monitoring Program in Baden-Württemberg, Germany, Journal of Chromatography. 2001.

[4] C. Zawadsky and A. Tiehm. PCR-detection of viruses in waste water and surface water of the Jordan valley. Proceedings "15th International Symposium on Health-Related Water Microbiology", 31st May – 5th June 2009, Naxos Island, Greece.

[5] P. Lipp, H. Groß and A. Tiehm. Improved elimination of organic micropollutants by a process combination of membrane bioreactor (MBR) and powdered activated carbon (PAC). Presentation at the MDIW 2010 conference in Trondheim. (to be published in Desalination and Water Treatment).

[6] WHO – World Health Organization: Guidelines for drinking-water quality. ISBN 978 92 4 154761 1 (WEB version). 2008.

[7] A. Tiehm, N. Schmidt, M. Stieber, F. Sacher, L. Wolf and H. Hoetzl. Biodegradation of pharmaceutical compounds and their occurrence in the Jordan Valley, Water Resources Management. 2010

[8] P. Lipp, K. Kreißel, S. Meuler, F. Bischof and A. Tiehm. Influencing parameters for the operation of an MBR with respect to the removal of persistent organic pollutants, Desalination and Water Treatment. 2009.

[9] N. Vieno, T. Tuhkanen and L. Kronberg. Elimination of pharmaceuticals in sewage treatment plants in Finland, Water Research. 2007.

▫Circular Economy and Closed Water Cycles: Sustainable Utilisation of waste water in Arid Areas

Prof. Dr. Peter Heck, Institute of Applied Material Flow Management (IfaS), Environmental-Campus Birkenfeld, P.O Box 1380, 55761 Birkenfeld, Germany

Abstract

The real challenge for the cities and regions of the future is the efficient management of scarcer and less reliable water resources to satisfy the water needs and manage waste water disposal without creating environmental, social or economic damage. In order to implement Zero-Emission-Strategies and closed water cycles, we need a paradigm shift in urban and rural water management. In order to develop a long-term strategy towards a water efficient society encompassing the agricultural sector, new management, technology and agricultural cultivation and crop strategies are demanded. Material Flow Management (MFM) defined as the holistic and efficient influencing of material flows in systems, is offering a suitable tool-kit to create a blueprint for a water efficient society, a circular economy society. Within two projects in Egypt and Morocco new technologies and approaches for the problem "Effluent" as irrigation water for agricultural areas will be tested and investigated as well as the adaption and development of existent technologies and methods towards an adjusted all-round solution. The SUWAAR as well as the SWORD project are designed and expected to deliver tangible results useful for decision maker to decide upon new protection strategies for water resources by increasing utilization of waste water in agriculture. It aims to provide a repeatable concept to balance water, energy and agricultural issues.

Keywords

Material Flow Management, Nutrient Recovery, Water Re-Use, Circular Economy, Zero-Emission, Carbon cycle, Climate Protection,

1. Introduction

Worldwide around 1.1 billion people do not have access to improved water supply and around 40 percent of the world's population has no sanitation access (WHO, 2002). The most affected are rural communities in developing countries, resulting that almost 82% of them are lacking of access to the adequate sanitation services (Massoud, Tarhini, & Nasr, 2009). Besides adverse effects on the environment, the lack of the sanitation and adequate wastewater treatment has an adverse impact to the general health, resulting that 2.1 million people die each year from diarrheal diseases (WHO, 2002).

Today, conventional urban water management (UWM) cycles are characterized by clear inefficiencies and irrational treatment practices, including high quality drinking water for all domestic purposes, large quantities of drinking water to transport human excreta, and loss of useful nutrients. The fast population growth and higher water demand per capita, higher industrial consumption and increased chemical load of the wastewater are no longer 'naturally' compensated and thus, sometimes result in severe ecological damage. In the developed countries the costs for drinking water supply and 'end-of-pipe' wastewater treatment are ever increasing. The predominant approach in developed countries, that human excreta are wasted without nutrient recovery or secondary utilization is a modern misconception and a clear burden for a sustainable ecological and economical development. It has led to the development of so-called "drop and store" or "flush and forget" sanitation solutions, where precious drinking water is used to transport excreta into the water cycle, misusing our rivers, oceans and aquifers as sinks for untreated wastewater. In addition, the existing high-technology wastewater treatment systems are directly requiring more and more energy, both operational and embedded. Indirectly, valuable nutrients and chemicals (such as nitrate and phosphate) are not recovered and produced with high environmental burdens and energy. Instead of re-using these nutrients, the agriculture sector is using chemical fertilizers, leading to soil degradation and ground water pollution. These conceptions and strategies are not suitable for developing or transition countries. In particular for those countries with water shortages and (semi-) arid areas, new treatment and management concepts and technologies are urgently needed, combining nutrient recovery, energy efficiency and water re-utilization aspects. In semi-arid to arid countries such as Egypt and Morocco, the agricultural usage of surface and ground water for the irrigation intensive cultivation of acreage led to acute water scarcity affecting strongly the security of supply of potable water.

Agriculture is by far the most intense water consumer, with about 70% of the global fresh water withdrawals (Lazarova et al., 2008), but in the same time by far the most important sector for the wastewater re-use. However, the wastewater re-use in agriculture accounts only for only 1% of agricultural water use (Jimenez & Asano, 2008). The large scale substitution of the fresh water in agriculture with treated wastewater with an aim of limiting water stress, preventing desertification and re-utilizing (semi-) arid areas for the production of food and energy crops as well as renewable raw materials is a challenge of the modern society. Circular economy (CE) and its integral part – material flow management (MFM) – emphasize on energy and material efficiency strategies, while activating the regional potentials and creating regional added value. CE promotes the idea of nutrient recover from wastewater and its usage in agriculture in order to generate added value products, such as energetic and/or pharmaceutical raw materials. Thus, CE and MFM can be considered as suitable tools and strategies to solve global sanitation problems and achieve the goals of Integrated Water Resource Management (IWRM).

2. Implementation of CE towards closing water cycles

The overall objective of Circular Economy (CE) approach in water management sector is to efficiently re-use wastewater (and embedded nutrients) in a sustainable manner with manageable environmental risks. The definition and implementation of ''Zero effluent discharge'' strategies for single industries, industrial clusters as well as entire cities or regions are playing a key role on the way towards a CE. Zero Discharge is a vision, a management concept in terms of a continuous improvement process of water management. The fundamental idea is continuous efficiency upgrading to attain complete closure of water and nutrient cycles.

For a long time, wastewater was identified as a problem. Besides hygienic hazards, macronutrients nitrogen (N), phosphorus (P) and potassium (P) were representing the main issue of concern when treating wastewater. Globally, sanitation has been mostly solved by applying concepts of ''drop and store'' and ''flush and forget''. Conventional sanitation systems have more disadvantages then advantages. Some of them are: over-exploitation of the limited potable water resources, adverse effects to the environments (e.g. soil and groundwater pollution) and as most important nutrient loss.

The implementation of CE approach and re-use of wastewater for irrigation results in protecting scarce fresh water resources, providing source of nutrients, environmental protection and creation of regional added value. The approach of nutrient recover from the wastewater is very important due to the fact it is estimated that for example phosphorus resources will be economically feasible to extract in next 100 years (Langergraber & Meullegger, 2005). On an annual base, an average person produces ca. 500 l of urine which is enough to fertilise 300-400m^2/a of crops and 50 kg of faeces which will be enough to fertilise 20-200 m^2/a of crops (EcoSanRes, 2008).

Closing water cycles requires intelligent and efficient management and organisation schemes and the process of evaluation and selection the most suitable wastewater treatment solution should consider the entire life cycle of wastewater generation. Solutions should be economically affordable, environmental sustainable and socially acceptable. In arid systems for example it is highly questionable to use water as a flushing device. Vacuum, air transport systems seem to be much more feasible in those environments. Irrigation management is next important driver and must be adapted to the local conditions. Flooding and sprinklers irrigation are cheap and less complex irrigation method, but however the water use efficiency is very low. For arid regions drip irrigation is most efficient but needs a proper pre-treatment in order not to block the irrigation pipes.

4. Show case SUWAAR in Qena, Egypt

Egypt is one of those developing countries currently facing chronically water stress. Currently, the daily wastewater re-use in Egypt is just 1.920.000 m^3/day accounting for only 1% of the total water demand (Jimenez & Asano, 2008). Energy efficient wastewater treatment and re-use are not commonly practiced in Egypt. Even existing centralized wastewater treatment plants (WWTP) are often ineffective.

The currently operating WWTP in the city of Qena based on the trickling filter technology is operating inefficiently and endangering the ground water sources. Instead of the projected 25.000 m^3/day the WWTP is receiving up to 100% more influent. As a result, almost untreated wastewater is discharged to the marginal desert soils, causing over fertilization of the soil and contamination of ground water sources resulting in closure of many fresh water wells. Hence, the ineffective WWTP is even increasing the water stress rather than contributing to the solution. New solutions ensuring and restoring the ground water quality are urgently needed in order to maintain public health and to limit the water stress. Like in Qena, where a new WWTP is currently under construction and will start the operation in mid 2011, the same situation and environmental problems could be observed in various other Egyptian cities with ineffective primary WWTP´s. For those primary WWTP´s, in particular in small scale communities and cities, the designated SUWAAR project[1] aim to develop an adapted and appropriate technology and management solution. The goal of the project is to prove that the following paradigm could be realized: "Provision instead of Disposal" and "Circular Flow instead of Throughput". Thus,

[1] The project ''Sustainable Utilization of wastewater in Arid Agricultural Regions - SUWAAR" is funded in the scope of German-Egyptian bilateral research cooperation and has been ongoing since 2009. The project is promoted by the BMBF [1] and the MHESR [2].

wastewater shall be treated decentralised energy extensive and the embedded nutrients shall be re-used. Due to the project complexity, it is divided into two phases. The currently on-going first phase of the SUWAAR project aims on devolving MFM based wastewater management concept. Besides sustainable wastewater treatment, the MFM concept will take into consideration maximal nutrients recovery and their re-use for production of an added value crops such as Jojoba or Jatropha. Hence an important cycle of matter can be closed, a ''problematic'' residue of the wastewater (e.g. phosphorus and nitrogen) is transformed into a profitable organic fertilizer and the level of hygiene will be improved. Concerning the development of new agrarian areas, desert areas which is up to now unused, free of competing utilization pressures and abundantly available, will be used for the production of potential renewal energy and/or pharmaceutical crops.

The overall aim of the SUWAAR II project phase is the justification, demonstration and commercialization of the entire SUWAAR project results. Based on the results and conceptional work of the first phase the demonstration unit consisting of an adapted constructed wetland system and a solar desalination unit will be set up as a supplemental treatment option for primary WWTP´s. The treated waste water will be used to irrigate the test field with intercropping. Various energetic and pharmaceutical as well as phyto-chemical utilization possibilities for the crops are explored in order to determine the value-adding potential and define an agricultural code of best practice. A continuous monitoring of the water and soil conditions shall help to predict the long-term impact on local ground water resources as well as the long-term development of soil conditions. Hence, the proposed project is a subsequent step of the ongoing SUWAAR project, with a practical implementation of the demonstration plants based on the gained results from the previous research. The investigated technology combination will be adapted towards Egyptian conditions. The network of Egyptian and German partner will establish a joint business unit to commercialize the results and disseminated the technology package as well as water management and agro-industrial consultancy services.

4.2 Show case SWORD in Layounne, Morocco

In this second case study IfaS is trying to prove that under special conditions there is no need for expensive and energy consuming treatment of waste water. Wastewater can be used as an irrigation source with little pre-treatment (sedimentation and sand filter) for energy and pharmaceutical crops in arid areas. The current situation in Layounne consists of the use of a dry valley (Qued) as a sewer channel. The untreated effluent of 200.000

people (21.000 m^3 waste water per day) is drained into the Qued (picture 3) creating at least odor problems. The planned conventional solution was a waste water treatment with lagoon systems. This takes a lot of space, needs high investment and still does not solve the odor problem entirely. And in addition this approach would not make use of the water and the incorporated nutrients.

IfaS proposed to analyze reuse options with a minimum pretreatment of the wastewater. This means the treatment is shifted to the use in Jatropha and Jojoba plantations. Aside from solving the treatment problem this would lead to alternative income options for the region. The required space to "treat" the existing amount of wastewater through an economic plant production would be around 1.000 – 1.500 ha calculated on base of 6.000 to 8.000 m^3 water demand per ha and year for irrigation. Based on the current amount of waste water around 4.000 liter of plant oil could be produced per year and ha and create regional added value to finance and upgrade the wastewater pretreatment.

5. Conclusion

The conventional end-of-pipe treatment of wastewater creates, if properly done and operated, in best case an outflow which is no longer a threat to the environment and the wellbeing of people. Unfortunately this kind of water management is expensive and creates continuous and increasing maintenance and operation costs. Thus, together with high initial investment the conventional end-of-pipe are resulting in high economic burdens and financial struggles neglecting the economic values of the nutrient and water content of the wastewater. In particular developing or transition countries with insufficient funds such conventional end-of-pipe are more likely to lead to structural poverty than to sustainable economic development. Furthermore, the production of sewage sludge, the increasing fresh water demand, the need for chemical fertilizer and the energy demand are creating new environmental burdens. Hence, the problem of wastewater is just shifted into other dimensions rather than adequately solved.

Although wastewater may contain heavy metals and pathogens, it could be seen as a highly valuable resource for the cultivation of energetic and pharmaceutical raw materials in (semi-) arid areas. Closing water cycles, creating new value adding option by re-utilizing wastewater using phyto-chemical concepts such as biocascading could be a suitable approach towards a CE in water management. Adapted energy efficient wastewater treatment options adjusted to local conditions are needed to limited the water stress and ensure an improvement on public health. In best case such "sustainable technology and management packages" can be re-financed and operated by income generated at the new to be cultivated areas. Instead of continuous deficit spending new value adding sectors could be developed and the financial pressure on developing or transition countries minimized. The previous presented aspects of waste water treatment are subject of some pilot projects which are still ongoing in Morocco and Egypt. Of course, there are still lots of problems to be solved, e.g. in arid areas the increasing amounts of salt may create problems and the long term accumulation of heavy metals in soil and underground need to be monitored. But the main issue to be dealt with is to convince the local people as well as the responsible authorities that for economic and environmental reasons new way of handling waste water are necessary.

List of Literature

[1] EcoSanRes: Stockholm Environemntal Institute. (2008). Guidlines on the Use of Urine and Faeces in Crop Production. Stockholm: EcoSanRes

[2] Jimenez, B., & Asano, T. (2008). Water reclamation and re-use around the world. In Water re-use - An International Survey of current practice, issues and needs (S. 648). London, UK: IWA Publishing.

[3] Langergraber, G., & Meullegger, E. (2005). Ecological sanitation - a way to slove global sanitation problems? Environment International 31 , 433-444.

[4] Massoud, M. A., Tarhini, A., & Nasr, J. A. (2009). Decentralised approaches to wastewater tretament and management: Application in developing countries. Environmental Management , 652-659.

[5] Parkinson, Jonathan N, Goldenfum, Joel A., Tucci, Carlos E.M.: Integrated urban Water Management: Humid Tropics, Urban Watr Series – UNESCO IHP, Volume 6, Leiden 2010.

[6] Qadir, M., Wichelns, D., Sally-Raschi, L., McCornick, P., Drechsel, P., Bahri, A., et al. (2009). The challenges of wastewater irrigation in developing countries. Agricultural Water Management.

[7] Raschid-Sally, L., & Jayakody, P. (2007). Understanding the drivers of wastewater agriculture in developing countries - result from a global assessment. Comprehensive Assment Resarch Series

[8] WHO - World Health Organosation. (2002). Environmental Helath. Eastern Mediteranean Regional Center for Environmental Health.

Funding organizations of the SUWAAR project:

[9] 1] BMBF, German Federal Ministry of Education and Research; Project number EGY 08/041

[10] 2] MHESR, Ministry of Higher and Scientific Research of the Arab Republic of Egypt, STDF-project-ID-670

Effects of long-term reclaimed water irrigation on soils of the suburban area of Beijing, China.

Meng Qingyi Sun Weihong Xu cuiping Ma Fusheng

Beijing Hydraulic Research Institute, No.21, Chegongzhuang west road, Beijing, 100048 China, Email: mqy@bwsti.com, Phone:+86(10)88416484

Abstract

The utilization of reclaimed water is one of the most effective measures to increase the amount of available water for crop irrigation. The objective of this study was to assess changes in soil chemistry and the related physical changes from long-term use of reclaimed water compared with soils irrigated with bore water. The result showed that the contents of Na^+, Mg^{2+} and K^+ generally increased in 0~100 cm of soil column, but the increase in 0~40cm depth was not significant, while in 40~80cm was significant. Differences in soil EC and pH due to irrigation with reclaimed water and bore water were not significant. Sodium adsorption ratio (SAR) slightly increased but maintained in a suitable level. This study will help assess whether reclaimed water irrigation in Beijing Suburban is sustainable in the long-term.

Key words

reclaimed water irrigation, winter wheat, summer maize, soil salinity

1. Introduction

In Beijing, the average water resource per capita is less than 200 m^3, the water shortage is a bottle-neck restricted the development of socio-economy. Therefore, the utilization of reclaimed water is an important measurement to alleviate the water stress situation of Beijing. Based on government development plan, reclaimed water utilization will account for 20 % of total water supply at 2010, and the amount of reclaimed water used for irrigation will reach 0.3 billon m^3, the area of reclaimed water Irrigation area will reach 40000 hm^2 in Beijing (Liu et al., 2006). There are two big irrigated area in Beijng i.e., the Nanhongmen reclaimed water irrigation area in Daxing District and the Xinhe reclaimed water irrigation area in Tongzhou District, and the reclaimed water utilization has been up to 0.29 billion m^3 in 2009. With the expansion of reclaimed water irrigation, the impact of reclaimed water irrigation on soil, crop and groundwater has been given great concern, particularly, the impacts on the crop quality and quantity remain the key issue. Numerous studies have proposed that reclaimed water irrigation helps the crop absorb the nutrient and salt accumulation in the soil without hazard to crop growth (Polglase et al., 1995; Falkiner and Polglase, 1997). In China, numerous experiments have been carried out (Ma et al., 2007; Huang et al., 2007; Li, 2007; Wan and Qi, 2007). However, the researches of the long term effects of reclaimed water application in irrigation on crop quality and quantity, soil salinity and alkalinity are seldom. In this study, the effects of the reclaimed water irrigation on soil salinity and alkalinity are evaluated during winter wheat and summer maize growing season in North China Plain.

2. Methods and Material

2.1 Experimental design

The experiment was carried out in the center irrigation experimental station of Beijing (39.33 N, 114.33 E, 12 m a.s.l.). The long-term average annual precipitation and water surface evaporation are 565mm and 1140mm, annual average temperature is 11.5℃, annual average frost-free period is 185 days. The experimental pits are bottomless with length and width of 3×2m. In order to avoid the influence lateral soil water movement, pits were isolated by geotextiles for about 1m depth. The soil texture is silt loam. The groundwater recharge effect is negligible due to the depth of water table is more than 8m depth.

For the study of reclaimed water irrigation impact on soil salinization, field experiments were conducted at Beijing central irrigation experimental station from October 2000 to June 2006. Four different irrigation schedules were applied for the winter wheat and summer maize, i.e., continuous reclaimed water irrigation for 6 years, bore water irrigation for 3 years and followed by 3 years reclaimed water irrigation, reclaimed water irrigation for 4 years and followed by 2 years bore water irrigation, continuous bore water irrigation for 6 years. The soil samples taken from each treatment, and the soil salt-ion, SAR value, EC value were measured for comparison study.

2.2 Water quality of reclaimed water

The reclaimed water used in the experiment produced by Gaobeidian sewage treatment plant, the water quality of reclaimed water measured before applied. The measured index includes TDS, SS, BOD_5, COD_{cr}, nutrition, heavy metals and pH. The water quality index of reclaimed water meets the national standards of irrigation water quality (GB5084-2005). The quality of irrigation water was listed in Table 1.

Table 1 The irrigation water quality during experiment

No.	Water quality index	Unit	Reclaimed water	Bore water
1	BOD5	mg/L	3.9~45.6	8.91
2	CODcr	mg/L	39.2~89.3	15.6
3	SS	mg/L	1.21~32	0.2
4	TN	mg/L	16.2~32	1.75~8.45
5	TP	mg/L	0.67~4.2	0.02~0.42
6	pH		7.3~8.46	7.5~8.18
7	TCa	mg/L	0.00003	0.000005
8	Chromium (VI)	mg/L	0.011~0.079	0.014~0.017
9	Fecal Coliform	/100ml	102~988	0
10	Ascaris Eggs	/L	0	0
11	Total Salt	mg/L	880~1050	610~760
12	Chloride	mg/L	148~300	49.3~96.2
13	Sulfide	mg/L	0.18~0.19	‹0.004

2.3 Analysis of soil samples

Soil samples were taken approximate at 10cm intervals from surface to the depth of 100cm. EC was measured by Conductivity Meter, pH was measured by pH Meter. Ca^{2+}, Mg^{2+}, K^+ and Na^+ were measured by Ion Mass Spectrometry, Sodium adsorption ratio of soil extract (SAR) with unit as $(mmol/L)^{0.5}$, calculated by the following formula.

$$SAR = \frac{[Na^+]}{\left[\left([Ca^{2+}]+[Mg^{2+}]\right)/2\right]^{1/2}}$$

Where, the concentration of Na^+, Ca^{2+} and Mg^{2+} were expressed in mmol/L.

3. Results and Analysis

3.1 Soil EC

As shown in Figure 1, with the increase of soil depth, the EC value showed a trend of first increase and then decrease for all of the treatments. The maximum EC occurs at 40cm depth for 3 years reclaimed water irrigation and 60cm depth for 4 years reclaimed water irrigation. It indicated that soluble salt may move downward with the elongation of reclaimed water irrigation.

The experimental field located in the North China Plain with typical monsoon climate, the soil salt accumulation controlled by both cumulative evaporation and precipitation leaching process. The precipitation events occurred at May 26th, 27th, and June 6th, 7th for about half month before the measurement of soil EC. The intensive precipitation and consecutive wet weather reduced soil evaporation and led significant soil salt leaching at topsoil, which results in lower salt content at upper layer (approximate 0~30 cm) and little differences of salt content among different treatments. However, for the bottom layer, soil physical characteristics contributed to low EC that relatively low clay content under 60cm depth is conducive to salt leaching.

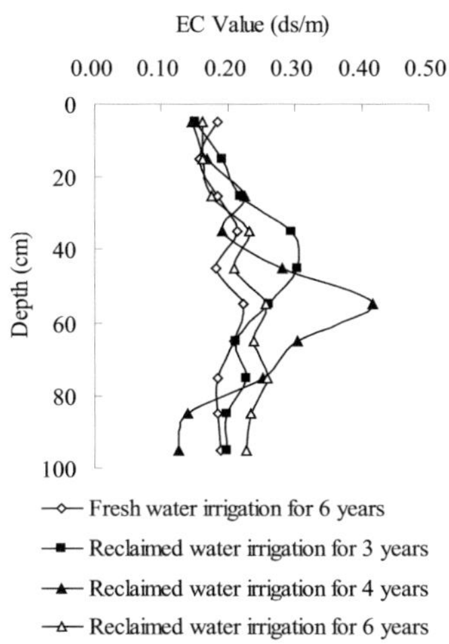

Fig.1 Soil EC versus soil depth for various treatments

Compared with bore water irrigation, soil salinity in soil profile (0~100 cm) increased slightly due to reclaimed water irrigation. But two-tailed t-test showed that there is no significant difference between bore water irrigation and reclaimed water irrigation for 3 years, 4 years, and 6 years in salt content (the significance level of 0.95). It indicated short-term reclaimed water irrigation does not cause soil salt accumulation due to precipitation leaching and not very high salt content in reclaimed water (compared with groundwater).

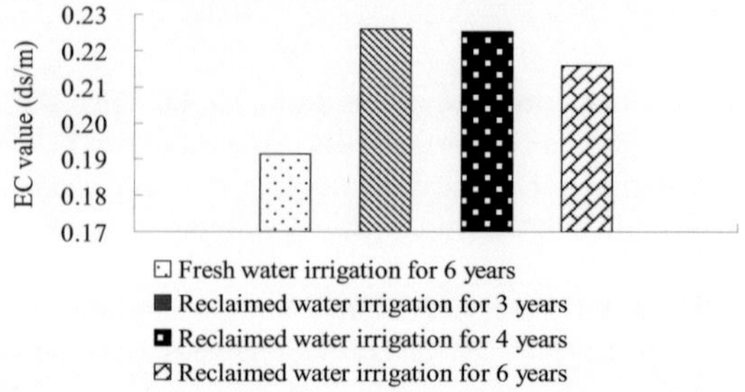

Fig.2 Average EC value of soil profile for different treatments

3.2 Soil pH

As shown in Fig 3, there was little difference of soil pH for different layers, and the soil pH for different treatment remained similar. Except for surface (0~10cm), 50~60cm, and 70~80cm, pH in other soil layers under reclaimed water irrigation are lower than bore water irrigation. This is due to the relative high organic matter content in reclaimed water. Organic acids from organic matter decomposition could lower the soil pH value. Due to the buffering capacity of the soil, the pH for different treatments varied slightly. Two-tail t-test showed that no significant difference (the significance level of 0.95) of profile 0~100cm between bore water irrigation and reclaimed water irrigation.

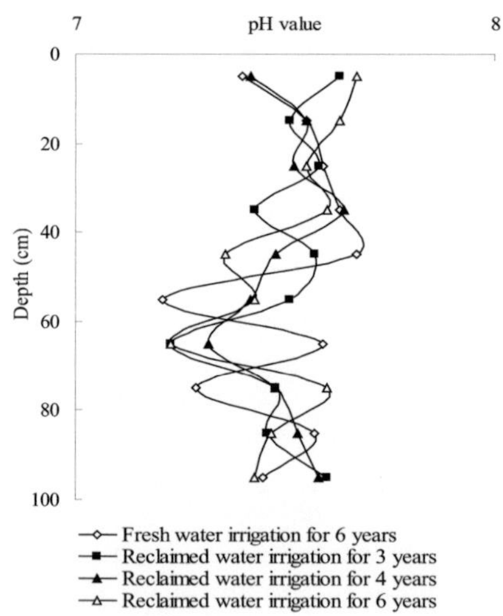

Fig.3 Variation of soil pH value versus soil depth for various treatments

3.3 Soil SAR

As Figure 4 showed, the soil SAR of reclaimed water irrigation for 3 years, 4 years, and 6 years is higher than bore water irrigation for all soil layers. For 6-year reclaimed water irrigation, the soil SAR values were higher than bore water irrigation except for the layer 20cm~30cm, with the increase of 30% to 136%. For 4-year reclaimed water irrigation, the SAR values were obviously higher in the layers 0~10cm and 40~90cm, with the growth rate of 33.7% to 104%. The soil SAR variation for 3-year reclaimed water irrigation was similar to that of 6-year reclaimed water irrigation, with differences from 3.8% ~ 151%. The characteristic of SAR for 4-year reclaimed water irrigation is different from those of 3-year and 6-year's, because 4-year reclaimed water irrigation was followed by 2-year bore water irrigation, which gradually decreased the influence of reclaimed water irrigation, the SAR approached the level of bore water irrigation. Although soil SAR for reclaimed water irrigation was higher than bore water irrigation with maximum value of $1.13(mmol/L)^{0.5}$, it is much lower than the critical soil salinazition SAR of 13 $(mmol / L)^{0.5}$ 、.

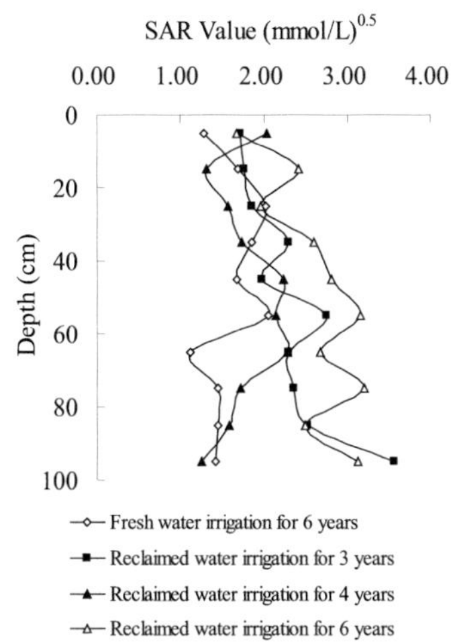

Fig 4 Variation of soil SAR value as soil depth for various treatments

Based on the above analysis, the content of Na^+ would be increased due to long-term reclaimed water irrigation, which could change the structure of soil aggregates. But after bore water leaching, the soil quality can be improved to some extent. Under the present experimental conditions, three to six years reclaimed water irrigation have not caused soil salinization or alkalization. This result is consistent with pioneer conclusion (Wang et al., 2007). However, the risk of soil salinization and alkalization for long term reclaimed water irrigation still exists.

3.4 Soil Ca^{2+}, Mg^{2+}, Na^+ and K^+ concentration

During the irrigation, ion exchange takes place between Na^+, K^+ and Ca^{2+}, Mg^{2+}, thus the various salt ion concentration in soil profile changes significantly for different treatments. Fig 5 showed the variation of Ca^{2+}、Mg^{2+}、Na^+ and K^+ concentration are 1%~3%, -3%~35%, 17%~61%, and 5%~38% of 0~100cm profile with different reclaimed water irrigation. There was no significant change for Ca^{2+}, Mg^{2+}, Na^+ concentration at 0~40cm, but significant accumulation at the layer of 40~80cm. This is because precipitation leaching caused salt moving downward and accumulated at bottom of root zone.

The Na^+ concentration for 3-year, 4-year and 6-year reclaimed water irrigation were about 52%, 17%, and 61% higher than bore water irrigation respectively. It shows that long term reclaimed water irrigation could result in Na^+ accumulation in the soil column, but certain fresh water leaching can reduce the risk of Na^+ accumulation. Two-tail t-test showed that K^+ concentration for 3-year, 4-year reclaimed water irrigation shows no significant difference with bore water irrigation (the significant level of 0.95). However, K^+ concentration for 6-year reclaimed water irrigation was significantly higher than bore water irrigation (the significant level of 0.95). It can be concluded that the K^+ accumulation is slower than Na^+ accumulation.

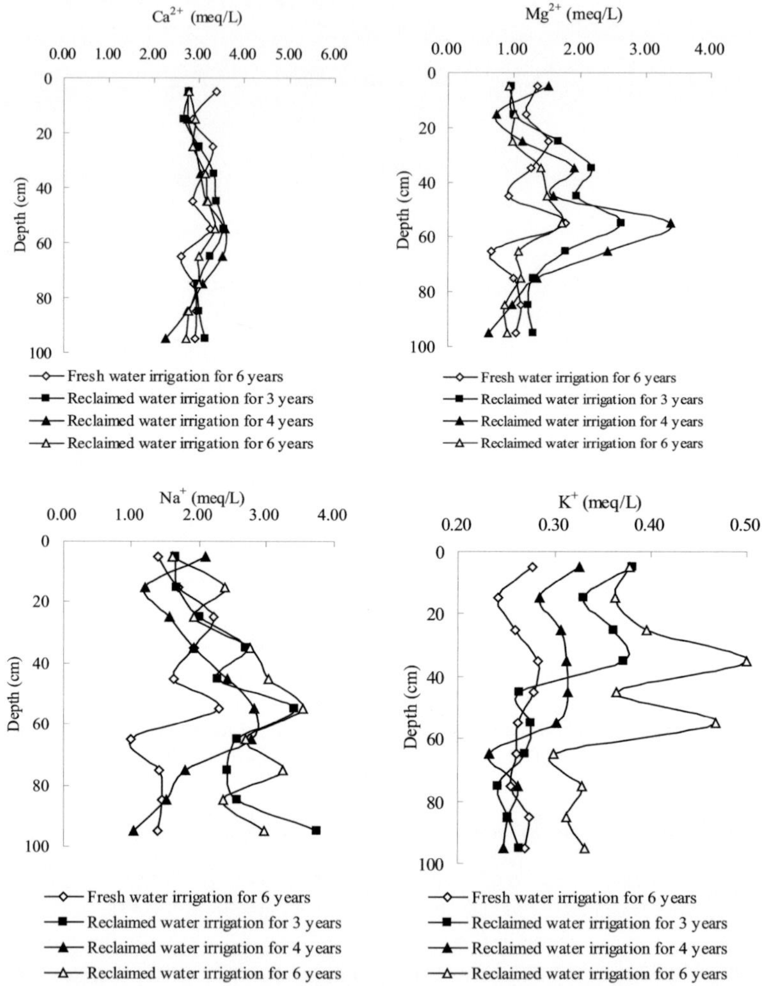

Fig 4 Effects of different treatments on averaged Ca^{2+}, Mg^{2+}, Na^+ and K^+ concentrations in soil profile

According to the experiment, long-term reclaimed water irrigation could lead to the increase of Na^+, Mg^{2+}, K^+ concentration in 0~100cm soil column. of which, Na^+, Mg^{2+} accumulates slightly in the soil depth 0~40cm and increased significantly in the soil depth 40~80cm. But the annual precipitation for the experiment period (1999-2006) only accounts for 70% of the long-term average, which weakened the effects of deep leaching. Therefore, in a normal precipitation year, the deep leaching is much more intensive and the risk of salt accumulation in soil column can be lowered. Furthermore, reasonable irrigation schedule, e.g., rotate irrigation of reclaimed water and bore water, could reduce the negative impact of reclaimed water irrigation on soil characteristics.

3. Conclusion

Reclaimed water irrigation for 3 years, 4 years and 6 years, resulted in the increase of soil salt, Na^{+}, Mg^{2+}, and K^{+} in depth of 0~100cm. Of which, the increase of Na^{+}, Mg^{2+} was not significant in depth of 0~40cm, while in depth of 40~80cm was relatively obvious. It showed that soil salt moved downward under precipitation leaching.

The SAR increased as the duration of reclaimed water irrigation, while pH showed no significant differences among treatments.

Under the experiment conditions, soil EC and SAR were maintained in an appropriate level with a tendency of decrease. Considering the mean annual precipitation for experiment period (1999--2006), which only accounts for 70% of the long-term average, deep leaching function was not fully realized. Therefore, the risk of salt accumulation due to reclaimed water irrigation is low in North China Plain.

However, it is important to establish the scientific irrigation schedule according to the precipitation and soil characteristics in different areas to avoid the negative influence with the long-term reclaimed water irrigation.

REFERENCES

1. Lazarova, V.. Role of water reuse in enhancing integrated water management in Europe[J]. Final, Report of the EU project CatchWater, ONDEO, Paris, France, 2001, p. 708.

2. Liu Changming, Chen Zhikai, The study of assessment of actuality water resourse and Supply-Demand development[M], China WaterPower Press Beijing, 2001 p.165~166.

3. Liu honglu, Wu wenyong, Shi yanwu, et al, Reclaimed wastewater use potential and collocation scheme for different industries in Beijing [J], Transactions of the CSAE, 2006, 22 (2) 289-291.

4. Polglase, P.J.; Tompkins, D.; Stewart, L.G.; Falkiner, R.A. Mineralization and leaching of nitrogen in an effluent-irrigated pine plantation[J]. Journal of Environmental Quality, v.24, p.911-920, 1995.

5. Falkiner R.A.; Polglase P.J. Transport of phosphorus through soil in an effluent-irrigated tree plantation. Australian Journal of Soil Research[J] , 1997, v.35, p.385-397.

6. Ma Min, Huang Zhanbin, Jiao zhihua, et al.. Effects of reclaimed wastewater irrigation on the quality of maize and soybean[J]. Transactions of the CSAE,, 2007,23(5):p.47~50.

7. Huang Zhanbin, Miao zhanxia, Hou liwei, et al.. Effect of Irrigation Time and Mode with Reclaimed Water on Growth and Quality of Crops[J]. Journal of Agro-Environment Science, 2007,26(6):p.2257-2261.

Dynamics of soil organic carbon and microbial activity in treated wastewater irrigated agricultural soils along soil profiles

Elisabeth Jueschke[1]
B. Marschner[1], Y. Chen[2], J. Tarchitzky[2]
[1] *Department of Soil Science/Soil Ecology, Ruhr-University Bochum, Universitaetsstr. 150, 44780 Bochum, Germany*
(E-mail: Elisabeth.Jueschke@rub.de, Bernd.Marschner@rub.de), Telephone: +49 (0)234-32-28683 [2] *Faculty of Agriculture, The Hebrew University, POB 12, Rehovot, 76100, Israel*

Abstract

Treated wastewater (TWW) is an important source for irrigation water in arid and semiarid regions and already serves as an important water source in Jordan, the Palestinian Territories and Israel. Reclaimed water still contains organic matter (OM) and various compounds that may affect microbial activity and soil quality [1]. Natural soil organic carbon (SOC) may be altered by interactions between these compounds and the soil microorganisms. This study evaluates the effects of TWW irrigation on the quality, dynamics and microbial transformations of natural SOC. Priming effects (PE) and SOC mineralization were determined to estimate the influence of TWW irrigation on SOC along soil profiles of agricultural soils in Israel and the Westbank. Clear effects for the stimulation of microbial activity could be detected and the turnover of SOC is influenced by using TWW for agricultural irrigation.

Keywords

Treated wastewater; irrigation; soil organic carbon; microbial activity; priming effects

1 Introduction

The application of treated wastewater as alternative irrigation water in agriculture is common in many regions of the world because of increasing water scarcity. Due to its high content of nutrients TWW can reduce fertilizer inputs. TWW also contains high amounts of organic matter which may stimulate microbial activity in the irrigated soils. In our study we investigated the effect of TWW irrigation on the dynamics of the soil organic matter and the microbial activity. This was investigated at three different sampling sites along the soil profiles down to 1 m depth.

As shown in previous studies, the addition of easily degradable substances can cause an increase of the microbial activity and the mineralisation of SOM will be enhanced [2, 3, 4]. This is the so-called priming effect (PE). Different mechanisms may play an important role in the activation of microorganisms. One possible mechanism, as reviewed by Kuzyakov et al. [5] is co-metabolism. Due to the availability of substrates the previous energy limitation of the microorganisms is abolished. Subsequently they are able to produce more enzymes and possibly energetically more expensive enzymes capable to degrade the soil organic matter. Thus the activity and the mineralisation of the microorganisms are obviously enhanced. Dissolved organic matter (DOM) may be the most important C-source in soils since all microbial uptake mechanisms require an aqueous environment [6]. Experiments on rhizosphere priming effects [7] and with model root exudates [8] indicate that dissolved organic substrates are important for inducing priming effects.

By irrigation with treated wastewater a considerable amount of DOM and POM (particulate organic matter) is added to the soil. Nelson et al. [9] proposed that the amount and composition of DOM in the soil solution strongly influences the microbial activity. If DOM is continuously entering the soil by effluent irrigation it will affect the soil-DOM and the microbial activity to a high extent. Priming effects may have a considerable effect on the carbon budget of soils irrigated with reclaimed wastewater.

Laboratory incubation experiments with additions of ^{14}C-labelled compounds to the soils showed that microbial activity in freshwater irrigated soils was much more stimulated by sugars or amino acids than in TWW irrigated soils. The lack of such "priming effects" [3, 4] in the TWW irrigated soils indicates that here the microorganisms are already operating at their optimal metabolic activity due to the continuous substrate inputs with soluble organic compounds from the TWW. The fact that PE are triggered continuously due to TWW irrigation may result in a decrease of SOC over long term irrigation. This could be detected at some agricultural fields by SOC measurements [10]. Therefore attention has to be drawn especially on the carbon content and quality of the used TWW for irrigation purposes.

2 Materials and Methods

The soils derived from three different sampling sites allocated in Israel and the Palestinian Authority. Soil samples were taken always from TWW irrigated sites and adjacent control plots from 6 different depths (0-10, 10-20, 20-30, 30-50, 50-70, 70-100 cm).

The soils in Israel were sampled in two orchards (Bazra grapefruit and Acco avocado) in January 2008. The third sampling site was in Wadi Faria close to Nablus on an agricultural field. The samples were bulked from 4 replicate plots in each case. Freshwater irrigated soils were taken as controls. The soils were air-dried directly after sampling, mixed and sieved to < 2 mm. Physical and chemical soil properties can be seen in Tab. 1. The total carbon was determined in a C/N-analyzer (elementar vario EL, Hanau) and the inorganic carbon in a C-analyzer (Eltra, analytikjenaAG).

	pH	EC		texture		organic C	inorganic C
			sand	silt	clay		
		dS m^{-1}	(%)	(%)	(%)	(%)	(%)
Acco	7.4	0.61	22	25	52	1.0	3.2
Bazra	7.5	0.8	82	5	12	0.5	2.4
Wadi Faria	nd	nd	11	23	66	2.2	4.5

Tab. 1: Soil properties

Soil incubations for the determination of SOM degradability were carried out with and without the addition of ^{14}C-labelled L-alanine and D-fructose (Amersham Pharmacia Biotech, Little Chalfont, England) mixed with respective unlabelled substrates D-fructose (Acros Organics, New Jersey, USA) and L-alanine (J.T. Baker B.V., Deventer, Netherlands) to obtain the required carbon concentration for the incubation experiment in order to determine the effects of these substrates on microbial activity and C-turnover. These two substrates were chosen as surrogates for easily degradable compounds present in effluent water, and had induced a priming effect in former experiments [3]. Before incubation the soil was adjusted to 60% water holding capacity and preincubated two weeks at 15°C in the dark. The soils were incubated in a Respicond apparatus (Nordgren Innovations, Bygdeå, Sweden) at 25°C in the dark and the respiration was recorded hourly by the changes of electric conductivity in 10 mL of 0.6 m KOH solution where the CO_2 is trapped [11]. The amount of $^{14}CO_2$ evolved was determined in the KOH by a liquid scintillation counting (Beckmann LS 6000 TA, Fullerton, USA).

3 Results and Discussion

The soil organic carbon (SOC) content decreases naturally with soil depth (Fig. 1). The SOC content is significantly lower in the FW irrigated soil in Bazra only in 0-10, 20-30 and 30-50 cm soil depth. In Acco the FW soil contains lower amounts of SOC in 0-10 and 70-100 cm depth. The lower SOC values for rain fed soils as reference site compared to TWW irrigated ones are more pronounced in the soil of Wadi Faria. The data show an increase of the SOC content due to TWW irrigation with depth compared to FW irrigated or rain fed soils.

Microbiological parameters as microbial activities, microbial biomass, soil organic carbon mineralization and priming effects (PE) were also investigated at all three sites and in all soil depth. Table 2 shows the significant differences between FW and TWW irrigated soils by a plus or a minus sign for each sampling site. The plus sign demonstrates a higher and the minus sign a lower value for the respective parameter in the TWW irrigated soil compared to the FW irrigated site. In the topsoils of TWW irrigated fields the general microbial activity is increased compared to the topsoils irrigated with FW. The soil respiration as well as the soil organic carbon mineralization showed a clear stimulation of the soil microbes in TWW soils. Therefore the turnover of soil organic material is affected by TWW irrigation.

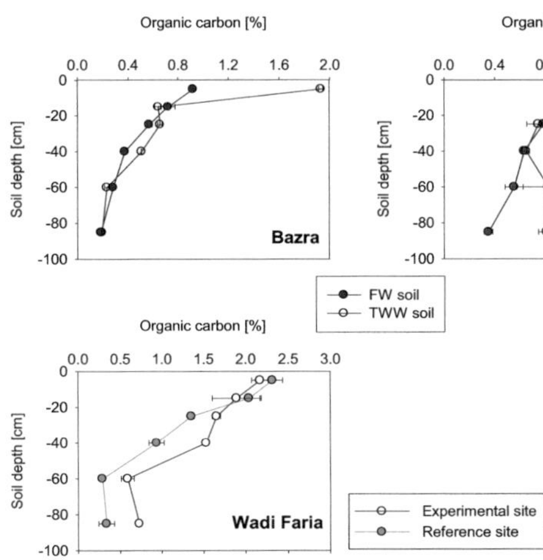

Fig. 1: Soil organic carbon content (%)

Laboratory incubation experiments with additions of ^{14}C-labelled compounds to the soils (alanine and fructose) showed that microbial activity in FW irrigated soils was much more stimulated by sugars or amino acids than in TWW irrigated soils. The lack of such "priming effects" [5, 6] in the TWW irrigated soils indicates that here the microorganisms are already operating at their optimal metabolic activity due to the continuous substrate inputs with soluble organic compounds from the TWW. The fact that PE are triggered continuously due to TWW irrigation may result in a decrease of SOC over long term irrigation.

In SOC measurements in this study no clear effects on the SOC content could be detected which might by due to heterogeneity of the soils. In former investigations by our group such a decrease was detected at some agricultural fields [10]. Therefore attention has to be drawn especially on the carbon content and quality of the used TWW for irrigation purposes.

Soil depth [cm]	C_{mic}	C_{mic}/C_{org}	$q\,CO_2$	acc. CO_2	SOC_{min}	SOC_{min} alanine	SOC_{min} fructose	PE alanine	PE fructose
0-10	+	+ -		++	++ -	++ -	++ -	+ -	-
10-20	++	+ -		+++	+++	+	++ -	- -	- - -
20-30		+	+	+	+ -	++	+ -		-
30-50	+ -	-		+++	+	- - -	- -	(2)	- (2)
50-70	+ -	-			+ -	+ -	+		+
70-100	++ -	- -	+ - -	+ -	-	-	- -	(2)	- (2)

Bazra, Acco and Wadi Fara

	no difference between TWW and FW
- - -	TWW < FW (three sampling sites)
- -	TWW < FW (two sampling sites)
-	TWW < FW (one sampling site)
+ - -	TWW < FW (two sampling sites) and TWW > FW (one sampling site)
+ -	TWW < FW (one sampling site) and TWW > FW (one sampling site)
++ -	TWW < FW (one sampling site) and TWW > FW (two sampling sites)
+	TWW > FW (one sampling site)
++	TWW > FW (two sampling sites)
+++	TWW > FW (three sampling sites)

(2) - only two of three sampling sites showed significant differences between control and treated sample, therefore only these values were calculated

Tab. 2: Overview of soil microbiological parameters at all three sampling sites.

4 Conclusions

From this study the importance of the quality of TWW for irrigation purpose is drawn to possible changes in soil organic carbon content and quality, which has a high impact on soil fertility and sustainability of the soil for agricultural use. Therefore it needs to be a priority of wastewater treatment to reduce the organic matter content if TWW should be used for irrigation of agricultural fields.

List of Literature

[1] Feigin, A., Ravina, I. and J. Shalhevet. Irrigation with treated sewage effluent. Berlin. Springer-Verlag. 1991. Comprehensive comparison of efficiency and CO2 emissions between biomass energy conversion technologies-position of supercritical water gasification in biomass technologies, Biomass and Bioenergy. 2003.

[2] De Nobili, M., Contin, M., Mondini, C. and P.C. Brookes. Soil microbial biomass is triggered into activity by trace amounts of substrate. Soil Biology and Biochemistry. 33. 1163-1170. 2001.

[3] Hamer, U. and B. Marschner. Priming effects in soils after combined and repeated substrate additions. Geoderma. 128. 38-51. 2005 a.

[4] Hamer, U. and B. Marschner. Priming effects in different soil types after addition of fructose, alanine, oxalic acid or catechol. Soil Biolology and Biochemistry. 37. 445-454. 2005 b.

[5] Kuzyakov, Y., Friedel, J.K. and K. Stahr. Review of mechanisms and quantification of priming effects. Soil Biology & Biochemistry. 32. 1485-1498. 2000.

[6] Metting, F.B. Structure and physiological ecology of soil microbial communities. Soil Microbial Ecology. New York. Marcel Dekker. 3-25. 1993.

[7] Kuzyakov, Y. Separating microbial respiration of exudates from root respiration in non-sterile soils: a comparison of four methods. Soil Biology and Biochemistry. 34. 1621-1631. 2002.

[8] Traoré, O., Groleau-Renaud, V., Plantureux, S., Tubeileh, A. and V. Bœuf-Tremblay. Effect of root mucilage and modelled root exudates on soil structure. European Journal of Soil Science. 51. 575-581. 2000.

[9] Nelson, P.N., Dictor, M.-C. and G. Soulas. Availability of organic carbon in soluble and particle-size fractions from a soil profile. Soil Biology and Biochemistry. 26. 1549-1555. 1994.

[10] Jüschke, E. Effluent Irrigation and Agricultural Soils: Effects on the Dynamics of Organic Carbon and Microbial Activity in Agricultural Soils in Israel. Schriftenreihe agrarwissenschaftliche Forschungsergebnisse, Bd. 35. Hamburg. Verlag Dr. Kovač. pp. 336. 2009.

[11] Nordgren, A. Apparatus for the continuous, long-term monitoring of soil respiration rate in large numbers of samples. Soil Biolology and Biochemistry. 20. 955-957. 1988.

Material flow management as an approach for reuse of wastewater in agriculture in arid and semi arid areas

Dr. Michael C. Birkle
Rolf Ebner
Gesellschaft für Meß- und Filtertechnik mbH, 76229 Karlsruhe, Germany,
E-mail: birkle@gmf-mbh.de, Telefon: +49(0)721/468791

Abstract

One way to reduce water scarcity in arid and semi arid areas is to use wastewater to irrigate plants. Doing this it is necessary to have look to the ingredients of the wastewater. On the one hand, some of the ingredients are plant nutrients, but if these nutrients are not in the right combination and quantity, the yield will decrease or discharge will happen. One the other hand some of the ingredients are harmful like pathogenic germs, high salts freight, residues of pharmaceuticals and heavy metals. To develop a sustainable agricultural land use with wastewater irrigation it is necessary that the input and the output of substances are in reasonable equilibrium. This means no accumulation of harmful substance. Material flow management is an approach to quantify these problems. In a first step a balance of the input and the output is considered for these substances. This approach will be tested in Egypt with the wastewater of Qena.

Catchwords

Wastewater reuse, agriculture, salinity, material flow, balance of water ingredients, nutrient flows

1 Introduction

The reuse of wastewater is of increasing importance. Because of a growing world population and an increasing economic status, the need for new methods of ecological and economic reuse of wastewater increases, in particular in countries with water scarcity problems. Connected with these problems is the fact of increasing wastewater production. Previous solutions have largely neglected either the ecological or the economic aspect. Wastewater reuse usually requires a more or less expensive processing of wastewater. The requirements of agricultural and health risk prevention should therefore be optimised together with the at least necessary wastewater treatment processes. An often intended solution of this optimisation process is the reuse of partially treated wastewater in irrigated agriculture for countries with water scarcity on one side and wastewater remediation problems on the other side. The aim is to recycle wastewater profitably and to conserve natural water resources.

To reuse – of primarily domestic – wastewater in agriculture for irrigation in arid regions, the material flows in wastewater (eg. organic carbon, nitrate and phosphate) and in agriculture have to be analysed respectively evaluated exactly and changed if necessary. In principle, wastewater contains all substances that plants need to grow, but these substances are usually not available in adequate concentrations and ratios. Thus wastewater has to be treated not as good as possible but as much as necessary, in order to be used sustainably in agriculture. This conditioning of the wastewater is primarily dependent on the type of agricultural use. On the one hand - especially from a hygienic point of view – it is important to distinguish between food or feed production and non-food-biomass for energy production. On the other hand a plant related consideration is necessary because the plant should consume all the nutrients supplied, in order to avoid an accumulation of nutrients in soil or groundwater.

Basically, only those substances should be added that are used by the plant and removed with the harvest. However, a short-termed, targeted over-supply, which is necessary for the building of a organic soil layer should be excluded from these considerations. So the conditioning of agricultural wastewater is directly dependent on the kind of plant and their cultivation, that are used in agriculture on the specific side.

In order to explore possible solutions to the mentioned aspects, since the summer of 2009 a project entitled "**S**ustainable **U**tilization of **Wastewater** in **A**rid **A**griculture **R**egions (SUWAAR)" is part of an international cooperation between Egypt and Germany in the field of wastewater management. The project partner from Germany are the Institute for Applied Material Flow Management (IfaS) University of Applied Sciences Trier and the two SMEs areal GmbH and G.M.F. mbH, and on the Egyptian side the Qena Company for Water & Wastewater (QCWW) and the South Valley University of Qena (SVU). In addition, the Holding Company for Water & Wastewater Egypt (HCWW), which is based in Cairo and responsible for a total of 13 governorates in Egypt, and the GTZ office Egypt are involved. The project is promoted by the BMBF [1] and the MHESR [2].

The project itself is carried out in Qena, a large city of around 300,000 inhabitants in Upper Egypt, which is located about 60 km downstream the Nile from the historically significant city of Luxor (Theben). In the wastewater treatment plant of Qena city about 55,000 m^3 of wastewater per day are treated, with increasing quantity. The development of a detailed regional material flow management for the reuse of wastewater for irrigation is actual in process.

2 Investigation of material flows

To balance material flows it is necessary to investigate the involved material flows. In the case of the wastewater of Qena and the use of this water in agriculture for irrigation the main components of balance are the input of substances with wastewater and corresponding output yield. To develop a fertile soil in desert it is important to build up organic soil matter. Organic soil matter will help to store nutrients and water in soil for the plants and protects soil from erosion. Therefore it is necessary to leave or to bring organic matter to the fields. The organic carbon will be converted to organic soil matter and to CO_2. It is unknown if there will be a level of too much organic soil matter, but it is known that if there is not enough organic soil matter many problems like erosion etc. will appear. Further on the degradation processes of wastewater organics and the production processes of soil organic matter in desert regions are not know exactly. The organic carbon flow in treated wastewater, respectively the soil organic matter, measured as total organic carbon, is therefore not included in the material flow management considerations. In first approximation there is no need eliminate the organic carbon in the wastewater treatment plant for agricultural application.

The wastewater of Qena is treated in the QCWW-plant in two steps. In the “preliminary step” the sludge and the floating material is mechanically separated. After passing the “trickling filters” in the second step of the plant the wastewater is actually pumped in ponds and in a forest. To make a material flow management, the ingredients of the wastewater after the “trickling filters” have to be investigated. Because of an overload of the treatment plant (planned for 25000m³/d, actually feeded with 55000 m³/d) the cleaning efficiency is not known. For some parameters there are existing reliable analyses from the QCWW. Unavailable parameters were replaced with values measured in raw water in Germany according to [3]. The parameters and the values (on molecular basis) are shown in Tab. 1 as the input concentrations.

Parameter	Source	Concentration [mg/l]
N	QCWW	33,4
P	QCWW	12,7
K	[3]	19,0
Mg	[3]	2,0
Ca	[3]	5,1
S	QCWW	11,2
Na	[3]	80,0
Cl	[3]	85,0

Tab. 1: Ingredients of Wastewater

To quantify the output material flow of the harvest information about the content and the quantity of the harvest is needed. Normally the quantity of the harvest is depending on the site conditions, the fertilization and the quantity of water. In a first step it has to be calculated how much water is needed for a yield of 100%. This was calculated with the program CROPWAT 8.0 [4]. In this program the evapotranspiration is calculated with the Penman-Monteith-formula. The input data and the evapotranspiration rate are shown in Tab 2. The temperatures, the relativ humidity and the wind speed are the average values of the years 2001 – 2003 [5]. The rainfall data are average rates in the time between 1935-1994 [5]. The sunshine hours are linear interpolated from data of [6] in [7].

According to the program CROPWAT [4] the irrigation requirement for spring-wheat in a sandy loam at 70 % irrigation efficiency is between 561 mm (planting date 01/11) and 1565 mm (planting date 01/05). The vegetation time is about 4 month. According to the federal statistical office of Germany the average yield of spring-wheat grain in Germany in the years from 2003-2007 was 5494 kg/ha. According to [8] this leads to a yield of straw of 4395 kg/ha. Most of the concentrations of the ingredients of the yield parts are taken from [8]. Only the concentration of N is taken from [9] and the concentration of Na in the grain is taken from [10]. The removal of nutrients (output) with the yield is shown in Tab. 3 under the assumtion that the dry substance is 86% of the harvestable yield [9].

	Jan	Feb	Mar	Apr	May	June	July	Aug	Sep	Oct	Nov	Dec
Tmax (°C)	23,1	24,9	31,3	35,2	39,3	42,2	42,1	42,1	40,2	35,4	29,4	26,7
Tmin (°C)	7,4	8,9	14,6	17,8	22,5	24,7	26,5	26,1	23,4	19,9	14,0	8,7
Relativ humidity (%)	49,4	41,0	35,0	26,3	23,3	25,6	28,2	28,5	30,2	37,7	46,2	50,9
Wind Speed (m/s)	4,5	5,2	6,0	6,1	6,7	6,7	6,7	7,3	6,7	5,7	4,6	4,8
Rainfall (mm)	0,2	0,3	0,4	0,1	0,2	0,0	0,0	0,0	0,1	0,8	0,7	1,1
Sunshine hours	10,6	11,3	12,0	12,7	13,4	13,8	13,6	13,0	12,3	11,6	10,8	10,5
Evapotranspiration (mm/d)	4,60	5,95	8,83	11,25	13,70	14,54	14,41	14,61	12,83	9,36	6,14	5,16

Tab. 2: Clima Data Qena (Altitude 78 m, Latitude 26,18 N, Longtude 32,73 E)

	grain [kg/ha]	straw [kg/ha]	total [kg/ha]
N			160,7
P	20,1	4,3	24,4
K	45,1	35,9	81,0
Mg	5,7	3,4	9,1
Ca	2,8	9,1	11,9
S	9,4	3,8	13,2
Na	0,4	2,6	3,0
Cl	4,7	7,6	12,3

Tab. 3 : Removal of nutrients with the yield (output)

3 Balance of Input and Output

The correlation of the concentration of nutrients in the wastewater with the irrigation requirement leads to total amount of nutrients taken to the field, called input. The balance between input and output for two different irrigation requirements is shown in Tab. 4. The ratio between input and output is a measure for the need to remove these nutrients from the wastewater. It is obvious that the most critical substances are Na and Cl. These substance should be removed nearly completely from wastewater if it will be used as irrigation water in a sustainable way and under the condition that there will be no salination of groundwater due to leaching. Because the need of irrigation water and nutrients during vegetation is not equally spread over time a layer of organic soil matter is needed to buffer at least the nutrients for the plants.

	Input (I)	Output (O)	Irrigation with 561 mm/m²			Irrigation with 1565 mm/m²		
			Input	Delta (I-O)	Ratio (I/O)	Input	Delta (I-O)	Ratio (I/O)
	[mg/l]	[kg/ha]	[kg/ha]	[kg/ha]		[kg/ha]	[kg/ha]	
N	33,4	121,1	187,4	66,3	1,5	522,7	401,6	4,3
P	12,7	24,4	71,4	47,0	2,9	199,2	174,8	8,2
K	19,0	81,0	106,6	25,6	1,3	297,4	216,4	3,7
Mg	2,0	9,1	11,2	2,1	1,2	31,3	22,2	3,4
Ca	5,1	11,9	28,6	16,7	2,4	79,8	67,9	6,7
S	11,2	13,2	62,8	49,6	4,8	175,3	162,1	13,3
Na	80,0	3,0	448,8	445,8	149,6	1252,0	1249,0	417,3
Cl	85,0	12,3	476,9	464,6	38,8	1330,3	1318,0	108,2

Tab. 4 : Balance between input with irrigation water and output with yield

4 Conclusion

The material flows quantified in Tab. 4 are founded on many assumptions and different publications. The investigation of the real material flows and there dependency on time is a substantial part of the project [1]. In general plants are able to adapt to different ranges of nutrients and different water availability leading to different output with the yield. Therefore the material flow calculations has to be done for each kind of supply and each sort of plant to be cultivated. Another substantial part of the work will be the development of optimized wastewater treatment methods to reach the aim of sustainable irrigation water from wastewater. This includes that some salts as well as hazardous components like heavy metals, pathogenic germs etc. have to be extracted from the irrigation water or have be destroyed. Thus an advanced, adequate (low cost) cleaning process is needed. In exclusive domestic wastewater heavy metals may be neglected. Essential will be primarily the removal of Na and Cl and the water disinfection. Not included in this paper is the use of the mechanically separated slugde and its content of soil particles, nutrients and organic carbon as a second input. Its inclusion in the material flows for reuse will be another substantial part in the project. The material flow management approach shows a way to calculated under what conditions wastewater reuse in agriculture can be sustainable to prevent soil and groundwater from an accumulation of substances that are not needed by plants.

Fig.1, List of Tables

List of Literature

[1] BMBF, German Federal Ministry of Education and Research; International Office, Project number EGY 08/041

[2] MHESR, Ministry of Higher and Scientific Research of the Arab Republic of Egypt, STDF-project-ID-670

[3] Koppe, P., Stozek, A., Kommunales Abwasser : Seine Inhaltsstoffe nach Herkunft, Zusammensetzung und Reaktionen im Reinigungsprozeß einschließlich Klärschlämme, 3. Auflage, Essen, 1993, Vulkan-Verlag

[4] CROPWAT 8.0, software, developed by the Land and Water Development Division of the Food and Agriculture Organization of the United Nations, downloaded from www.fao.org

[5] A. M. M. Mahmoud, Hydrogeological and hydrochemical evaluation of the area between Qena and Qus, Eastern Desert, Egypt, thesis for degree of M.Sc. in Geology at the South Valley University, Faculty of Science (Qena)

[6] Doorenbos, J. Pruitt, W. O., 1975, Guidelines for predicting crop water requirements. Irrigation and Drainage Paper 24, Food and Agriculture Organization of the United Nations, Rome

[7] Achtnich, W., Bewässerungslandbau: Agrotechnische Grundlagen der Bewässerungswirtschaft, Stuttgart, 1980, Eugen Ulmer GmbH&Co.

[8] Schilling, G., Pflanzenernährung und Düngung, Stuttgart, 2000, Ulmer Verlag

[9] Internet, http://www.lfl.bayern.de/iab/duengung/mineralisch/10536/linkurl_0_10_0_0.pdf

[10] Souci, S.W., Fachmann, W., Kraut, H., Kirchhoff, E., Der kleine Souci/Fachmann/Kraut. Lebensmitteltabelle für die Praxis, 4.Auflage, Stuttgart, 2009, Wissenschaftliche Verlagsgesellschaft mbH

The challenges of Integrated Urban Water Management – Case Study of San Luis Potosí, Central Mexico

Stefanie Kralisch,[1]
Sandra Martínez[2], Leif Wolf[3],Oscar Escolero[4]

1 Institute for Applied Geosciences (AGW), Karlsruhe Institute of Technology, kralisch@kit.edu
3) UNESCO-IHP for Latin America and the Caribbean. Montevideo
2) AGW, KIT / CSIRO Land & Water, Australia, leif.wolf@csiro.au
4) Instituto de Geología, UNAM, Mexico-City

Abstract

Intensive groundwater exploitation has permitted and maintained an accelerated urban and economic growth in (semi-)arid northern and central Mexico. Nevertheless, today abstraction outweighs recharge in most of the urban centers. This imbalance, in many cases aggravated by water quality deterioration, has become a severe limitation to further development. A promising way forward to ease pressure on conventional water sources is the adoption of a multiple resources concept with diversified supply options, leading to increased supply security considering potential climate change. However, the current lack of readily applicable systems analysis tools and the knowledge gaps especially concerning optimum utilization of subsurface water resources poses major barriers to the successful adoption of a multi-resource concept. The paper presents results of an urban water balance study for San Luis Potosi and reports on the requirement analysis with the local stakeholders. Specifically, the applicability of urban groundwater models to support the innovative water rights management in San Luis Potosi is demonstrated as an example for other areas in arid or semi-arid contexts.

Keywords

Integrated Urban Water Management, Total Water Cycle Analysis, semi-arid areas

1 Introduction

Economic growth in Mexico has been supported by indiscriminate groundwater withdrawal, especially in the dryer central and northern parts of the country. Nationwide, more than 120 aquifers are declared as overexploited. Though agriculture still is the most important water user (around 70%), in many of these aquifers urban water use holds the main share in abstraction. The urban setting, eventually with peri-urban agriculture offers a variety of management options to ease pressure on the groundwater reserves. The question is how to tap alternative water sources, in which sectors groundwater demand can be reduced or replaced, and how to optimize water allocation in this complex situation of multiple uses and multiple resources. While water scarcity and quality problems have been recognized by decision makers, uptake of new concepts to mitigate the situation is slow. Tools are needed to depict and analyze the extended urban water balance – to identify potentials and to evaluate the effects of different management scenarios on the underlying resources.

2 The urban water management challenge – Example San Luis Potosí

The Metropolitan area of San Luis Potosí is an urban agglomeration of 1.1 Mio inhabitants in the arid central part of Mexico. It constitutes an important center of regional development and concentrates economic and demographic growth in the region. San Luis Potosí eventually faces all of the complex water related problems that are or in the near future will be of relevance in many other Mexican cities. Drinking water supply is the main water user (97 of 143 Hm3/y) and depends almost solely on the underlying deep aquifer. The intensive exploitation doubles recharge and has lead to declining piezometric levels (up to 4 m/year in some city wells, 0.5-1.0 m/year in the surrounding agricultural area), with the side effects of differential land subsidence and infrastructural damage, increased pumping costs and elevated F concentrations in deep wells. On the contrary, high water losses in the drinking water supply network result in close-to-surface water levels in the shallow aquifer, causing problems with flooded basements. This shallow aquifer layer is barely exploited (5-10 Hm3/y) due to quality deterioration by leakage from sewers, the presence of industrial contaminants in some areas, and the effects of return flows from wastewater irrigation. Stormwater is a threat as well as a potential resource, as extreme rain events tend to overload the mixed drainage system. Since decades, (untreated) wastewater has been used for irrigation in peri-urban agriculture (~ 55 Hm3/y), but nowadays, with a treatment rate of already 85%, cost recovery is an issue and competition for the reclaimed water is arising from other sectors.

3 Management Potentials from a total water cycle perspective

A variety of promising measures have been proposed to tap additional water resources in SLP and to improve water use efficiency and allocation. (Tab. 1) These strategies include water imports, new reservoirs, artificial recharge, reduction of mains leakage, etc. Nevertheless, management is handled by various institutions with competing interests, approaches are often partial, information is disperse, and so far solutions have been facing considerable implementation problems.

The potential benefit of integrated strategies with effective allocation of all urban water resources in a quality-fit-for-purpose approach is high in SLP: Good quality water from intermediate aquifer layers could be reserved for drinking water use, but to be sufficient conjunctive use with surface water would have to be improved. The use of shallow groundwater and treated wastewater in the industry and service sector can be increased, to ease pressure on the deep aquifer. Therefore, the shallow and deep aquifer would have to be decoupled in local regulation, and wastewater treatment would have to be closely adapted to reuse demand. Decentralized infiltration schemes or separate sewers could be implemented to use runoff from the intensive rain events during summer for managed aquifer recharge, reducing at the same the risks of inundations that annually strike the city.

Approach	Implementation	Potential
Problem: Losses in Distribution system		
Network rehabilitation	Slow and cost intensive, current losses ~36%	Priorisation with respect to a concept of shallow aquifer use, role of urban recharge within the whole water balance
sectorisation	Planned for 2012	basis for improving monitoring and control, E-efficiency
Problem: overexploitation of deep aquifer, drawdown of aquifer levels		
Spatial optimization of wells	temporal well shut-down in case of deep abstraction cones / high F-concentrations	Model effects of different abstraciton scenarios, improved monitoring
Control & regulation	Despite of ban on well drilling continuously increasing abstraction volume	Participative approach, local regulation as instrument for allocation following fit-for-purpose criteria
Enhance use of surface water	Planned water import from adjacent basin	Control indicators: additional source to replace GW use
	Enhanced treatment capacity	Seasonal optimization of reservoir operation and GW use
	Rehabilitation of local dams foreseen,	Demonstrate synergies between drinking water supply

	not prioritized	and flood protection
Use of shallow urban aquifer	Low abstraction, lack of control & data, willingness to regulate	Model dynamics and quality. Regulative framework: decoupling of aquifers. Mixing with deep well water with high F concentrations after prior treatment
Reuse of treated WW	Reuse mainly in agriculture with low efficiency, lacking cost recovery	More efficient irrigation to liberate volumes for other uses. Infiltration of treated WW in rainy season (water banking)
Fit-for-purpose approach	Detailed database on surface and GW-rights and discharge permits (REPDA)	Local regulation system: differentiated access to water sources according to type of use and quality requirements
Water saving	Adjusted tariffs, water meters, campaigns	Setting incentives, transfer of adapted technologies (greywater, rainwater)
Problem: rising Fluoride concentrations in deep wells		
mixing	Mixing groundwater from different wells, concentrations often exceed standard	Mixing with (treated) shallow aquifer water. Potential of directed artificial recharge?
pretreatment	Pilot plants at single wells, high costs	Selective pretreatment closer to end user
tap water ≠ drinking water	due to monitoring (UASLP) lower F-concentrations in bottled water	Socio-economic assessment: drinking water from the tap? → social effects & cost-benefit analysis
Problem: flooding		
Seperate sewers, infiltration	Few stormwater sewers in critical areas. Pilot projects for artificial recharge	potential of decentralized infiltration? Set incentives selection of artificial recharge sites based on modelling
Protect upper catchments	Despite of regional planning increasing construction and sealing of piedmont zone	Protection of recharge areas = flood protection, PES

Tab. 1: Potential of Integrated Urban Water Management in the San Luis Potosí Valley.

4 Tools to tap the potential

New tools are needed, which are able to estimate the impacts of different management scenarios on the whole urban water cycle. The specific requirements of such tools have been evaluated from a technical and user perspective (Kralisch, 2009). Stakeholders (incl. the water provider, local water authority, GW-user association) seek support for their decision processes, and have well included options like aquifer recharge, stormwater management, or shallow aquifer use on their agenda. An appropriate tool should be able to facilitate integrative scenario analysis and trigger conjunctive use of multiple resources.

A groundwater flow model (Modflow) has been set up for the SLP-area by Kohn-Ledesma (2009), to estimate the impact of different abstraction scenarios on the aquifer. Fig. 1 shows on the left side the abstraction cones in a 2015 scenario, and on the right the results of a well relocation (R1-R4) towards the recharge areas - leading to less drawdown given the same overall pumping volume.Though numerical GW-modeling is the basis for evaluating the spatially differentiated effects of GW-use, the modeling attempt failed in including land use and recharge with sufficient detail. Urban recharge depends on management options in other components of the urban water cycle, like leakage reduction, stormwater management, artificial recharge with various water types, etc.

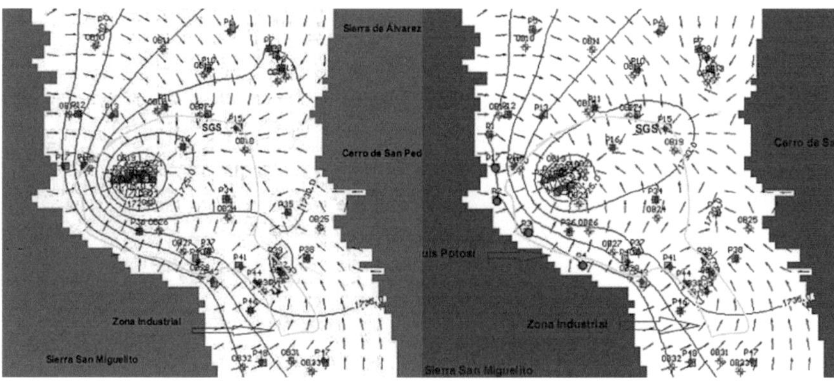

Fig. 1: Modflow results of well relocation (modified from Kohn-Ledesma, 2009)

Possible tools for integrated urban water management like Mike-Urban, UVQ, Hydroplanner, WEAP or Watercress range from process-oriented water balance and flow routing to more focused on water demand management and allocation, and differ in scale and detail (see e.g. Mitchell et al., 2007).

A first urban water balance has been set up for SLP-city, using the model UVQ (Urban Water and Quality Model, Martinez 2009). UVQ () is a conceptual water balance model with daily time steps that estimates contaminant loads and flow volumes in the urban area from sources to discharge points (Mitchell & Diaper, 2006). Different scenarios of water supply, stormwater management and sanitation were modeled. Results provided an estimation of the potential benefits from water network rehabilitation, shallow groundwater use and decentralized rainwater infiltration on the overall water balance (**Fig. 2**). While leakage reduction and shallow aquifer use lead to substantial decrease in deep well abstraction, an infiltration scenario limited to industrial and service sector neighborhoods showed little influence on the water balance.

This water balance is nevertheless limited to the urban area, not including important flows like runoff from the surrounding mountain areas on the source side (just urban runoff is estimated in 30,25 Hm3/y) and the peri-urban agriculture as important water user of ground- and wastewater and recharge component for the aquifer.

The urban water balance has to be further improved on the process level (e.g. interaction between aquifer and urban networks, stormwater volumes), but also in terms of water allocation scenarios, to close the gaps between demand and supply of different water sources at a temporally and spatially adequate scale.

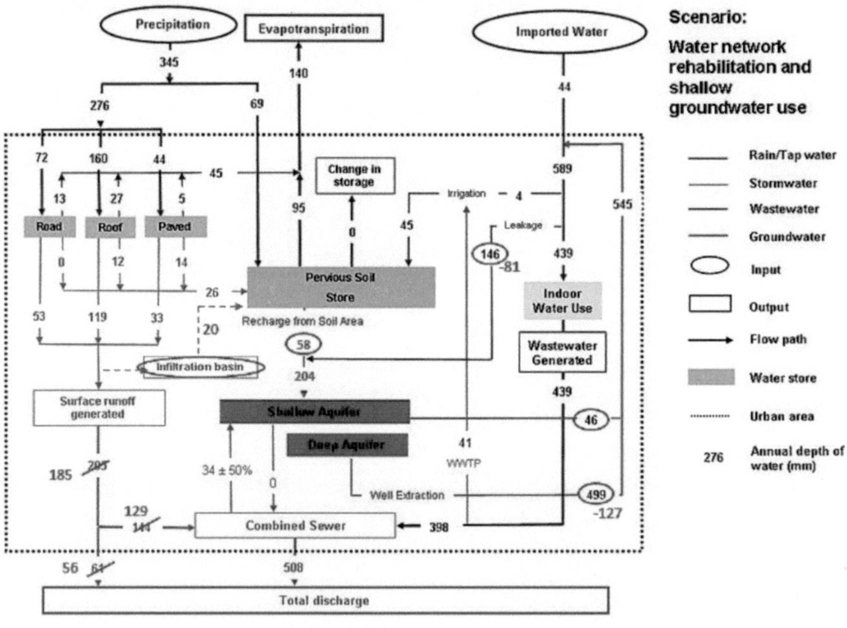

Fig. 2: Urban Water Balance modelled with UVQ by Martínez (2009). The Red circles show the changes to the baseline scenario.

Mexico has an advanced water rights system, including the respective institutional structures at national, regional and aquifer/basin scale (with active user participation), and also mechanisms for the trade of water rights among users and sectors. GW and surface water concessions as well as WW disposal permits are all included in the central water rights registry (REPDA). A tool for water allocation based on a total water cycle analysis should be able to extract data directly from this database. The aim is to convert the water rights administration system into a policy instrument for effective water allocation in a fit-for-purpose approach and to develop sustainability indicators to base water transfer decisions and incentives on the estimated impacts on the underlying resources. This goes along with the recent developments of establishing water banking schemes in Mexico. Due to the restriction of water rights trading within aquifer boundaries, these have to be made operable at local to regional scales.

5 Conclusion and Forecast

In conclusion, promising strategies and the adequate regulative framework for integrated urban water management do exist in Mexico. Now, adequate tools and monitoring concepts need to be developed. Some advances to model the urban water balance have been made in the SLP case, which show the potential of multi-resource strategies. Further work will be done in adapting and combining existing tools to improve the estimation of processes (like spatially distributed urban recharge), scaling up to include peri-urban areas and recharge zones, but also to allow for coupling resource availability and demand using water allocation models. The final goal is to provide a toolbox to (i) develop integrated and common management strategies in group environments (facilitate interinstitutional and participative planning), (ii) to evaluate the expected impacts of these strategies on the underlying water resources and on regional economy, and (iii) to be able to monitor, predict and eventually steer water transfers between multiple resources, according to environmental and development goals.

List of Illustrations

List of Tables

List of Literature

[1] Mitchell, V., Duncan, H., Inman, M., Rahilly, M., Stewart, J., Vieritz, A., Holt, P., Grant, A., Fletcher, T., Coleman, J., Maheepala, S., Sharma, A., Deletic, A., and Breen, P. State of the art review of integrated urban water models. NOVATECH 2007 - 6th International Conference on Sustainable Techniques and Strategies in Urban Water Management, Lyon, France. 2007

[2] Martínez, S., Escolero, O., Wolf, L. Quantifying the total urban water cycle as tool to integrated management. Results to the Area of San Luis Potosí. Submitted to Water Resources Management. 2009.

[3] Kralisch, S. Análisis de requerimientos para un sistema de soporte de decisiones para la gestión integrada de agua subterránea: Estudio de caso Valle de San Luis Potosí. Master Thesis in the Postgraduate Program of Earth Sciences at the Universidad Nacional Autónoma de México, UNAM. 2009

[4] Kohn-Ledesma, I.A.: Modelo de Flujo del Acuifero de San Luis Potosí, SLP. Unpublished Masterthesis. Instituto de Geofísica, Universidad Nacional Autónoma de México. 2009.

[5] Mitchell, V.G., Diaper, C.: Simulating the urban water and contaminant cycle. Environmental Modelling & Software, Volume 21, Issue 1, January 2006, Pages 129-134. 2006.

Abstract for IWRM Congress, Karlsruhe, Nov. 24-25, 2010

Institutional Prerequisites for Decentralized Wastewater Treatment and Reuse – a Case Study of Jordan

Ines Dombrowsky

Since the mid 1990s a paradigm shift has been advocated to move from centralized to a decentralized wastewater treatment and reuse. Such decentralized approaches are expected to serve remote areas at less cost, to close local cycles of water, energy and biomass flows, and to enhance adaptability under conditions of global change. However, a number of studies show that it proves difficult to implement wastewater reuse in general, and there are indications that decentralized solutions are even less prevalent. It is argued that institutional constraints are an important factor inhibiting wastewater reuse. However, there are only few studies addressing governance issues in detail and even less are specifically concerned with decentralized wastewater treatment and reuse (WWT&R). This begs the question what the institutional prerequisites for decentralized approaches are: Do 'decentralized' physical solutions also require a decentralized decision-making and management approach?

In order to address this question a case study of Jordan is presented. Jordan is a developing country, suffering from extreme water scarcity with a centralized water governance system in place. Water reuse has long been advocated as a major management option, but almost no decentralized approaches have been implemented so far. Therefore, it was analyzed whether decentralized WWT&R is principally institutionally feasible and whether potential alternative operators have incentives to pursue decentralized WWT&R. We assumed that decentralized plants can be set up and operated sustainably, if potential operating entities, such as the water authority, regional utilities, private companies, municipalities, municipality-owned utilities, NGOs or community-based initiatives have the legal competences to initiate and such plants and to recover their costs. Methodologically the case study included (1) an assessment of the formal institutional framework conditions in form of a review of legal and policy documents, (2) a review of the current practice of wastewater reuse in Jordan and (3) an analysis of stakeholder perceptions on decentralized WWT&R.

The paper finds that in Jordan in principal decentralized WWT&R has always been possible, but it did not happen under a strictly centralized water governance regime that was in place before the year 2001. A potential explanation is that from an economic point of view it was rational to connect large urban conglomerations first, and to do so in a centralized manner. Furthermore Jordan's Wastewater Management Policy of 1998 explicitly stated that centralized WWTPs would be pursued, even in rural areas.

Modest steps towards privatization and decentralization and legal reforms in the year 2001 improved the institutional feasibility of decentralized WWT&R to some extent. But still, the decision-making process remains cumbersome, as any transfer of operational responsibilities to entities other than the central water authority and any flexible pricing mechanism require a cabinet decision. One consequence of the current decision-making and financing competences is that none of the potential operating entities has particular incentives to pursue decentralized WWT&R: Property owners not connected to the central sewage system are affected by overflowing cesspits or high tanker transportation costs, but already pay part of their property tax for wastewater services and expect the service from the government. Municipalities are also affected from the nuisances of overflowing cesspits, but are financially weak, have limited capacity and no formal responsibility. The concept of municipality-owned utilities does not exist in Jordan so far. Private companies need to be granted permission for operation by the cabinet, and have little leeway to influence the tariffication process for cost recovery and profit generation. NGOs are not active in the wastewater sector in Jordan. Regional utilities may play a role, but are by and large still under creation. For the central water authority, it has been rational to serve urban areas first. Hence, so far, the institutional setting in Jordan did not particularly encourage affected actors to take action and de facto lacked flexible pricing arrangements for cost-recovery. In that sense, the centralized governance regime was not conducive towards decentralized WWT&R.

Still, institutional reforms envisioned in Jordan's 2009 Water Strategy possibly provide a window of opportunity to improve the conditions for decentralized WWT&R. First, the strategy reflects a drastic policy shift from centralized to decentralized approaches in semi-urban and rural areas. Second, the strategy envisions far reaching institutional reforms including the establishment of a regulatory authority and transformation of governorate water administrations into utilities. Third, it pushes for the implementation of full cost recovery and for regionally differentiated prices, and envisions that private and public operators shall be able to set tariffs for their customers, approved by the regulatory authority. If implemented, the envisioned reforms of the water sector could contribute towards a significant simplification of procedures. One open question remains whether and to what extend municipalities would be able to initiate projects for decentralized WWT&R as they usually have the most intimate knowledge of the local conditions, but no formal decision-making competences with respect to wastewater to date.

Community based approach towards water quality restoration and pollution prevention of the Flores Creek (Uruguay)

C.T. Avellán*
D. Rousseau**, P. Lens**
** Ludwig Maximilian Universität, Department of Geography, Chair of Geography and remote sensing, Luisenstr. 37, 80333 Munich, Germany, E-mail: t.avellan@iggf.geo.uni-muenchen.de, Telephone: +49(0)89 2180-6689*
***UNESCO-IHE, Department of Environmental Resources, Westvest 7, Delft, The Netherlands, Email:d.rousseau@unesco-ihe.org, p.lens@uensco-ihe.org, Telephone: +31(0)15 215-1783*

Abstract, Kurzfassung
The micro-river basin of the Flores Creek is a typical example of the surface water conditions within the dairy farm belt around the capital of Montevideo. It serves simultaneously as the drinking water supply of the mid-sized town of Libertad and irrigation source of the surrounding rural area while receiving raw or primary treated municipal and dairy sewage thus suffering from chronic water quality and quantity deviations.
To mitigate this issue, an integrated river basin project based on technical and community approaches was initiated in 2008, raising awareness of the problems to relevant stakeholders, identifying and executing the practically possible solutions.
The main implemented measures were:
a) design and construction of a constructed wetland for one of the larger dairy farms as a measure to improve the water quality of the effluent,
b) active participation of the local rural school children in the water quality/level monitoring,
c) integration of the collected data into GIS based software as to enhance the visualization of the problems and to facilitate its integration into modeling programs,
d) increased monitoring of the water quality, the flow volume and the biodiversity.
This study represents an initial approach to Integrated River Basin Management in Uruguay where the key emphasis was laid on the active inclusion of the rural and urban population in the process in order to raise awareness of water and ecosystem problems in the area. Apart from enhancing the water quality in certain areas of the creek the main result of this project was the variety of stakeholders that was reached and that took a serious and, hopefully, long lasting interest in the matter of surface water pollution and ecological remediation.

Keywords, Stichworte
water quality, constructed wetlands, SWAT, dairy farms, Santa Lucia River

1 Introduction, Einleitung

Integrated water resource management (IWRM) and its principles were formulated in Dublin and Rio de Janeiro in 1992 and further improved at the World Summit on Sustainable Development (1992) and the Millennium Assembly (2000) and are defined by the following items:

1. Freshwater is a finite and vulnerable resource, essential to sustain life, development and the environment.

2. Water development and management should be based on a participatory approach involving users, planners and policy makers at all levels.
3. Women play a central part in the provision, management and safeguarding of water.
4. Water is a public good and has a social and economic value in all its competing uses.

Since then, a vast amount of projects on small and even large areas (often catchments) have been implemented trying to improve the livelihoods and ensure sustainability within this new geopolitical unit of a catchment (Molle 2009). Central to the principles is the aspect of participation of stakeholders, as well as finding the balance of competing needs of social, economic and environmental issues.

In Uruguay, one of the smaller countries in South America (176.000 km^2; population 3 million people), the new government entity DINASA (Direccion nacional de aguas y saneamiento) was created in 2007, in part, to implement a top-down design for the future implementation of river basin committees (pers. com. with Mr. Genta in 2008, director of DINASA).

So far IWRM principles have been put into practice in the coastal marine regions in the last few years as a joint effort between the Universidad de la Republica and several NGO's (Trimble et al. 2010). The Rio de la Plata basin has a long history of trans-boundary management between Uruguay and Argentina (i.e. the EcoPlata project) - lately being the cause of great tumult as a litigation was carried out in front of the international court of The Hague concerning the establishment of a paper pulp mill on the Uruguayan side of the Rio Uruguay. The Guarani aquifer has also been the target of transnational scientific and economic cooperation (Uruguay, Paraguay, Argentina and Brazil), since it is the largest aquifer in South America. However, none of the inland waters, which are used for urban and rural needs, have so far been targeted as study objects.

Presently, we intended to implement IWRM/IRBM measures on a typical small catchment in the dairy belt region that surrounds the capital Montevideo. The dairy sector plays an important economic role in Uruguay and a trend of increased milk production has been occurring for the past decade (1980: 600000 cows- 2008: < 750000, MGAP 2009). This goes hand in hand with the increased use of fertilizers, pesticides, raw sewage disposal and irrigation (Conaprole, DINAMA, 2007). At the same time, these creeks often lay within reach of small urban settlements that use the surface waters as dumping grounds (pers. observation) and the subsurface waters as drinking water source. Although almost 50 % of the inland urban population is connected to a public sewer system (OSE 2010) these often lack maintenance and thus do not serve the purpose of protecting from (sub)surface water pollution. Therefore, a typical creek in southwestern Uruguay will be exposed to various types of pressure, mainly excess of nutrients from rural and urban sources and fluctuating water levels due to irrigation.

Our goal was to explore the feasibility of the implementation of IWRM principle in a small catchment at the rural-urban interface, where opposing needs and viewpoints are present (i.e. water for drinking vs. irrigation vs. ecological minimal flow).

2 Text, Textkörper

Study area:

The Flores Creek watershed (ca. 80 km^2) is a characteristic example of the ones found within the dairy belt, being lined with 6 medium sized dairy farms (up to 200 cows in milking) as well as the town of Libertad (10000 inhabitants) at its headwater. The creek is situated in the Southwestern part of Uruguay within the department of San José, 60 kms north of Uruguay's capital Montevideo. Moreover, it will shortly become part of the Santa Lucia Wetlands National Park (HSL 2009), where water and ecosystem quality will be a priority.

Moreover, the creek and its adjacent wetland areas are an important recharge element for the underlying Raigon aquifer since the infiltration rates through the surrounding limestone soil is slow and the aquifer is largely rainwater fed (RLA 2005). High nitrogen rates have already been confirmed in the aquifer and correlate strongly with anthropogenic pollution sources (RLA 2005).

Water quality sampling since 2003 show that both the rural as well as the urban areas impact the surface water negatively, reducing oxygen levels to 0 in some areas and thus affecting aquatic life (Fig. 1) (Avellan 2004).

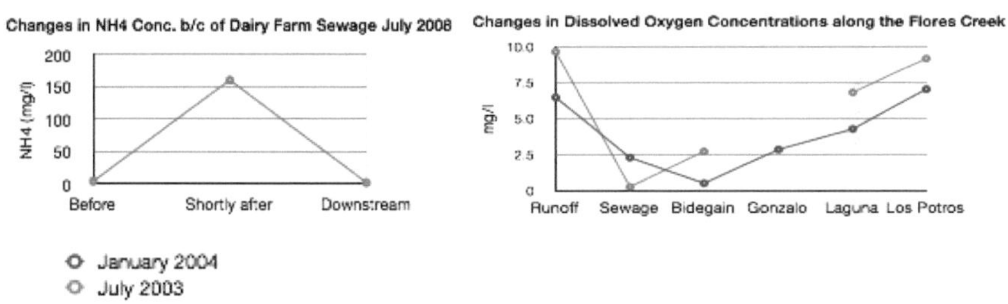

Fig. 1, Abb. 1: Right: Influence of untreated dairy farm sewage on the ammonium concentration of the Canada Clara (July 2008) and Left: Dissolved Oxygen concentrations along the Flores Creek in 2003 and 2004

Materials & Methods:

Stakeholder identification

From November 2007 onwards, we actively engaged in dialogues with potential interest groups. In a first approach, we contacted the dairy farmers of the creek unearthing their main concerns in view of the environmental degradation and potential solutions they had thought of, heard of or implemented.

Subsequent to this first meeting, we produced a list of potential stakeholders, using the information gathered from the farmers, from NGO's (Friends of the Earth, the local Rotary Club) and from a consultant working for the German-Uruguayan chamber of commerce in the field of ecological management (Johannes Frommann). We then started to address the identified stakeholders to sense their interest for the matters of integrated catchment management and water pollution at the rural-urban interface.

Problem identification and solving

Several other meetings that were open to the general public followed throughout 2008 (in the town hall of Libertad), in which the problems were identified and potential solutions were presented and explained.

After thorough discussions and feasibility studies, the following 4 issues were addressed:

- innovative and sustainable solution for the dairy farm waste waters --> pilot constructed wetland for one of the dairy parlors
- ongoing water quality and ecological monitoring
- integration of the locals --> environmental education with the school children
- use of IT tools to aid in the visualization of the problems and model future scenarios

Results& discussion:

Stakeholder identification

We produced an extensive list (Table 1) of interest groups and identified their potential input (kind of help) that could be expected.

Tab. 1: List of potential stakeholders and interest groups

Country of origin	Type of institution	Name of insititution	Kind of help
Uruguay	University	UdelR Limnology	water analysis equipment
			Laboratory space for water analysis
		UdelR Ingenieria quimica	design and installation of bioreactors
			potentially water analysis
		UdelR Ing. Micorbiologica	carry out microbiological soil analysis
			lab equipement for bacterial analysis (soil &water?)
		UdelR Agronomia	realtionship crop-soil-water
		UdelR Veterinaria	sanitary hazards
		UdelR Ing. Civil	design and installation of primary sewage treatment facilities, lagoons
		RedTema	aid in environmental education, contacts
	Scientific institute	Clemente Estable	carry out microbiological soil analysis
		INIA	Experience with agricultural experimental work
	NGO	CEUTA	know how in wetland design for low income single family homes
		Rotary International	small scale projects (i.e. environmental education)
		Red de Amigos de la Tierra	environmental education programs
	Governmental institution	APL	contact to farmers
			experience with installation of traditional sewage systems for dairy farms
			experience with governmental funding agencies
		IMSJ	water quality equipment, political pressure
		MGAP-RENARE	Water and soil quality testing; in charge of irrigation permits
		OSE	responsible of sewage treatment facility Libertad
		DINAMA	project of sewage treatment facilities of dairy farms with Conaprole
		DINASA	IWRM implementing gov. body
		Conaprole	responsible of Libertad milk farmers
		SNAP	In charge of protected areas
	Funding agency	CSIC	funding for university projects
		PPR	funding for farmers, conservation of biodiversity
	Farmers	Milk Farmers	Sewage, space, unskilled &skilled workers, minor funds
	Press	Diario local de libertad	Awareness raising, Publication of interesting results etc.
		Radio El espectador	Awareness raising, most critical radio station,

Germany	University	UFZ	molecular ecology (Argentina BMBF fund)
		ILZ	aid in finding funding agencies
	Company	Subterra	know how in constructed wetlands design and construction in the world
			interest in implementing techniques in Uruguay with paying clients
Chile	Company	Wetlands SA	know how in wetland design in Chile
Italy	NGO	Rotary International	know how in sewage treatment facilities
			funds
USA		Engineers w/o frontiers	know how in sewage treatment facilities
	NGO	WWMD	equipment for kids water quality testing
Netherlands	Company	Ecofyt	design and installation of wetlands, low cost, high;y enthusiastic

From the previously identified stakeholders all showed interests in working with us. Many applications for work agreements were undertaken (i.e. with Rotary International using contacts in Italy, Germany and Uruguay; or with Engineers without Frontiers in the US), however, ultimately only three concrete working assignments were affirmed.

a) The local municipality (Intendencia de San Jose)
   - ongoing water quality monitoring of the creek
   - use of political pressure in order to force the water company to enhance the municipal sewage treatment facility.

b) The Ministry for Agriculture, (Water and Soil section and their laboratory),
   - water quality test for the compliance with irrigation standards
   - soil humidity and water absorption capacity of the surrounding soils.

c) The Universidad de la Republica (Seccion de limnologia)
   - macro - invertebrate species composition change in the past 5 years
   - monitor further changes which may arise due to the implementation of measures that shall enhance the water quality.

Lastly, we were also able to integrate some environmental education campaigns due to the voluntary help by staff from Red de Amigos de la Tierra as well as through donations of water quality kits by the World Water Monitoring Campaign.

Examples of problem solving

Constructed wetland for one of the larger dairy farms

The Bidegain family is one of the six mid sized dairy farms of the Flores Creek catchment and has traditionally milked cows for the past 100 years here. In their dairy parlor they milk approx. 200 cows twice a day and use roughly 16 m^2 of freshwater a day for washing the cows and rinsing the parlor. About ten years ago they had constructed a double lagoon system as a means of treating the waste water before it entered the Cañada Clara, a tributary to the Flores Creek. However, this system had filled up recently due to the lack of maintenance and they therefore decided to build a new system.

Using an online calculation tool developed by local civil engineers and agronomists we were able to calculate the dimensions needed for a sediment trap as well as a double lagoon system (MGAP PPR 2009).

Following this, we designed a free water surface constructed wetland of 736 m^2 planted with *Scirpus americanus* (see figure 2) (according to methods proposed in Kadlec&Knight, 1996).

Fig. 2, Abb. 2: Images of the construction of the wetland and the vegetated areas.

Environmental education campaigns

School children, especially from the rural school within the watershed, were actively engaged in water quality and ecology testing primarily through the water quality kits from the World Water Monitoring Day Campaign from WEF and IWA (figure 3). But also water quantity measures were carried out daily by a group of students on their way to school.

Fig. 3, Abb. 3: Left: Workshop on waste management with the children of 3rd grade of a school in Libertad in cowork with REDES- Friends of the Earth; Right: Water quality monitoring with the rural school children using the donated kits of the WWMD campaign.

3 Charts, Tables, Tabellen

4 Formulas, Mathematik-Modus

5 Conclusion and Forecast, Ergebnis und Ausblick

This project is one of the first intents in Uruguay to apply integrated river basin management techniques on inland freshwater streams, bringing together rural and urban populations. We were able to bring farmers, NGOs, academics and governmental officials to the same table and successfully carry out innovative and integrative solutions to some of the issues. Nevertheless, coordinated work between these diverse stakeholders was often tiresome and unproductive, leaving many problems unsolved, such as the pressing issue of surface water pollution due to urban sewage. We hope that this work will serve as a foundation for future works in that area.

List of Illustrations, Abbildungsverzeichnis

List of tables, Tabellenverzeichnis

List of literature, Literaturverzeichnis

[1] Avellan, T. (2004). Developmental instability in Polygonum punctatum and Sagittaria montevidensis in a sewage impacted stream. Biological Sciences. Detroit, Wayne State University. Master of Sciences.
[2] CONAPROLE, DINAMA, IMFIA, (April 2008) Evaluación de los sistemas de tratamiento de tambos
[3] HSL (Junio 2009), Propuesta de proyecto de selección y delimitación del área de humedales del Santa Lucía para su ingerso al sistema nacional de áreas protegidas, SNAP DINAMA
[4] JICA (2005). Gestion de calidad del agua. MVOTMA. Montevideo: 4.
[5] Kadlec&Knight, (1996) Treatment Wetlands, CRC press, pp. 555, 643
[6] MGAP PPR (2009) (http://www.cebra.com.uy/presponsable/adjuntos/2008/06/manual-efluentes.pdf), (http://www.chipsadventure.com/SCMET/scmetest.html (accessed April 2009)
[7] MGAP. (2009) From http://www.mgap.gub.uy/portal/hgxpp001.aspx?7,5,85,O,S,0,MNU;E;27;5;MNU;, (accessed December 2009)
[8] Molle, F. (2009). "River-basin planning and management: The social life of a concept." Geoforum 40(3): 484-494.
[9] OSE (2010) From http://www.ose.com.uy/s_indicadores_cobertura.html; accessed August 2010
[10] RLA (2005) Informe final sobre el proyecto de cooperación técnica "Manejo integrado y sostenible de las aguas subterráneas en América Latina" – "Gestión sostenible del Acuífero Raigón, Uruguay", MGAP-RENARE
[11] Trimble, M.; Rios, M.; Passadore, C. (2010) Ecosistemas costeros uruguayos: Un guia para su conocimiento; Proyecto Arenas, Aver Aves, Cetaceos Uruguay, Karumbe

Optimal Water Allocation in the Middle Olifants Sub-Basin of South Africa: How Can Water Markets Increase Water Use Efficiencies?

Teresa Walter
UN-Water Decade Programme on Capacity Development (UNW-DPC), UN Campus, 53113 Bonn, Germany, e-mail: walter@unwater.unu.edu, tel.: +49 2288150663

Abstract

This study aims to investigate possible benefit gains from water trade between alternative sectors in the Middle Olifants sub-basin of South Africa. The research area is characterized by inelastic water supply, increasing competition for scarce water resources and a poor water demand management, leading to an imbalance of water supply and demand. Accordingly, this study seeks to estimate water demands of the two production sectors irrigation and mining. The determined inverse water demand functions of the two sectors together with water supply figures represent the basic elements in the water allocation model, where benefits from water use are maximized. Three different scenarios are modeled and results are analyzed. By reallocating water from low-value uses to high-value uses benefits are maximized and efficiencies of water use are improved. In the third scenario trade is allowed between all tertiary catchments in the study area, leading to one equilibrium price in the whole region of 1.819. Model results give necessary insights to sectoral water demands and economic aspects in the study area, and they can be of support in the decision-making of appropriate policies.

Keywords

IWRM, Middle Olifants, water allocation, water efficiency, agriculture, mining

1 Introduction

Many countries of the world have to face the challenge of rapidly growing water demands. As South Africa is one of the various countries in the world experiencing water shortages, this research focuses on an extremely water scarce region in the north east of the country; the Middle Olifants sub-basin. The research area is one of the four sub-basins of the Olifants Water Management Area, which is the third most water stressed basin in South Africa. The Middle Olifants is characterized by its high water scarcity (500mm/year), high poverty and population rates, as well as unequal water distribution. A big part of the former homelands[1] is located in the Middle Olifants and many people still do not have access to water and different users are competing for the scarce water resources, forcing the South African Government to replace the emphasis on increasing water supply by a strategy of demand management [2]. Accordingly determining water demand in different water using sectors has become a major task in developing an Integrated Water Resources Management (IWRM). Although main water users include households, agriculture and the mining sector, this study will only focus on water uses in the production process, that is the agricultural and the mining sectors.[2] Results of the study will give useful information on the implications of water allocation practices and will be further used in an optimization model also including rural and urban households.

2 Model Formulation

2.1 Water Demand Functions and Elasticities

Using primary data water demand functions and their according inverse water demand functions in the agricultural and the mining sectors were estimated for each of the five tertiary catchments.[3] In irrigation a linear logarithmic function and in the mining sector a double-log linear function were found to show the best fit. Thereby, the various coefficients vary for each tertiary catchment. Using equation 1 water price elasticities are calculated with Q_w as the water quantity and P_w as the water price [1]:

$$\eta_{WW} = \frac{\partial Q_W}{\partial P_W} * \frac{P_W}{Q_W} \qquad [1]$$

Derived water demand functions in different uses show varying water price elasticities with the most inelastic water demand for irrigation. Tab. 1 shows the respective water price elasticities in alternate uses for each tertiary catchment.

---

[1]Homeland was territory set aside for black inhabitants of South Africa as part of the apartheid policy.

[2]Since the study is embedded in a project framework it only represents one part of the whole water demand system and water demand on the household level is analyzed in another study.

[3] Detailed calculations and methodologies used are available from the author by request.

<table>
<tr><th>Sector</th><th>B31</th><th>B32</th><th>B51</th><th>B52</th><th>B71</th></tr>
<tr><td>Irrigation</td><td>-0.198</td><td>-0.192</td><td>-0.181</td><td></td><td></td></tr>
<tr><td rowspan="2">Mining</td><td rowspan="2">-0.954</td><td rowspan="2"></td><td rowspan="2"></td><td>-0.774</td><td>-0.897</td></tr>
<tr><td>-0.872</td><td>-0.767</td></tr>
</table>

Tab. 1: Own water price elasticities per sector and catchment

For irrigation elasticities are calculated per tertiary catchment whereas in the mining sector, water price elasticities are calculated for each mine (one mine is located in catchment B31 and two mines are located in catchments B52 and B71 each). For all sectors, elasticities are calculated at average consumption levels and prices.

2.2 The Water Allocation Model

Typically prices reflect market scarcity as well as the equilibrium between demand and transportation and supply costs [7]. If this is not the case, there is market imperfection and that impedes the market from adjusting to changes in quantities demanded or to changes in costs of supply. Hence, political and institutional mechanisms must be established to ensure an efficient water use. Although the National Water Act (NWA) provides the constitutional framework for water markets and technical water research has received priority in the past, little is known about the economics of water use and impacts of alternative water policies [5]. With a nonlinear model the potential of reallocation of water rights is simulated by deterministic mathematical optimization.

The optimization model is programmed in GAMS, where the different sets and subsets for catchments q (B31, B32, B51, B52, B71), and water uses u consisting of irrigation (i) and mining (m) are formulated. Further sets include the parameters of the inverse water demand functions. The objective function maximizes total benefits TB as the sum of the areas under the inverse demand curves [4, 6]. Accordingly, considering agricultural and water demands in the mining sector, the model can be formulated as:

$$\max TB = \sum_{q} \sum_{u} \left[\int_{LL}^{UL} F_{qu}(Q_w) dQ_w + CP_{qu} Q_{\min qu} \right] \qquad [2]$$

where F_{qu} (Q_w) are the inverse functions for water consumption Q_w. Benefits B from each water usage u in each catchment q can be calculated as the area under the inverse demand functions from the upper limit Q_w to the minimum quantity demanded Q_{min} at the choke price CP.[4] Q_w is a decision variable in the model, whereas the minimum quantity demanded is a parameter calculated with the water demand function and the choke price inserted.

Equations 3 and 4 respectively show the benefit functions for the agricultural and mining sectors with 'a' and 'b' as the coefficients of the inverse water demand functions.

$$\max B_{qi} = \left[-a_{qi} * e^{(b_{qi} - Q_{wqi})/a_{qi}} \right] - \left[-a_{qi} * e^{(b_{qi} - Q_{\min qi})/a_{qi}} \right] + CP_{qi} * Q_{\min qi} \qquad [3]$$

$$\max B_{qm} = a_{qm}^{1/b_{qm}} \left[Q_{wqm}^{1-1/b_{qm}} - Q_{\min qm}^{1-1/b_{qm}} \right] / [1 - 1/b_{qm}] + CP_{qm} * Q_{\min qm} \qquad [4]$$

Since water is not endlessly available, total water supply per catchment S_q represents the major restriction in the model.

$$\sum_q \sum_u Q_{wqu} \leq \sum_q S_q \qquad [5]$$

3 Results

The model is estimated for the three Scenarios: (1) the status quo situation, (2) intraregional trade and (3) interregional trade. In the status quo situation water demands are fixed to current consumption levels, implying that water is used unsustainably, and according benefits from water use are determined. In the second scenario trade between agriculture and mining is only allowed within tertiary catchments[5], while water can be traded in the whole study area in the third scenario.

[4] The choke price defines the backstop price, at which quantity demanded becomes exactly zero or approximately zero.

[5] Based on hydrological features, the Middle Olifants basin is sub-divided into five tertiary catchments; B31, B32, B51, B52 and B71.

In Scenarios 2 and 3 available water quantities are reduced according to the NWA, which states that the ecological reserve and human consumption have to be considered and given priority before water is allocated to agriculture and industries. Total water supply decreased by approximately 52% with highest reductions in catchment B31. From Tab. 2 it can be seen that water allocation decreases by around 53% in agriculture while in the mining sector water allocation is only decreased by 19% in Scenario 2. Fig. 1 shows that total benefits per year decrease from 2582 Million Rand in Scenario 1 to 2344 Million Rand in Scenario 2 – a decrease of 9%. Compared to high water supply decreases this decrease in benefits is relatively low, strengthening the assumption of positive impacts of water trade on water use efficiencies.

In Scenario 3 total water supply is limited to the sum of available water amounts in each tertiary catchment. The equilibrium price is with 1.819 Rand/m^3 relatively high, compared to current water prices. The high decrease of water allocation in the mining sector has its reason in different water supply quantities. In tertiary catchments where water is mostly used for mining, supply levels are relatively high with resulting low water values, whereas especially in catchment B31 water is very scarce and its value is therefore relatively high. When trade is allowed between catchments water is allocated from low value uses to high value uses, hence especially from catchment B71 to B31.

Water Use per Catchment		Scenario 1		Scenario 2		Scenario 3	
		P_w	Q_w (in ’00 000)	P_w	Q_w (in ’00 000)	P_w	Q_w (in ’00 000)
Irrigation							
B31		0.07	1769	7.563	401	1.819	817
B32		0.07	1664	0.567	1119	1.819	819
B51		0.07	506	0.895	324	1.819	273
Mining							
B31		0.966	3	7.563	0.44	1.819	1.7
B52	Mine1	0.008	67	0.086	9	1.819	0.69
	Mine2	0.821	9	0.086	51	1.819	5
B71	Mine1	0.891	11	0.661	14	1.819	6
	Mine2	0.35	8	0.661	5	1.819	2

Tab. 2: Equilibrium Prices in Rand/m^3 and Quantities in m^3/year per Tertiary Catchment for Trade Scenarios

In Scenario 3 total benefits shown in Fig. 1 are 2470 Million Rand – a 4% decrease from Scenario 1 and 5% rise from Scenario 2, implying that decreased water supplies can best be compensated when interregional trade is allowed.

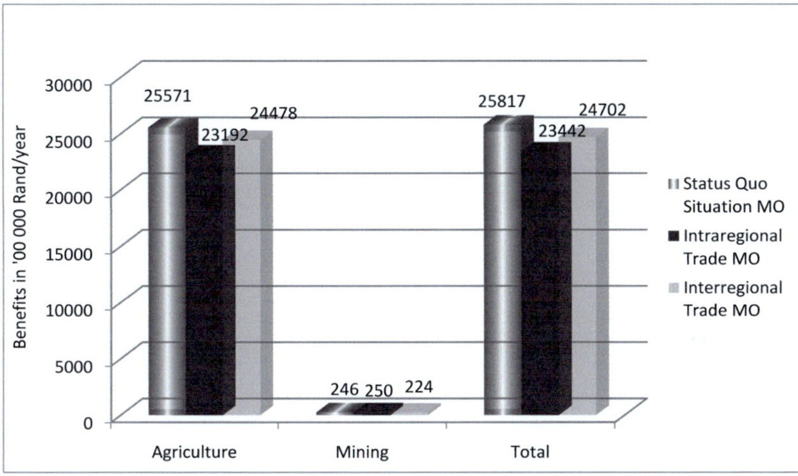

Fig. 1: Benefits from Water Use for Trade Scenarios

Results show that water scarcity can partly be compensated by allowing water trade and that water markets could be a response to water scarcity under appropriate institutional conditions.

4 Conclusion and Outlook

The economic benefits for off-stream water uses, provided in this article encompass the agricultural and the mining sector. The estimates provide a basis for policy decisions affecting water users as well as water resources. Benefits from water trade were analyzed for different Scenarios. Water use efficiencies increase by reallocating water from agriculture to mining in Scenario 2. Trade scenarios show benefits increases, which can be assigned to intra- and interregional trade. Benefits only decrease by 9% and 4% in Scenario 2 and 3 compared to Scenario 1, although total water supply decreased by 52%. These results show that according market mechanisms can counteract or even offset water scarcity and that through appropriate water demand management an optimal water allocation can be reached and efficient water prices can be determined.

However, politicians should consider that the costs of moving to market pricing may include up-front administrative and compliance investments and recurrent costs borne in all years. Such institutional and environmental costs, associated with a change of water allocation policy, are not included in the model and further research is necessary.

List of Figures

List of Tables

References

[1] Berndt, E R, and D.O. Wood. "Technology, Prices and the Derived Demand for Energy." The Review of Economics and Statistics, 1975: 259-268.

[2] Conradie, B. The Value of Water in the Fish-Sundays Scheme of the Eastern Cape. Report, WRC, Pretoria: WRC, 2002.

[3] DWAF. Olifants Water Management Area. Pretoria: WRC, 2003.

[4] Mahan, R. C., Th. M. Horbulyk, and J. G. Rowse. "Market Mechanisms and the Efficient Allocation of Surface Water Resources in Southern Alberta." Socio-Economic Planning Sciences, 2002: 25-49.

[5] Nieuwoudt, W. L. Water Markets Institutions In South Africa, Lessons from Colorado. Pietermaritzburg, South Africa: Department of Agricultural Economics, University of Natal, n.a.

[6] Nieuwoudt, W.L., G.R. Backeberg, and H.M. du Plessis. "The value of water in the South African Economy - A review." Draft Paper, Pretoria, 2004.

[7] Spulber, N., and A. Sabbaghi. Economics of Water Resources: From Regulation to Privatization. USA: Kluwer Academic Publishers, 1998.

Development and optimization of measures against the contamination of the food chain by arsenic from polluted groundwater resources in rural areas of the Red River Plain in Vietnam

Norra, S.[1,2]
Kunert, H.J.[3]; Bahr, C.[4]; Berner, Z.[1]; Bich, P.T.N.[5]; Blömecke, B.[6]; Boie, I.[1]; Cat, L.V.[5]; Chieu, L.V.[3]; Driehaus, W.[4]; Eiche, E.[1]; Kellermeier, E.[1]; Jekel, M.[7]; Lan, V.T.M.[8], Vogt, J.[9]; Long, D.D.[10]; Lübken, M.[11], Sperlich, A.[7]; Wegner, A.[9]; Wichern, M.[11]; Wiesmüller, G.A.[12] Ngoc, D.T.M.[13]; Norra, C.[14]; Trang, P.T.K.[8]; Viet, P.H.[8]

[1]: Institute of Mineralogy and Geochemistry, Karlsruhe Institute of Technology, 76128 Karlsruhe, email: stefan.norra@kit.edu, phone: +49-(0)721-6087245
[2]: Institute of Geography and Geoecology, Karlsruhe Institute of Technology
[3]: AHG Allgemeine Hospitalgesellschaft AG, Düsseldorf
[4]: GEH Wasserchemie GmbH Co KG
[5]: Institute of Chemistry, Vietnamese Academy of Science and Technology
[6]: Ecotoxicology/Toxicology and Institute for Biogeoanalytics & Environmental Specimen Bank, Universität Trier
[7]: Chair for Water Quality Control, Technische Universität Berlin
[8]: Research Center for Environmental Technology and Sustainable Development, Hanoi University of Science
[9]: Institute of Regional Science, Karlsruhe Institute of Technology
[10]: Genetic Department, Hanoi University of Science
[11]: Chair for Sanitary Environmental Engineering and Environmental Technology, Ruhr Universität Bochum
[12]: Health Authorities, Department for Sanitation and Environmental Medicine, Cologne
[13]: National Institute of Occupational and Environmental Health, Hanoi
[14]: Psychiatry Psychotherapy Psychosomatics and Preventative Medicine, Ruhr Universität Bochum

Abstract

In Vietnam, the health of 10 million people living in rural areas of the Red River Delta (RRD) is acute endangered by Arsenic contaminated water. This calamity is caused by the establishment of numerous tube wells during the last century. These wells show arsenic concentrations from 100 µg/L up to more than 400 µg/L, which is far too high in regard to the WHO limit value of 10 µg/L and in regard to the recommended daily human uptake of 2 µg/kg body weight. The uptake of high arsenic amounts can cause general health problems such as black foot disease and cancer but a high incidence of mental diseases, too. Many Vietnamese people already show symptoms and it will become more severe in the near future.

This project aims to investigate the various aspects of the uptake of Arsenic by human beings and the toxicological consequences as well as the development and establishment of adapted mitigation techniques and quality control methods. Therefore an interdisciplinary team was formed covering subjects from environmental medicine to water management and filter technology. In its first phase from 2008 to 2010, this program is funded as initial project[1] by the International Bureau of the German Ministry of Education and Research (BMBF), the German Research Foundation (DFG) and the Vietnamese Ministry of Science and Technology (MOST).

Catchwords

Arsenic, food chain, toxicology, neurotoxicology, groundwater pollution, drinking water quality, filter techniques, mental disorder, Vietnam, Red River Delta

1 Background and Challenge

The contamination of groundwater with arsenic from geogenic sources is a severe problem throughout the world. Using this groundwater as drinking water can cause arsenicosis as already early reported from countries like Taiwan [1], Chile [2], Mexico [3] or Argentina [4]. These casualties have been known for a long time, but only a limited number of people were affected. Following the discovery of the arsenic calamity in the densely populated Bengal Delta [5] the topic came into the focus of the scientific world leading to intensive investigations. Other regions with elevated arsenic levels in groundwater have since been identified, primarily in relatively young alluvial deposits, such as the densely populated deltas of Mekong and Red River in Cambodia and Vietnam, where numerous tube wells have been established during the green revolution [6, 7, 8, 9, 10, 11].

In Vietnam, the health of 10 million people living in farms of the Red River Delta (RRD) is acute endangered by Arsenic contaminated water [6, 7, 11]. The release of Arsenic is caused by chemical redox processes in the underground from where the water is extracted. During the last decades, numerous new wells have been drilled resulting in an intensive use of groundwater for drinking and irrigation purposes (fig. 1). Those waters contain often more than 100 up to more than 400 µg/L Arsenic. The current limiting

[1] Name of the initial project: VIGERAS - Initiation of a Vietnamese-German network of experts for the assessment of Arsenic in the food chain and on the development, optimisation and implementation of filter techniques to remove Arsenic from contaminated groundwater in rural areas of the Red River Delta, Vietnam.

value proposed by the WHO for drinking water is 10 μg/L and for Arsenic uptake by food is 2 μg/kg bodyweight. Uptake of higher amounts of Arsenic by human beings causes illnesses such as skin cancer, black food disease and mental disorders. Many Vietnamese in the affected areas in the RRD show symptoms of those illnesses and many more will show those symptoms in future if no counter measures are taken. However, only little is known about the health effects on the neurological system [12] and of arsenic in levels of subtoxic concentrations, which could be manifested in mental disorders or an impaired development during childhood, e.g. of intelligence. Furthermore, arsenic toxicity can be influenced by genetic polymorphisms, nutrition and general environmental pollution. These are all factors still not sufficiently investigated.

Figure 1: left side – self made filter system at Van Phuc; right side – sewage pond at Maidong

A major pathway of Arsenic uptake by humans is by drinking water. Although in the affected areas self made filter systems using sand are sometimes established, they can not guarantee Arsenic free water. In the particular case of the shown filter system the resulting concentrations of Arsenic in drinking water are still more than twice the WHO threshold value [13]. These filter systems are made mainly to extract dissolved Iron from the groundwater. However, groundwater is also intensively used for the irrigation of rice and wheat fields, which are the main crops cultivated in that area. The volume of water used for the irrigation of a specific crop varies considerably depending not only on climatic factors, but also on the permeability of soil. The water demand of rice is particularly high [14]. Though the potential risk is evident, studies on the impact of irrigation with high Arsenic groundwater on soil and crop attracted some attention only during the last couple of years [15, 16, 17].

Over the years, various processes have been postulated in order to explain high arsenic concentrations in the aquifers and they are still subject of dispute. The reductive dissolution of different iron oxides, which are common in sedimentary environments, is widely accepted as a key process for the release of arsenic into groundwater [18, 19, 20, 21]. However, the reduction of iron oxides alone cannot explain the wide range of groundwater arsenic concentrations encountered in similarly reducing aquifers [22]. What is clear is that the microbially driven decomposition of organic material plays an important role for the onset and the maintenance of reducing conditions in aquifers [23, 24, 25, 26]. A characteristic feature of Vietnamese villages in the Red River Delta are the numberless sewage ponds through which organic compounds enter environmental systems (fig. 1). Despite its importance, not enough is know about the nature and the origin of organic carbon [25]. Different sources of organic carbon have been proposed over the years: peat layers or confining sediment layers rich in total organic carbon (TOC) [24, 27, 28], recharge from ponds and rivers commonly high in dissolved organic carbon (DOC) and anthropogenic sources [19, 27, 29]. Further processes under discussion, which could have an influence on the arsenic concentration in groundwater are competition with other dissolved ions like phosphate [39] or bicarbonate [19, 31] oxidation of pyrite [32] or precipitation and dissolution of secondary mineral phases (e.g. siderite, magnetite, amorphous phases incorporating arsenic) [33, 34, 35]. It was also suggested that arsenic released in the surface soil by redox cycling could be transported downwards towards sandy aquifer [36].

There is still much disagreement about underlying causes of the patchy arsenic distribution commonly observed in affected areas. Pronounced differences in arsenic levels can be found within distances of 100 m [37, 38, 39]. Recent studies in portions of the Red River Delta have also revealed significant differences even within short distances of 10 - 20 m [9]. Several explanations have been proposed for the complex spatial distribution of arsenic, including differences in the subsurface lithology, mineralogy, geochemistry, local hydrology, and the abundance of organic material [22, 40, 41]. Considerable uncertainty remains, however, and too little is known to predict with confidence how arsenic concentrations will evolve over time and to what extent aquifers currently providing potable water can be relied on in the future [42]. However, this patchy and till now not predictable distribution of high Arsenic concentrations in groundwater resources is highly endangering its usability as potable water and irrigation water by the rural population of Vietnam.

2 Objectives

The overall aim of this program is to place clean and healthy water at the disposal for the rural population of the Red River Delta in Vietnam. From this, following project milestones can be defined:

- Comprehensive capacity building and scientific exchange in the various fields concerning arsenic contamination of water resources and food chain. Training of Vietnamese partners on assessing health consequences of Arsenic contamination of the food chain.
- Investigation of distribution processes of Arsenic in the environment, the food chain and in selected human tissue in dependency of different Arsenic concentrations in groundwater as depicted in figure 2. Identification of major transfer pathways of Arsenic from groundwater into the human body.
- Selection, development and establishment of optimized filter and treatment techniques (fig. 2) and adaptation of those techniques to the specific economic and cultural situation in Vietnam.
- Development and establishment of medical indicators for control of success of reduction of Arsenic exposition of human beings.
- Development and establishment of a sustainable water management system for the rural areas of the Red River Delta.

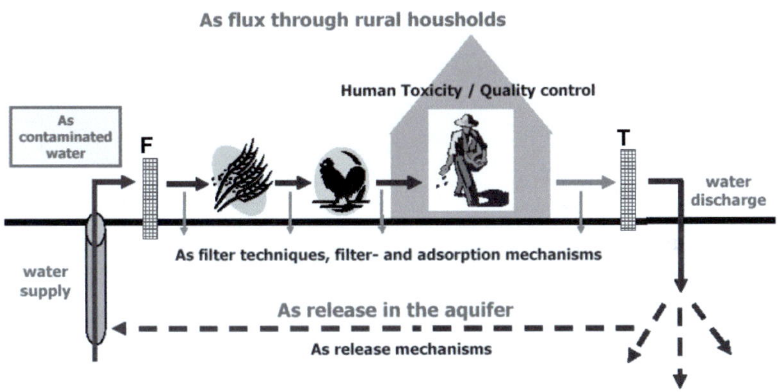

Figure 2: Arsenic flux through rural households of the Red River Delta, Vietnam. F is representing adapted filter systems for arsenic and T is representing a treatment system for waste water. The development of F and T are aims of the presented program.

The final results of this program shall be made available to the local population at a pilot farm in the rural region of the Red River Delta, where specific filter systems will be installed as demonstration projects, help in water supply issues will be offered and health support will be provided. This general concept is shown in figure 3.

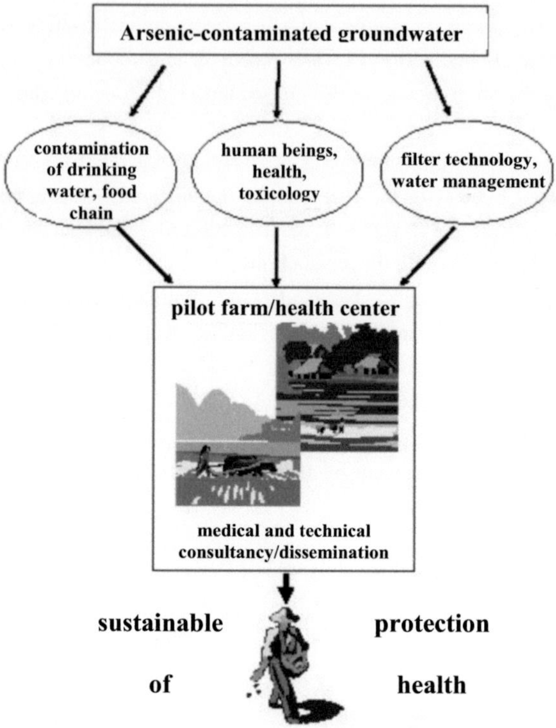

Figure 3: General project concept with the objective to establish a pilot farm where necessary information will be made available for the local population.

Figure 1 depicts the necessary technological inventions to filter (F) Arsenic from groundwater resources for various uses as irrigation and drinking water and to clean (T) the waste water to minimize the dispersion of dissolved organic carbon compounds, which have the potential to contribute to reducing conditions in the aquifer.

The results of this initial project will be compiled in a final report acting as basis for the design of the future program phases. The health centers of the rural villages in the Red River Delta shall be used for dissemination of the results. In future phases of the program, the local water authorities will be involved in the establishment of a water management system providing clean water for the population.

List of Illustrations

Figurc 2: Arscnic flux through rural households of the Red River Delta, Vietnam.

Figure 3: General project concept with the objective to establish a pilot farm where necessary information will be made available for the local population.

List of Literature

[1] Tseng, W.P., Chu, H.M., How, S.W., Fong, J.M., Lin, C.S., Yeh, S., 1968. Prevalence of skin cancer in an endemic area of chronic arsenicism in Taiwan. J. Nat. Cancer Inst. 40, 239-254.

[2] Zaldivar, B.J., 1974. Arsenic contamination of drinking water and foodstuffs causing endemic chronic poisoning. Beiträge zur Pathologie 151, 384-400.

[3] Del Razo, L.M., Arellano, M.A., Cebrián, M.E., 1990. The oxidation states of arsenic in well-water from a chronic arsenicism area of northern Mexico. Environ. Pollut. 64, 143-153.

[4] Nicolli, H.B., Suriano, J.M., Peral., M.A.G., Ferpozzi, L.H., Baleani, O.A., 1989. Groundwater contamination with arsenic and other trace-elements in an area of the Pampa, province of Córdoba, Argentina. Environ. Geol. Wat. Sci. 14, 3-6.

[5] Das D, Samanta G, Mandal BK, Chowdhury TR, Chanda CR, Chowdhury PP, Basu GK, Chakraborti D., 1996. Arsenic in groundwater in six districts of West-Bengal, India. Environmental Geochemistry and Health 18, 5-15.

[6] Berg, M., Tran, H.C., Nguyen, T.C., Pham, H.V., Schertenleib, R., Giger, W., 2001. Arsenic contamination of groundwater and drinking water in Vietnam: A human health threat. Environ. Sci. Technol. 35, 2621-2626.

[7] Polya, D.A., Gault, A.G., Diebe, N., Feldmann, P., Rosenboom, J.W., Gilligan, E., Fredericks, D., Milton, A.H., Sampson, M., Rowland, H.A.L., Lythgoe, P.R., Jones, J.C., Middleton, C., Cooke, D.A., 2005. Arsenic hazard in shallow Cambodian groundwaters Mineralogical Magazine 69, 807-823.

[8] Buschmann, J., Berg, M., Stengel, C., Sampson M.L., 2007. Arsenic and manganese contamination of drinking water resources in Cambodia: coincidence of risk areas with low relief topography. Environ. Sci. Technol. 41, 2146-2152.

[9] Berg, M., Stengel, C., Trang, P.T.K., Viet, P.H., Sampson, M.L., Leng, M., Samreth, S., Fredericks, D., 2007. Magnitude of Arsenic Pollution in the Mekong and Red River Deltas – Cambodia and Vietnam. Sci. Total Environ. 372, 413-425.

[10] Buschmann, J., Berg M., Stengel, C., Winkel, L., Sampson, M.L., Trang, P.T.K., Viet, P.H., 2008. Contamination of drinking water resources in the Mekong delta floodplains: Arsenic and other trace metals pose serious health risks to population. Environ. International.

[11] Eiche E, Neumann T, Berg M, Weinman B, van Geen A, Norra S, Berner Z, Kim Trang PT, Viet PH, Stüben D 2008. Geochemcial processes underlying a sharp contrast in groundwater arsenic concentrations in a village on the Red River Delta, Vietnam. Applied Geochemistry 23, 3143-3154.

[12] Kapaj, S.; Peterson, H.; Liber, K.; Bhattacharya, P. 2006: Human health effects from chronic arsenic poisoning – a review. Journal of Environmental Science and Health part A, 41: 2399-2428.

[13] Boie, I. 2009: Arsenic contamination in rural areas of the Red River Delta in Vietnam: A case study considering geochemical and basic socio-economic parameters in exposure assessment and planning of mitigation measures. Unpublished Diploma Thesis, Institute of Mineralogy and Geochemistry, Karlsruhe Institue of Technology

[14] Gupta, R.K., Naresh, R.K., Hobbs, P.R., Ladha, J.K. 2002: Adopting conservation agriculture in rice-wheat systems of the Indo-Gangetic Plains – new opportunities for saving on water. Water wise rice production workshop, 5 – 10 April 2002, IRRI, Phillippines.

[15] Abedin Md. J., Cotter-Howells, J., Meharg A.A., 2002. Arsenic uptake and accumulation in rice (*Oryza sativa* L.) irrigated with contaminated water. Plant and Soil 240, 311–319.
[16] Alam, M.G.M., Snow, E.T., Tanaka, A., 2003. Arsenic and heavy metal contamination of vegetables grown in Samta village, Bangladesh. Sci. Tot. Environ. 308, 83–96.
[17] Norra, S.; Berner, Z.; Agarwala, P.; Wagner, F.; Chandrasekharam D.; Stüben, D. 2005. Impact of irrigation with As rich groundwater on soil and crops: a geochemical case study in Maldah District, West Bengal. Applied Geochemistry 20 (10): 1890-1906.
[18] Nickson, R.T., McArthur, J.M., Ravenscroft, P., Burgess, W.G., Ahmed, K.M., 2000. Mechanisms of arsenic release to groundwater, Bangladesh and West Bengal. Appl. Geochem. 15, 403-413.
[19] Harvey, C.F., Swartz, C.H., Badruzzaman, A.B.M., Keon-Blute, N., Yu, W., Ashraf Ali, M., Jay, J., Beckie, R., Niedan, V., Brabander, D., Oates, P.M., Ashfaque, K.N., Islam, S., Hemond, H.F., Ahmed, M.F., 2002. Arsenic mobility and groundwater extraction in Bangladesh. Science 298, 1602-1606.
[20] Stüben, D., Berner, Z., Chandrasekharam, D., Karmakar, J., 2003. Arsenic enrichment in groundwater of West Bengal, India: geochemical evidence for mobilization of As under reducing conditions. Appl. Geochem. 18, 1417-1434.
[21] Charlet, L., Polya, D.A., 2006. Arsenic in shallow, reducing groundwaters in Southern Asia: an environmental health disaster. Elements 2, 91-96.
[22] Stute, M., Zheng, Y., Schlosser, P., Horneman, A., Dhar, R.K., Datta, S., Hoque, M.A., Seddique, A.A., Shamsudduha, Ahmed, K.M., Van Geen, A., 2007. Hydrological control of As concentrations in Bangladesh groundwater. Water Resour. Res. 43, W09417.
[23] Lovley, D.R., 1992. Microbial Oxidation of Organic matter coupled to the reduction of Fe(III) and Mn(IV) oxides. Catena Suppl. 21, 101-114.
[24] Lovley, D.R., Chapelle, F.H., 1995. Deep subsurface microbial processes. Rev. Geophysics 33, 365-381.
[25] Rowland, H.A.L., Polya, D.A., Lloyd, J.R. and Pancost, R.D., 2006. Characterisation of organic matter in a shallow, reducing, arsenic-rich aquifer, West Bengal. Organic Geochemistry 37, 1101-1114.
[26] Rowland, H.A.L., Pederick, R.L., Polya, D.A., Pancost, R.A., van Dongen, B.E., Gault, A.G., Bryant, C., Anderson, B. Charnock, J.M., Vaughan, D.J. and Lloyd, J.R., 2007. Control of organic matter type of microbially mediated release of arsenic from contrasting shallow aquifer sediments from Cambodia. Geobiology 5, 281-292.
[27] McArthur, J.M., Ravenscroft, P., Safiulla, S., Thirlwall, M.F. (2001): Arsenic in Groundwater: Testing Pollution Mechanisms for Sedimentary Aquifers in Bangladesh. Water Resources Research 37, 1: 109-117.
[28] Zheng, Y., Stute, M., van Geen, A., Gavrieli, I., Dhar, R., Simpson, H.R., Schlosser, P., Ahmed, K.M., 2004. Redox control of arsenic mobilization in Bangladesh groundwater. Appl. Geochem. 19, 201-214.
[29] Buckau, G., Artinger, R., Geyer, S., Wolf, M., Fritz, P., Kim, J.I., 2000. Groundwater in-situ generation of aquatic humic and fulvic acids and the mineralization of sedimentary organic carbon. Appl. Geochem. 15, 819-832.
[30] Su, C., Pulse, R.W., 2001. Arsenate and arsenite removal by zerovalent iron: effects of phosphate, silicate, carbonate, borate, sulphate, chromate, molybdate and nitrate, relative to chloride. Environ. Sci. Technol. 35, 4562-4568.
[31] Apello, C.A.J., Van der Weiden, M.J.J., Tournassat, C., Charlet, L., 2002. Surface complexation of ferrous iron and carbonate on ferrihydrite and the mobilization of arsenic. Environ. Sci. Technol. 36, 3096-3103.
[32] Chowdhury, T.R., Basu, G.K., Mandal, B.K., Biswas, B.K., Samanta, G., Chowdhury, U.K., Chanda, C.R., Lodh, D., Roy, S.L., Saha, K.C., Roy, S., Kabir, S., Quamruzzaman, Q., Chakrabort, D., 1999. Arsenic poisoning in the Ganges Delta. Nature 401, 545-546.

[33] Sengupta, S., Mukherjee, P-K., Pal, T., Shome, S., 2004. Nature and origin of arsenic carriers in shallow aquifer sediments of Bengal Delta, India. Environmental Geology 45, 1071-1081.

[34] Swartz, C.H., Blute, N.K., Badruzzman, B., Ali, A., Barbander, D., Jay, J., Besancon, J., Islam, S., Hemond, H.F., Harvey, C.D., 2004. Mobility of arsenic in a Bangladesh aquifer: Inferences from geochemical profiles, leaching data and mineralogical characterisation. Geochim. Cosmochim. Acta 68, 4539-4557.

[35] Herbel, M., Fendorf, S., 2006. Biogeochemical processes controlling the speciation and transport of arsenic within iron coated sand. Chem. Geol. 228, 16-32.

[36] Polizzotto, M.L., Harvey, C.F., Li, G., Badruzzman, B., Ali, A., Newville, M., Sutton, S., Fendorf, S., 2006. Solid-phases and desorption processes of arsenic within Bangladesh sediments. Chem. Geol. 228, 97-111.

[37] BGS, DPHE, 2001. Arsenic contamination of groundwater in Bangladesh. In: Kinniburg, D.G., Smedley, P.L. (Eds.), Vol. 2, Final Report, BGS Technical Report WC/00/19. British Geological Survey, Keyworth, U.K.

[38] van Geen, A., Zheng, Y., Versteeg, R., Stute, M., Horneman, A, Dhar, R., Steckler, M., Gelman, A., Small, C., Ahsan, H., Graiano, J.H., Hussain, I., Ahmed, K.M., 2003. Spatial variability of arsenic in 6000 tube wells in a 25 km^2 area of Bangladesh. Water Resour. Res. 39, 1140.

[39] McAthur, J.M., Banerjee, D.M., Hudson-Edwards, K.A., Mishra, R., Purohit, R., Ravencroft, P., Cronin, A., Howarh, R.J., Chatterjee, A., Talukder, T., Lowry, D., Houghton, S., Chadha, D.K., 2004. Natural organic matter in sedimentary basins and its relation to arsenic in anoxic ground water: the examples of West Bengal and its worldwide implications. Appl. Geochem. 19, 1255-1293.

[40] Pal, T., Mukherjee, P.K., Sengupte, S., 2002. Nature of arsenic pollutants in groundwater of Bengal Delta – A case study from Baruipur area, West Bengal, India. Current Science 82, 554-561.

[41] van Geen, A., Zheng, Y., Cheng, Z., Aziz, Z., Horneman, A., Dhar, R.K., Mailloux, B., Stute, M., Weinman, B., Goodbred, S., Seddique, A.A., Hoque, M.A., Ahmed, K.M., 2006. A transect of groundwater and sediment properties in Araihazar, Bangladesh: Further evidence of decoupling between As and Fe mobilization. Chem. Geol. 228, 85-96.

[42] Zheng, Y., van Geen, A., Stute, M., Dhar, R., Mo, Z., Cheng, Z., Horneman, A., Gavrieli, I., Simpson, H.J., Versteeg, R., Steckler, M., Grazioli-Venier, A., Goodbred, S., Shahnewaz, M., Shamsudduha, M., Hoque, M.A., Ahmed, K.M., 2005. Geochemical and hydrogeological contrasts between shallow and deeper aquifers in two villages of Araihazar, Bangladesh: Implications for deeper aquifers as drinking water sources. Geochim. Cosmochim. Acta 69, 5203-5218.

Modelling network for analysing diffuse nitrogen leaching reduction in agriculture to meet the targets of the water framework directive in the Weser River Basin of Germany

Peter KREINS[a)], Horst GÖMANN[a)], Claudia HEIDECKE[a)*], Ulrike HIRT[b)], Frank WENDLAND[c)]

a) Institute of Rural Studies, Johann Heinrich von Thünen-Institute, Bundesallee 50, D-38116 Braunschweig, Germany,

b) Department of Lowland Rivers and Shallow Lakes, Institute of Freshwater Ecology and Inland Fisheries (IGB), Germany,

c) Research Centre Jülich, Germany,

**corresponding author: E-mail: claudia.heidecke@vti.bund.de, Phone: +49(0)531/5965519*

Abstract

The objective of this paper is to present a model based approach to identify the demand for measures for the reduction of nitrogen surpluses from agriculture and to compose a combination of possible measures with their associated costs to be able to meet the targets of the water framework directive (WFD). An interdisciplinary model network consisting of the regionalised agricultural economic model RAUMIS and two hydro(geo)logical modelling systems GROWA/WEKU and MONERIS is used to analyse the spatially differentiated impacts of nutrient reduction measures on the water quality of groundwater and surface water. The study region of the German Weser river basin is chosen that comprises 49,000 km^2 and features heterogeneous natural site conditions. In order to support the preparation of a management plan for the Weser river basin the AGRUM Weser project has been initiated. A combination of measures is specified for each county in the Weser basin depending on regional peculiarities and farming practices. Results show that about 100 million Euros per year are needed for agri-environmental measures to fulfil the objectives of the Water Framework Directive. This paper presents a regionally comprehensive and area wide modelling approach that is capable of linking hydrological and agricultural economic aspects in an interdisciplinary approach and which can easily be transferred to other river basins.

Catchwords: Water framework directive, diffuse pollution, agricultural economic and hydrological modelling, cost of nutrient reduction measures, Weser River Basin

1 Introduction

The fundamental objective of the EU-Water Framework Directive (WFD) (European Commission, 2000) is to secure a good chemical and ecological status or rather, potential, for all surface water bodies, and good chemical and quantity status for groundwater bodies by 2015 (2027 at the latest). To a large extent, the pollution originates from nutrient discharges of agriculturally used land. Appropriate nutrient reduction measures are needed in order to achieve the environmental objectives of the WFD. For the preparation of river basin wide coordinated management plans integrated analyses on the impacts of measures to reduce nutrient discharges have to be carried out. The Weser river basin was selected as study region because in addition to the existence of a water pollution problem it is characterized by a wide range of hydrological conditions and agricultural structures which ensures that the approach is transferable to other river basins. The objective of the paper is to present a model based approach to identify the demand for additional measures for the reduction of nitrogen surpluses. A combination of possible additional measures with their associated costs is set up enabling to meet the targets of the WFD for surface and groundwater. The paper presents the model network and its application to the Weser river basin, some major results for the status quo situation and the baseline projection for 2015 as well as a possible combination of measures to reach the environmental objectives of the WFD and their costs associated with their implementation.

2 Modeling network

Analyses in this paper are based on a model network applied in the AGRUM Weser project (Kreins et al., 2010), a project which analyzed current and future nitrogen inputs as well as measures to prevent diffuse nutrient leaching from agriculture in the Weser River Basin. The model network consists of the agricultural sector model RAUMIS (Henrichsmeyer et al., 1996), the hydro(geo)logical model GROWA/WEKU (Wendland et al., 2002, 2004) and the MONERIS model (Modelling Nutrient Emissions in River Systems, Behrendt et al., 1999; 2003). A detailed description of the results of surface water interaction and the application of MONERIS can be found in Hirt et al. (2008). The groundwater processes and the nitrogen loads into groundwater bodies are described in detail by Wendland et al. (2010). In this paper we concentrate on the use of the results of the hydrological models within the agricultural sector model RAUMIS and the calculation of additional measures and costs necessary to fulfill the WFD in 2015. For agricultural economic simulations and projections the Regionalised Agricultural and Environmental Information System RAUMIS is applied. The model consolidates various agricultural data sources as a framework of consistency and reflects the whole German agricultural sector with its sector linkages. The model has been successfully applied in the last two decades to a variety of agricultural policy analysis e.g. Kreins and Gömann (2008) and Gömann et al. (2008). In RAUMIS, a set of agri-environmental indicators is linked to agricultural production. Regarding diffuse water pollution, the indicator "nutrient surplus" is of particular importance. The model GROWA (Kunkel and Wendland, 2002) is used to carry out area differentiated water balance analyses. The mean long term total runoff is modelled as a function of the regional interaction of climate, soil, geology, topography and land use conditions. The model WEKU (Kunkel & Wendland, 1997; Wendland et al., 2004) simulates the reactive nitrate transport in groundwater. The MONERIS model is applied, which quantifies nitrogen (N) and phosphorus (P) emissions into river basins, via various point and diffuse pathways, as well as the retention and the nutrient load in rivers (Behrendt et al., 2003). Sub-catchments of river basins are the basis of the spatial resolution.

For the coupling and interaction of the models a central issue is the adjustment of the different spatial resolutions, i.e., administrative units in RAUMIS on the one hand and grids/raster cells in GROWA/WEKU or (sub)river basins in MONERIS on the other hand. In a first step, the spatial allocation of agricultural production is disaggregated from the RAUMIS "region farm" level (NUTS III) to community level (NUTS IV). With the model RAUMIS, regionally differentiated status quo and impact analyses of agricultural and agri-environmental measures are undertaken and nutrient balances are calculated representing the amount of nutrients that potentially leach into water bodies. The GROWA/WEKU model provides high resolution spatial results (i.e., 100x100 m grids) for the diffuse nutrient leaching in groundwater. The MONERIS model considers all relevant paths of nutrient leaching (point sources and diffuse leaching sources) for surface waters.

3 Results

Various factors drive the future development in agriculture and thus determine nutrient surpluses and the need for action in the field of diffuse water pollution. Using the model RAUMIS a baseline is projected until the year 2015 which is a milestone for the implementation process of the WFD. Important driving factors are developments of political framework conditions such as the Common Agricultural policy (CAP), EU and national environmental regulations, projections of prices and yields of agricultural products, and "conventional" measures to reduce nutrient surpluses.

In the first step, nitrogen surpluses and nitrogen concentrations are calculated for 2003 and for the baseline scenario of 2015. Under the assumption of the baseline scenario the average N-surplus (without atmospheric decomposition) declines from 70 kg N per ha UAA by around 9 kg per ha agriculturally used area. This leads to a reduction from 40 mg NO_3 per litre in 2003 to 30 mg NO_3 per litre in 2015 in groundwater bodies on average in the Weser river basin. However, model results also show that a lot of regions still face nitrogen concentration of over 50 mg NO_3 per litre in 2015. The nitrogen emission into surface water bodies in 2015 is about 75.700 t/a, and are thus reduced by 17 % in contrast to the situation in 2003. In the second step, the projected nitrogen surpluses and nitrogen leaching in the baseline scenario are evaluated with respect to the management objectives for a good ecological status of groundwater bodies of 50 mg/l of nitrate in groundwater (groundwater regulation, Richtlinie RL 2006/118/EWG). For surface water bodies the preliminary objective is 3 mg N/l at the gauge of Hemelingen (close to Bremen). Overall, the results show that the targets of the WFD cannot be achieved without additional efforts.

This means that to achieve the WFD targets an additional nitrogen reduction demand of 23,000 t per year is necessary until 2015. Within a comprehensive literature survey (Osterburg and Runge, 2007) the ecological impacts and the capability of technical and organisational water protection measures were recorded for several criteria. Table 1 provides an overview of selected measures used in the further analyses. The selected measures feature substantial differences in regard to their impact on water quality and related costs. Table 1 also depicts the calculated levels and costs of measures necessary to achieve the management targets for groundwater and the provisional target for surface water. First, the levels of measures to fulfil groundwater targets are determined. In total a measure combination that amounts to about 1.1 million hectares is necessary that would require a funding of about 74 million Euros. Second, the surface water targets are evaluated under the assumption of a good status of groundwater. Another 5,500 tons (5 %) of nitrogen on additional 273,000 hectares must be reduced to achieve the targets for surface water which would arouse costs of approximately 20 Million Euros.

			Groundwater		Surface Water	
	Impact on N-Surplus (kg N/hectare)	Costs (€/hectare)	Level ('000 hectare)	Total costs ('000 Euro)	Level ('000 hectare)	Total costs ('000 Euro)
No application of organic fertilizer after harvest	15	15	103	1,543	3	43
Intertillage	20	80	350	27,966	96	7,670
Groundwater protective application of dung	15	30	98	2,950	26	790
Extensive grassland production	30	100	86	655	19	764
Promotion of extensive farming	40	70	75	5,280	31	2,163
Reduced mineral fertilizer in cereal production	30	80	165	13,217	44	3,517
Cultivation of turnip rape	10	60	141	8,449	38	2,280
Organic farming	60	170	83	14,046	16	2,707
Total	/	/	1,101	74,106	273	19,934

Table 1: Total levels and costs of measures for the achievement of WFD objectives in the Weser River Basin (impacts and costs per hectares are based on Osterburg and Runge, 2007)

Adding up cost of ground and surface water, the measures proposed here would sum up to around one hundred million Euros. Regionally however, the reduction demand, land use and thus the combination of measures are very different (see Figure 1).

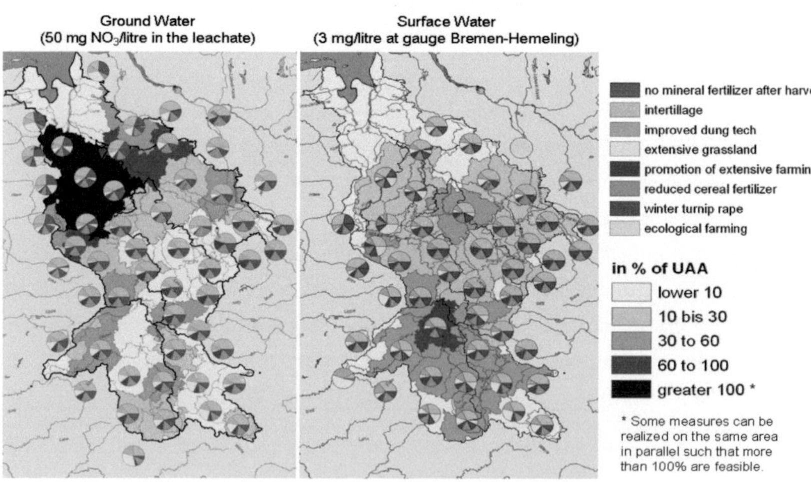

Figure 1: Levels and combination of measures to achieve WFD objectives (Kreins et al. 2010)

4 Conclusion

According to the model results diffuse nutrient pollution will decrease overall in the Weser river basin until the year 2015 due to the changes of framework conditions in agriculture. Even though agriculture will comply - by assumption - with best management practice in 2015 the environmental targets set by the water framework directive will not be achieved for all ground and surface water bodies in the river Weser basin. In order to achieve these targets additional measures have to be offered to farmers. Preliminary calculations for the expected costs of these measures result about 100 million Euros per year. These costs do not include the costs of counselling e.g. by agricultural extension services in the field of agricultural water protection. However, an intensive counselling is a crucial measure to help implement efficiently the changes of the administrative law in particular the manure regulation until 2015.

List of Literature

[1] Behrendt H, Bach M, Kunkel R, Opitz D, Pagenkopf W-G, Scholz G, Wendland, F. Quantifizierung der Nährstoffeinträge der Flussgebiete Deutschlands auf der Grundlage eines harmonisierten Vorgehens. UBA-Texte 82. 2003.

[2] Behrendt H, Huber P, Ley M, Opitz D, Schmoll O, Scholz G, Uebe R. Nährstoffbilanzierung der Flußgebiete Deutschlands. Berlin, UBA-Texte 79. 1999.

[3] European Comission. WFD (Water Framework Directive, European Commission Directive 2000/60/EC. Official Journal, 22. 2000.

[4] Gömann H, Heiden M, Kleinhanß W, Kreins P, von Ledebur O, Offermann F, Osterburg B, Salamon P. Health Check der EU-Agrarpolitik - Auswirkungen der Legislativvorschläge : Studie im Auftrag des BMELV. Braunschweig, Arbeitsberichte des Bereichs Agrarökonomie. 2008.

[5] Henrichsmeyer W, Cypris C, Löhe W, Meudt M, Sander R, von Sothen F, Isermeyer F, Schefski A, Schleef K-H, Neander E, Fasterding F, Helmcke B, Neumann M, Nieberg H, Manegold D, Meier T. Entwicklung eines gesamtdeutschen Agrarsektormodells RAUMIS96. Endbericht zum Kooperationsprojekt. Forschungsbericht für das BML (94 HS 021). 1996.

[6] Hirt U, Venohr M, Kreins P, Behrendt H. Modelling nutrient emissions and the impact of nutrient reduction measures in the Weser river basin, Germany. Water Science and Technologie 58/11: 2251-2258. 2008.

[7] Kreins P, Behrendt H, Gömann H, Heidecke C, Hirt U, Kunkel R, Seidel K, Tetzlaff B, Wendland F. Analyse von Agrar- und Umweltmaßnahmen im Bereich des landwirtschaftlichen Gewässerschutzes vor dem Hintergrund der EG-Wasserrahmenrichtlinie in der Flussgebietseinheit Weser. Braunschweig Landbauforschung - vTI agriculture and forestry research : Sonderheft 336. 2010.

[8] Kreins P, Gömann H. Modellgestützte Abschätzung regionaler landwirtschaftlichen Landnutzung und Produktion in Deutschland vor dem Hintergrund der „Gesundheitsüberprüfung" der GAP Agrarwirtschaft, Band 57, Heft 3-4: 195-206. 2008.

[9] Kunkel R, Wendland F. The GROWA98 model for water balance analysis in large river basins – the river Elbe case study. Journal of Hydrology 259, 152-162. 2002.

[10] Kunkel, R. & F. Wendland (1997): WEKU - A GIS-supported stochastic model of groundwater residence times in upper aquifers for the supraregional groundwater management. *Environmental Geology*, *30*, 1-9.

[11] Kunkel, R., H. Bogena, B. Tetzlaff & F. Wendland (2006): Digitale Grundwasserneubildungskarte von Niedersachsen, Nordrhein-Westfalen, Hamburg und Bremen: Erstellung und Auswertungsbeispiele. *Hydrologie und Wasserbewirtschaftung*, *50*, 212-219.

[12] Osterburg B, Runge T. (eds.). Maßnahmen zur Reduzierung von Stickstoffeinträgen in Gewässer – eine wasserschutzorientierte Landwirtschaft zur Umsetzung der Wasserrahmenrichtlinie. Landbauforschung Völkenrode, Sonderheft 307. 2007.

[13] Wendland F, Behrendt H, Hirt U, Kreins P, Kuhn U, Kuhr P, Kunkel R, Tetzlaff B. Analyse von Agrar- und Umweltmaßnahmen zur Reduktion der Stickstoffbelastung von Grundwasser und Oberflächengewässer in der Flussgebietsgemeinschaft Weser. Hydrologie und Wasserbewritschaftung 4: 231-243. 2010.

[14] Wendland F, Kunkel R, Grimvall A, Kronvang B, Müller-Wohlfeil D I. The SOIL-N/WEKU model system - a GIS-supported tool for the assessment and management of diffuse nitrogen leaching at the scale of river basins. Water Science and Technology 45 (9): 285-292. 2002.

[15] Wendland F, Kunkel R, Voigt H-J. Assessment of groundwater residence times in the pore aquifers of the River Elbe Basin. Environmental Geology 46: 1-9. 2004.

Grey Water Treatment and Reuse Experience in the Palestinian Rural Areas

Authors *
Abu-Rahma A.R* , Ghanem M** and Ali W.***

*Palestinian Hydrology Group (PHG)
**Birzeit University
***Karlsruhe University

Abstract

Wastewater is a very significant pollution source that has serious adverse impact both on the environment and local residents. In the Palestinian Territories raw wastewater is disposed in wadis or left to infiltrate through cesspits into the underlying vulnerable groundwater aquifers. These aquifers are the only water sources and suffer from overexploitation and contamination. The problem is further complicated by the poor economical conditions that prevent the construction of infrastructure and burden the locals with the costly expenses of wastewater disposal by collection tanks. This has motivated deferent interested local organizations, since 2000, to adopt low-cost grey water treatment and reuse technologies. These technologies were initially acceptable to the public but unfortunately later faced technical problems that led to insufficient treatment efficiency and bad odors. continuous efforts invested in the sector by piloting other more advanced treatment units such as a septic tank up-flow gravel filters. This intervention is not only a cheap wastewater disposal method but also a new water source used for agricultural purposes that can save considerable quantities of fresh water especially since grey water forms about 80% of the total water used at household level. At least 60% of grey wastewater can be recovered, treated and reused thus more than 150,000 liters can be saved annually per household. By introducing a reliable water source poor families will be encouraged to cultivate their home gardens and be engaged in different agricultural activities which will without doubt contribute in the food security of the household in addition to improving their monthly income. Local organizations had successfully managed to provide treatment units to schools and households both in the West Bank and Gaza that have an average treatment efficiency of about 80% with COD values that fall within the acceptable guidelines set by WHO and PWA. The implementation of these units was accompanied by awareness activities about wastewater, the impact of poor disposal practices and the anticipated solutions in addition to training about the operation and maintenance of the units.

Key words: Grey water, treatment, reuse

1. Introduction

The problem of water supply in the Palestinian territories is one of the most difficult problems facing Palestinian society. The lack of water resources and the competition between different uses i.e. domestic, agricultural and industrial is increasing with time. Indeed, the limitation of water resources for the Palestinians is mainly due to the Israeli occupation authorities' laws and practices. Israeli settlements control water resources, waste a lot of fresh water quantities, and produce a lot of wastewater which is disposed on Palestinian land contaminating the soil and the limited water resources available for Palestinians.

Cesspits used by Palestinians to dispose their wastewater are another source of pollution to water resources. These cesspits also form a large burden on the income of the Palestinian families, where some families spend about 20% of their monthly income to manage water and wastewater at house level.

For that the Palestinian NGOs, especially Palestinian Hydrology Group (PHG) and Palestinian Agricultural Relief Committees (PARC) have implemented about 500 of grey wastewater treatment units and reuse in the West Bank and Gaza Strip that serve about 650 families and 30 schools, treating an average of about 0.5 cubic meter per day per unit. The treated grey wastewater is reused for irrigating the home and school gardens, fruit trees and vegetables eaten cooked.

2. Objectives of Grey Wastewater Treatment (GWWT) & Reuse.

1. Reuse of grey wastewater to irrigate home and schools gardens.
2. To reduce water resources pollution.
3. To reduce environment pollution.
4. Substitute for cesspits and reduce pumping cost.
5. Reduce fresh water consumption used to irrigate the home garden.
6. Preserve health safety by reducing the odors on roads and around homes.
7. Enhancing the role of women in managing domestic wastewater.
8. Raising the awareness of the farmers about reusing grey wastewater by illustrating actual cases.

3. Palestinian NGOs experience in GWWT: (Case: PHG).

3.1 Palestinian Hydrology Group's (PHG) experience in GWWT.

PHG's history with GWWT began in 2000, PHG focused on this subject and used a septic tank up flow gravel filter followed by an aerobic filter system (see Figure 1) that was installed in households and schools. PHG implemented these projects aiming to create new sources of water to be used for irrigation in home gardens in addition to making valuable improvements to existing sanitation systems in households and schools, which are threatening fresh water sources and public health.

Grey wastewater forms about 80% of the total water used at household level. At least 60% of grey wastewater can be recovered, treated and reused. More than 150,000 liters of fresh drinking water are be annually saved per household, through implementing grey waste water treatment and reuse systems in 159 households and schools implemented by PHG. Large amounts of water are efficiently managed, treated and reused in irrigating home gardens.

The general shortage of water and the high grey wastewater generation from households requires careful consideration of solutions that can be locally applied, considering environmental elements, health issues and the socio-economic situation. Septic tank up flow gravel filter followed by aerobic filter system installed in households and schools showed enough satisfaction to these elements.

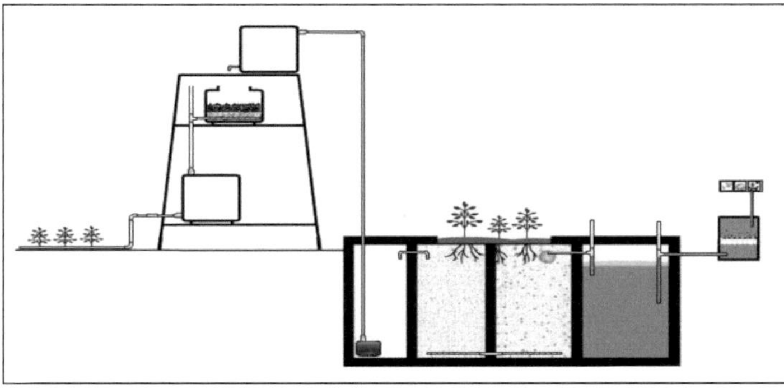

Figure (1): Grey waste water treatment plant scheme used by PHG.

4. GWWT Projects implemented by PHG.

PHG implemented a number of GWWT projects in different places in Palestine, which all had the main goal of introducing a new source of water for the irrigation of home gardens safely and preserving a safe environment. The following projects were implemented by PHG:

1. **Gaza Strip**: In the years 2004 and 2005 PHG's branch in Gaza implemented a project that included the construction of 14 grey water treatment units and the reuse of the treated grey water for agricultural purposes in Abbasan Al Kabeera and Abbasan Al Sagheera and Bani Suhaila in Khan Younis Governorate in order to benefit from the treated grey water by reusing it for growing plants and for it to pose as a model for other rural areas that are not served by sewage networks but rely on cesspits for disposing their wastewater (see Figure 2). Figure 2 illustrates the distribution of the units implemented in Gaza.

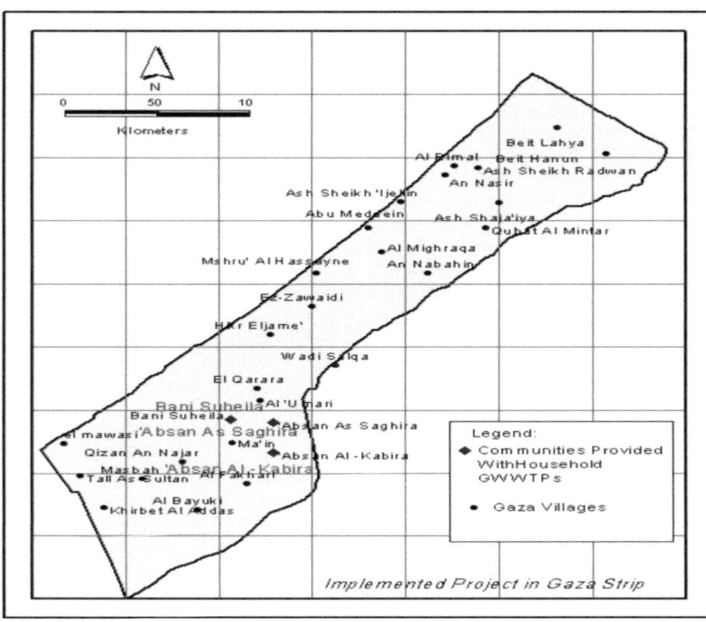

Figure (2): Implemented projects in Gaza Strip.

2. **Southern part of the West Bank**: Funded by ECHO and in cooperation with GVC association, in the year 2005 PHG's branch in Hebron implemented a project that included the construction of 10 grey water treatment units and the reuse of the treated grey water for agricultural purposes in schools in Hebron Governorate. In 2006 and through the Polish Humanitarian Organization (PHO) PHG's branch in Hebron also implemented a project that included the construction of 3 grey water treatment units and the reuse of the treated grey water for agricultural purposes in schools in Bethlehem Governorate.
3. **Northern part of the West Bank:** In 2002, PHG's branch in Nablus implemented a project that included the construction of a centralized grey water treatment plant and the reuse of the treated grey water for agricultural purposes in the village of Ijnisinya which has a population of 600 residents. This project served more than 70 families, and had the main goal of reducing pollution resulting from leaking wastewater cesspits to the main and only source of water which is the spring that supplies the community with drinking water. In addition to that, Nablus Branch implemented in 2002, 60 grey water treatment units in Seir, Meselyia, Al-Jdayidah, Tayaseer and Rabah, and 57 units in Sanour/Jenin District, where 117 families have benefited from that. Moreover, 14 units for grey water were in 14 schools; Aqqaba, Jenin, Kafr Thulth, Awarta, Jamal Abdel Naser (Nablus City), Sabastyia, Tallouza and Al-Badhan.
4. **Ramallah Governorate**: In the years 2002 and 2006 PHG's branch in Ramallah implemented a project that included the construction of 12 grey water treatment units and the reuse of the treated grey water for agricultural purposes in Bil'in, Ras Karkar, Deir Ibzea and Kharbatha Al Mousbah communities.

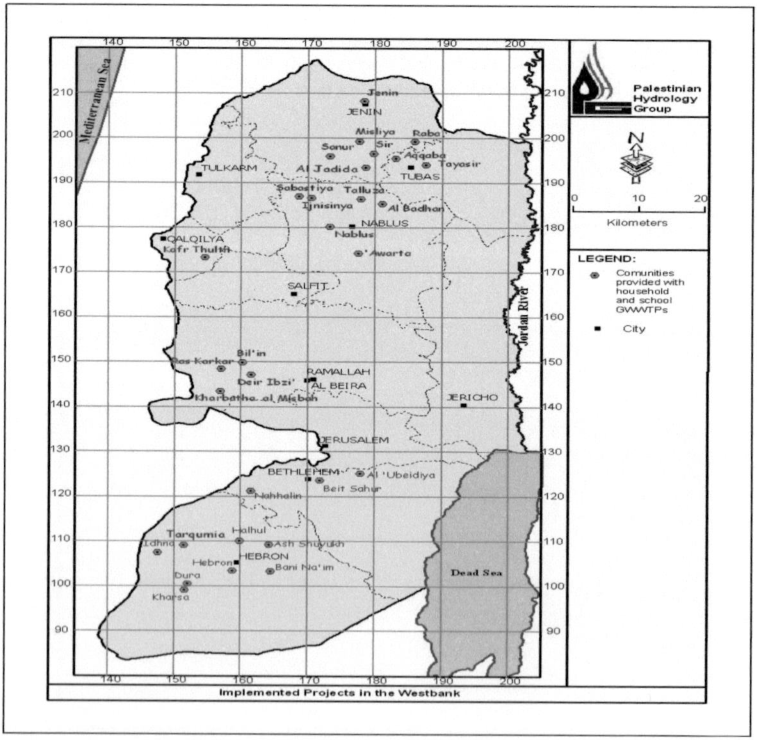

Figure (3): Implemented projects in the West Bank.

5. Sampling and analysis

Within the monitoring activities on the grey wastewater treatment units implemented by PHG samples were collected from the West Bank and Gaza areas in order to measure the efficiency of the GWWT system. Tables 1 & 2 illustrate the results. It is clear that most values fall within the acceptable guidelines set by WHO and PWA (COD= 150-200mg/l). However where they do exceed the permitted limits reasons are due to inadequate operation and lack of maintenance.

Table 1: Characteristics of raw and treated grey wastewater from different sites Ramallah, Bethlehem and Gaza Strip.

Sample #	Location	In-effluent COD(mg/l)	Effluent COD(mg/l)	Efficiency(%)
1	West Bank	462.1	176.1	62
2	West Bank	692.7	475.0	31
3	West Bank	933.1	166.4	82
4	West Bank	702.5	192.4	73
5	West Bank	50.0	30.0	40
6	Gaza	1500.0	270.0	82
7	Gaza	820.0	80.0	90
8	Gaza	1500.0	230.0	85
9	Gaza	590.0	170.0	71

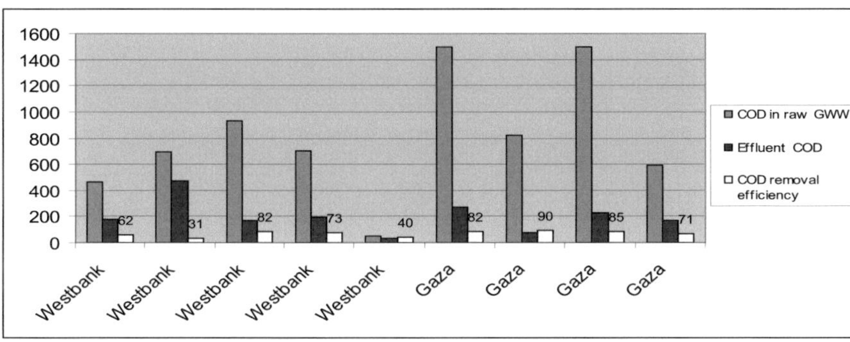

Chart 1: COD removal efficiency

6. Treated grey wastewater reuse in the Palestinian rural areas.

The Palestinian rural areas experience a high level of drinking water shortage. There is a lack of water resources for all purposes (industrial, agricultural, and residential uses), and the existing sources of water are controlled by the Israeli occupation. The quantities of available water for the Palestinian people are limited and less than the minimum recommended by World Health Organization (WHO) standards, where the average consumption of the Palestinian per capita is 76 L/day(water for life 2006, PHG).

The Palestinian communities experience a great weakness in their water and sanitary systems infrastructures where most of the water networks experience to deterioration and devastation, and most of the Palestinian communities are not connected to sanitary network and treatment plants.

A high percentage of Palestinian families in the villages especially the far-off ones live in poverty and low incomes. Most of these families depend on farming to ensure their basic needs of life. Because of the lack of water many use untreated grey wastewater to irrigate their farms and gardens without taking into consideration health and environmental risks especially to the children

and women who are the direct user of this kind of water in addition to the social problems with neighbors because of the foul odors and insects, which ultimately affects the social bonds.
Grey wastewater treatment and reuse in the irrigation of gardens and farms is acceptable by the Palestinian families which helps ensure a good and durable source of water for agricultural use ensuring their basic needs from vegetables which are planted in the house garden with an area of 150 to 200 square meter, in addition to the reduction in pollution and sanitary problems caused by untreated grey wastewater (UNEP 2003).
As earlier mentioned in this paper the Palestinian Hydrology Groups has implemented many grey wastewater treatment units in the Palestinian communities in addition to reclaim house gardens through providing the Palestinian families with irrigation networks and vegetable plants which can be irrigated with this kind of water according to the Palestinian and international standards. This helps in improving the healthy livelihood and conditions for poor Palestinian families.
However, reuse in Palestine remains to be very primitive and requires development mainly due to the lack of large scale infrastructure, social obstacles and public unacceptability. Farmers and households must be fully aware of the importance of reuse and how they can benefit from the treated wastewater by increasing the production and saving the costs that would in other case be spent on fertilizers and fresh water (ARIJ, 1994).

7. Conclusions & lessons learned:

1) Grey water can be a very significant water source for home gardens.
2) Treated grey wastewater is more acceptable for re-use than other kinds of treated water, due to religious and cultural beliefs.
3) Grey water treatment can contribute to improving the socio-economic conditions of poor families in rural areas by saving a considerable portion of their income spent on water supply and wastewater disposal.
4) Using treated grey water for agricultural activities protects the public from the risk of being infected with water borne diseases.
5) Grey water treatment also contributes in preserving the food security of these families by providing a reliable water source to produce a vegetable basket for the household.
6) Due to the poor condition of the sanitation infrastructure, the political situation and the deteriorating economical condition, PHG strongly believes that grey water treatment can be an effective solution to the wastewater management problem especially in the marginalized rural areas.
7) Through its experience, and that on a regional level, PHG has concluded that the up flow gravel filter is an acceptable technology that is low cost and easy to operate and maintain.
8) The concrete chambers developed by PHG are better than infiltrations ponds and parallel barrels previously used since:
    1. They are easier to control.
    2. They are more durable (have a longer life).
    3. More efficient since the volume is not limited.
    4. Environmentally friendly.
9) The mixed system consisting both aerobic and anaerobic processes overcomes many of the problems encountered in the previously implemented units especially due to the generated odors.

8. Recommendations*:*

It is recommended that:

1. Grey and black wastewaters are separated for the purpose of on-site treatment and reuse of grey wastewater.
2. On-site treatment through a septic tank and upflow gravel filter system (the system used by PHG) is provided for the reuse of grey wastewater for gardening.
3. Black wastewater must be directed to modified cesspit for safe and cheap disposal.
4. Alternative treatment options for the on-site reuse of wastewater should be investigated.
5. Such systems must be introduced in poor marginalized areas especially those classified as sensitive groundwater recharge zones.
6. Awareness and training activities related to the topic must be organized in these areas.

References

PHG, Water for Life 2006, WaSH Monitoring Program Report.

UNEP (2003), Desk Study on the Environment in the Occupied Palestinian Territories.
Palestinian Water Authority (2003), Rural Sanitation Report.

ARIJ, (1994). A Palestinian Vision of Environmental Protection and Sustainable Development.

Forecasting the effects of EU policy measures on the nitrate pollution of groundwater and surface waters

Ralf KUNKEL[a1], Peter KREINS[b], Björn TETZLAFF[a], Frank WENDLAND[a]

a: Institute of Chemistry and Dynamics of the Geosphere (ICG), Institute 4: Agrosphere, Research Centre Juelich, Germany

b: Institute of Rural Studies, Johann Heinrich von Thünen-Institute, Bundesallee 50, D-38116 Braunschweig, Germany.

1: corresponding author: E-mail: r.kunkel@fz-juelich.de

Abstract

We used the model network AGRUM to predict the actual nitrate concentration in percolation water at the scale of the Weser river basin (Germany) using an area differentiated approach. AGRUM combines the agro-economical model RAUMIS for estimating nitrogen surpluses and the hydrological models GROWA/DENUZ/WEKU and MONERIS for assessing the nitrate leaching from the soil. For areas showing predicted nitrate concentrations in percolation water above the EU groundwater quality standard of 50 mg NO_3/L, effective agri-environmental reduction measures need to be derived and implemented to improve groundwater and surface water quality by 2015. The effects of already implemented agricultural policy are quantified by a baseline scenario projecting the N-surpluses from agricultural sector to 2015. The AGRUM model is used to estimate the effects of this scenario concerning groundwater and surface water pollution by nitrate. From the results of the model analysis the needs for additional measures in terms of required additional N-surplus reduction and in terms of regional prioritization of measures can be derived. Research work will therefore directly support the implementation of the Water Framework Directive of the European Union in the Weser basin.

Catchwords

Catchment management; diffuse source pollution; mitigation methods; river basin management; Water Framework Directive

Introduction

The fundamental objectives of the European Union-Water Framework Directive and the EU Groundwater Directive are to attain a good status of water and groundwater resources in the member states of the EU by 2015. For water bodies, whose good status cannot be guaranteed by 2015, catchment wide operational plans and measurement programs have to be drafted and implemented until the end of 2009.

The river basin Weser comprises a catchment area of about 49,000 km^2 (see figure1). It is the only macro-scale 1st order river basin, which is completely located within Germany's territory. As the Weser basin stretches across various landscape types, different pedological, hydrological and hydrogeological properties occur. Hence, the different agricultural land use, production and farm types occurring in the basin vary substantially. In the northern and north-western part of the Weser basin land use is dominated by intensive land-independent animal husbandry with accordingly high livestock densities. The middle part of the Weser basin is characterized by intensive catch cropping. In the midlands in the South-Western and South-Eastern part extensive forms of land use are predominant.

In total, about 55% of the area of the Weser basin is used agriculturally. Consequently, the achievement of the good status of groundwater according to the EU Water Framework Directive is unclear or rather unlikely for 63% of the groundwater bodies [1]. Inputs from diffuse sources and most of all nitrate losses from agriculturally used land have been identified as the main reasons for failing the "good qualitative status" of groundwater.

In the AGRUM-WESER project [2] management options for the Weser river basin with respect to their environmental and economical relevance have been investigated and analyzed. As a target value for groundwater protection measures a nitrate concentration in percolation water of 50 mg/L has been defined. By backward calculation using the model the tolerable N-surpluses needed to meet the environmental target as well as the reduction requirements for N-surpluses from agriculture can be estimated. Based on the results of this analysis regionally

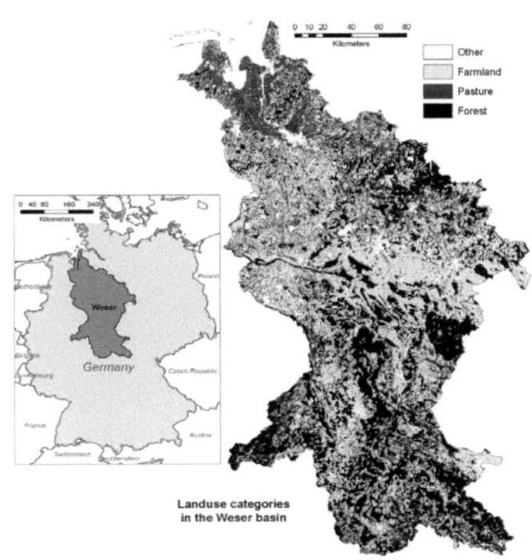

Figure 1: Location of and land use categories in the Weser basin

adapted agri-environmental reduction measures can be derived and implemented to improve groundwater and surface water quality by 2015, as requested by the Water Framework Directive.

Methods

The achievement of good qualitative status of groundwater bodies entails a particular challenge especially for large river basins as the complex ecological, hydrological, hydrogeological and agro-economic relationships have to be considered simultaneously. Integrated large scale agri-economic- hydrologic models are powerful tools to analyze the actual pollution loads and "hot spot" areas and to predict the temporal and spatial effects of reduction measures. We used the model network AGRUM [1] to predict the nitrogen intakes into groundwater and the nitrogen losses to surface waters by different pathways at the regional scale using an area differentiated approach. The model system combines the agro-economic model RAUMIS [3-5] for estimating nitrogen surpluses from agriculture and the hydrological models GROWA/DENUZ/WEKU [6-8] and MONERIS [9] for describing the reactive nitrate transport in the soil-groundwater system. Nitrogen transport by groundwater runoff, surface runoff, drainage runoff and natural interflow is considered.

The mean long-term nitrate concentration in percolation water is calculated from the long-term percolation water rates, annual nitrogen surpluses from agriculture and denitrification in the soil as input variables. Denitrification losses in the soil are calculated using the DENUZ model as a function of the diffuse N-surpluses, denitrification conditions and the residence time of percolation water in the soil according to a Michaelis-Menten kinetics. The reaction kinetic parameters were assessed from observed denitrification rates in German soils [10] according to the geological substrate, the influence of groundwater and perching water of the soils and the residence time of perching water in the soil [7, 11].

The displacement of N-surpluses into surface waters is coupled to the different runoff components. These are calculated by the water balance model GROWA [6]. It employs an empirical approach with a temporal resolution of one or more years. Annual averages of the main water balance components (percolation water rate, surface runoff, drainage runoff, (natural) interflow and groundwater runoff) are quantified as a function of climate, soil, geology, topography and land use conditions.

Results

In a first step the model was used to analyze the present situation using N-surpluses from agriculture for the year 2003 (see figure 2). In many regions of the Weser basin, particularly in the north-western part, characterized by high livestock densities, predicted nitrate concentrations in percolation water exceed the threshold value for groundwater of 50 mg NO_3/L by far. In parallel high nitrogen outputs to surface waters are calculated for these areas. Low nitrate concentrations in the leachate occur in the north-eastern loose rock areas as well as in regions of the southern part of the Weser basin used for extensive agriculture. With reference to a nitrate concentration of 50 mg NO_3/L in the leachate as an environmental target nitrogen, substantial reductions of the inputs from agriculture are required for many regions of the Weser basin [12].

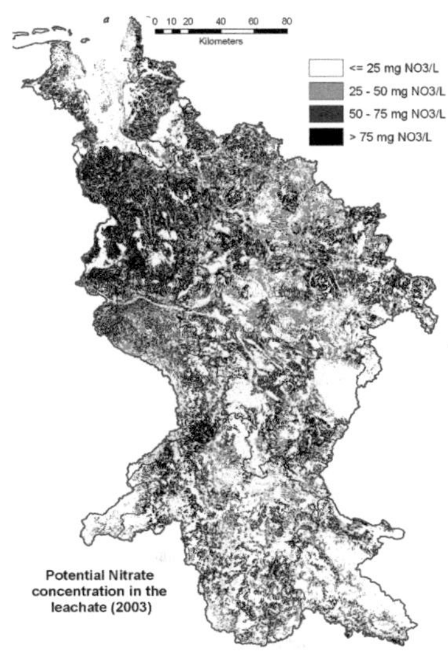

Figure 2: Potential nitrate concentration in the leachate of the Weser basin for the present state (2003).

It is important to distinguish between the effects of measures, which have already been implemented by current agricultural policy and measures which have to be implemented additionally. For this purpose a baseline scenario is developed, which projects the effects of modified general conditions of the agricultural sector on the nitrogen surpluses to the year 2015 [5, 13]. The left part of figure 3 shows the potential changes of N-surpluses from agriculture in 2015 with respect to the current situation (2003). For most regions of the Weser basin, a reduction of the N-surpluses from agriculture can be expected. Especially for the agriculturally intensive used regions the expected reduction of N-surpluses may amount to 40 kg ha^{-1} yr^{-1} or more. On average for the whole Weser basin, however, the N-surplus reduction is much smaller and expected to be about 10 kg ha^{-1} yr^{-1} or less.

The right part of figure 3 shows the calculated nitrate concentration in the leachate based on the N-surpluses from the baseline scenario. In comparison to the present situation already reduced nitrate concentrations in the leachate can be expected by 2015. A comparison shows that the potential nitrate concentration in the leachate will decrease in almost

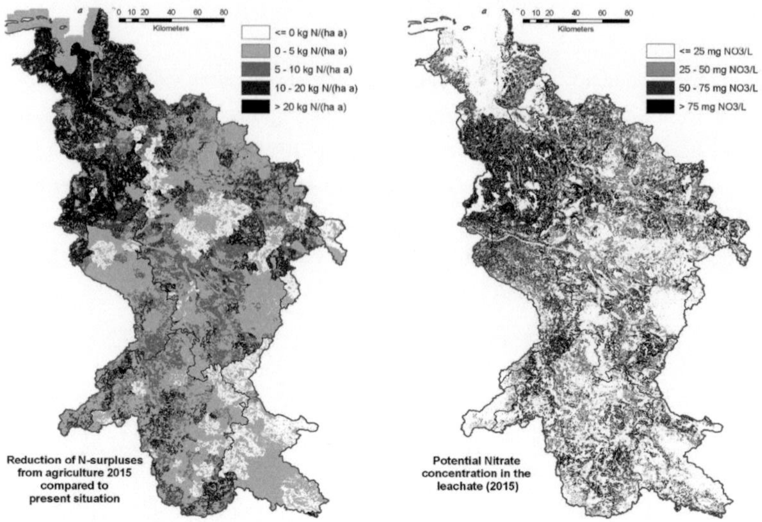

Figure 3: Reduction of N-surpluses from agriculture for 2015 according to the baseline scenario (left) and calculated nitrate concentration in the leachate for 2015 (right).

all regions of the Weser basin, mostly by about 10 mg NO_3/L. In the agricultural intensive used regions much higher reductions in the order of 40 mg NO_3/L may be expected. Consequently, reduced nitrogen outflows to surface waters via the different pathways are obtained.

On the other hand it becomes clear that additional reductions of N-surpluses from agriculture are still required for many regions of the Weser basin. The amount of the necessary reduction of N-surpluses can be quantified by backward calculation of the model. Figure 4 shows the necessary N-reduction from agriculture calculated under the assumption that 50 mg NO_3/L or less is to be realized below each agricultural acreage. The required N-reduction to reach 50 mg NO_3/l in the percolation water ranges between 0 and more than 75 kg ha^{-1} yr^{-1}.

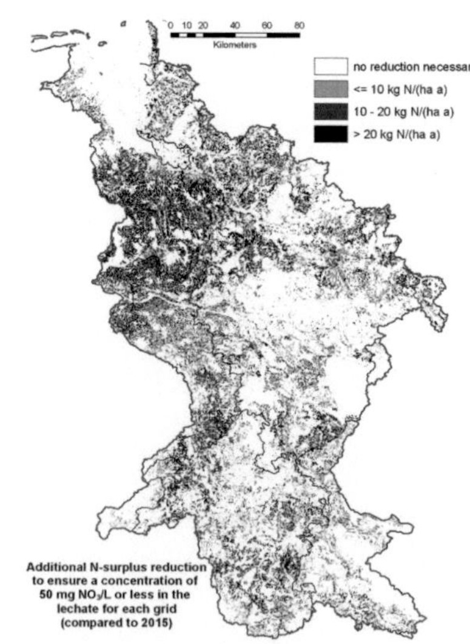

Figure 4: Additional N-reduction required reaching 50 mg NO_3/L in the percolation water per grid with respect to the baseline scenario (2015).

Especially in regions with area-independent animal production the necessary N-surplus reduction levels are high and affect vast areas. In these areas appropriate measures to achieve this extent of N reduction necessary are the prohibition to apply manure fertilizers after harvesting, improved techniques to apply liquid manure and (to a certain extent) the introduction of catch cropping systems.

Conclusions

The combined agro-economical/hydrological model network AGRUM has been used to assess mean long term nitrate concentrations in the leachate for the Weser basin. Model results showed large regional differences ranging between less than 10 mg NO_3/L and more than 100 mg NO_3/L. The modelled mean long term nitrate concentrations in the leachate have been used in a backward modelling approach to determine nitrogen reduction levels necessary to reach the groundwater quality target of at most 50 mg NO_3/L, i.e. the EU drinking water standard for nitrate in groundwater. Especially in the regions dominated by intensive animal husbandry, i.e. the north western part of the Weser basin, the necessary N-surplus reduction would be very high (> 50 kg ha^{-1} yr^{-1}). Most of all the north-western part of the Weser basin could be identified as a priority area for implementing nitrogen reduction measures.

The presented results of the project here will support the river Weser commission (FGG) and the Federal German States abutting the Weser to develop and implement the river Weser basin management plan, including proposals for the allocation of compensation payments to the regions in the Weser basin concerned by agricultural reduction measures.

List of Literature

[1.] Flussgebietsgemeinschaft Weser. *Bewirtschaftungsplan Flussgebietseinheit Weser - Bestandsaufnahme 2005.* 2.5.2008]; Available from: http://www.fgg-weser.de/wrrl/bericht_2005.html, 2005.

[2.] Kreins, P., H. Behrendt, H. Gömann , U. Hirt, R. Kunkel, K. Seidel, B. Tetzlaff, and F. Wendland, *Analyse von Agrar- und Umweltmaßnahmen im Bereich des landwirtschaftlichen Gewässerschutzes vor dem Hintergrund der EG-Wasserrahmenrichtlinie in der Flussgebietseinheit Weser.* Landbauforschung. Vol. Sonderheft 336. Braunschweig, Germany: vTI agriculture and forestry research, 2010.

[3.] Henrichsmeyer, W., C. Cypris, W. Löhe, M. Meudt, R. Sander, F. von Sothen, F. Isermeyer, A. Schefski, K.-H. Schleef, E. Neander, F. Fasterding, B. Helmcke, M. Neumann, H. Nieberg, D. Manegold, and T. Meier, *Entwicklung eines gesamtdeutschen Agrarsektormodells RAUMIS96. Endbericht zum Kooperationsprojekt.* Institut für Agrarpolitik, Marktforschung und Wirtschaftssoziologie der Universität Bonn, Institute für Betriebswirtschaft, Strukturforschung und Landwirtschaftliche Marktforschung der Bundesforschungseinrichtung für Landwirtschaft (FAL): Bonn, Braunschweig, Germany, 1996.

[4.] Kreins, P., H. Gömann, S. Herrmann, R. Kunkel, and F. Wendland, *Integrated agricultural and hydrological modelling within an intensive livestock region.* Advances in the Economics of Environmental Resources. **7**: p. 113-143, 2007.

[5.] Kreins, P., H. Gömann , C. Heidecke, and U. Hirt, *Modelling network for analysing diffuse nitrogen leaching reduction in agriculture to meet the targets of the water framework directive in the Weser River Basin of Germany*, in *Integrated Water Resources Management (IWRM) Karlsruhe 2010*: Karlsruhe, Germany, 2010.

[6.] Kunkel, R. and F. Wendland, *The GROWA98 model for water balance analysis in large river basins - the river Elbe case study.* Journal of Hydrology. **259**(1-4): p. 152-162, 2002.

[7.] Kunkel, R. and F. Wendland, *Diffuse Nitrateinträge in die Grund- und Oberflächengewässer von Rhein und Ems.* Schriften des Forschungszentrums Jülich, Reihe Umwelt/Environment. Jülich, Germany: Forschungszentrum Jülich GmbH, 2006.

[8.] Kunkel, R. and F. Wendland, *WEKU - A GIS-supported stochastic model of groundwater residence times in upper aquifers for the supraregional groundwater management.* Environmental Geology. **30**(1-2): p. 1-9, 1997.

[9.] Hirt, U., M. Venohr, P. Kreins, and H. Behrendt, *Modelling nutrient emissions and the impact of nutrient reduction measures in the Weser river basin, Germany.* Water Science and Technology **58**(11): p. 2251-2258, 2008.

[10.] NLfB, *Grundwasser: Methodenbeschreibung.* Niedersächsisches Landesamt für Bodenforschung, Niedersächsisches Landesamt für Ökologie, Bezirksregierung Hannover: Hannover, Germany, 2005.

[11.] Kunkel, R., P. Kreins, B. Tetzlaff, and F. Wendland, *Forecasting the effects of EU policy measures on the nitrate pollution of groundwater and surface waters.* Journal of Environmental Sciences. **22**(6): p. 872-877, 2010.

[12.] Wendland, F., H. Behrendt, H. Gömann, U. Hirt, P. Kreins, U. Kuhn, R. Kunkel, and B. Tetzlaff, *Determination of nitrogen reduction levels necessary to reach groundwater quality targets in large river basins: the Weser basin case study, Germany.* Nutrient Cycling in Agroecosystems. **85**(1): p. 63-78, 2009.

[13.] Gömann, H., P. Kreins, and F. Wendland, *Zukünftige Entwicklung der N-Bilanzen und geeignete Maßnahmenkombinationen zur Reduzierung diffuser N-Einträge im Einzugsgebiet der Weser*, in *Stoffströme in Flussgebieten. Von der Bilanzierung zur Bewirtschaftung*, S. Fuchs, S. Fach, and H.H. Hahn, Editors. Verlag Siedlungswasserwirtschaft Karlsruhe: Karlsruhe, Germany. p. 267-277, 2008.

Relevance and Importance of Groundwater Artificial Recharge in the Jordan Valley Area

Prof. Dr. Elias Salameh

Faculty of Science, Department of Geology, University of Jordan, Amman-Jordan

Abstract

The Jordan Valley is the base level for all the natural water courses of the northern highlands. It is built of recent clastic sediments forming local aquifers. These aquifers are not filled with water, but are even overexploited.
According to ongoing geologic and geophysical investigations rechargeable aquifers which can serve as storage reservoirs are found in the JV. They can accommodate several tens of millions of cubic meters of recharge water.
Flood waters of inter catchments and wadis downstream of dams in addition to treated waste water, especially during the rainy season still flow unutilized and can be used as recharge water.
Water /water and water /rock interactions show that the recharge water will not have any adverse impacts on the groundwater quality or on major dissolution precipitation processes in the aquifers.
Therefore, countries sharing the Jordan Valley area should utilize its recharge potential during the rainy season and in times when there is no need for water use, to store any excess water in the underground for times of need. This seems a more appropriate way of storing water than surface reservoirs with all their problems of evaporation, eutrophication and pollution.

Keywords: Jordan Valley, groundwater, artificial recharge, treated waste water, intercatchments.

1. Introduction and Problem Statement

Due to increasing urbanization, industrialization and improving living standards along the northwestern highlands of Jordan which host the main cities and around 3/4 of Jordan's population, increasing water sources have been, and are planned to be imported into the area from other areas to satisfy the increasing demand (MOWI 2010).
As a result, increasing amounts of treated effluents are produced, which together with the flood flows of un-dammed wadis require storage, especially during the wet season when water use for irrigation is low (RWC 2008).

Surface storage facilities such as weirs and dams can surely solve the storage problem of the unused water until needed. But, surface storage of treated waste water and/ or a mixture of it with flood water suffer, in the arid and semi arid climatic zones of:

- High losses to evaporation
- Eutrophication
- Salinization due to evaporation

The Jordan Valley is the base level for all the natural water courses of the northern highlands. It is built of recent clastic sediments forming local aquifers. These aquifers are not filled with water, but are overexploited (Salameh 2002).

Therefore, it lies at hand to use these aquifers, which waters are generally used for irrigation, for the storage of water by recharging the excess surface water into them.

Such a program will have the advantages:

- Storing excess water for times of need.
- Saving water from evaporation and salinity increases.
- Guaranteeing further self-purification of treated effluents during the recharge and underground flow processes.
- Halting saline water encroachments, caused by aquifer over exploitation.

The relevance of such artificial recharge schemes will be explained on examples from the Jordan Valley area, illustrating the recharge potentials and the water/groundwater chemical reactions.

2. Topography

The Jordan Valley (JV) extends from Lake Tiberias at an elevation of 212 m below Sea Level (bsl) southward to the Dead Sea at an elevation in 2009 of 422 m bsl (Potash Co.). The width of the JV, just to the north of the Dead Sea is around 20 km, in its northern part around 10km and in its central part around 4 km. On the east side of the Jordan River this width varies between 2 and 10 km.

The length of the JV is 105 km. It is bordered on both sides; east and west by high, steep escarpments with differences in elevations between the valley floor and the surrounding mountains of 1200 m going up to 1700 m.

The lower Jordan River flows longitudinally through the valley. It takes its water from Lake Tiberias and from the side wadis to its east and west. The river has eroded its way through the valley to form a secondary trough, some 60m deeper than the JV.

The river meanders in very large loops in the alluvial plain. The difference in elevation between Lake Tiberias and the Dead Sea is around 200 m. The overall slope of the valley is 1.81 %. Due to meandering, the length of the river itself is around 220 km with an average gradient of 0.86 % or about half the slope of the valley.

This river gradient is still a steep one with erosion capabilities. The strong and violent floods of the river erode along its entire course from the north to the south.
The main features of the JV area are the relatively flat terraces on both sides of the River Jordan, which constitutes the bed of the valley.
The area under consideration extends from the Jordan River in the West to the mountain foot hills in the East. It is a relatively flat area in its western parts; in the

JV proper and hilly with deep incised wadis in its eastern parts at the mountain foot hills.

3. Climate

The prevailing climate in the Jordan Valley area is unique. It differs totally from the climate of its surroundings, which to the east and west lie at elevations of more than 1000m rising within a few kilometers from the valley floor (at elevations of 200 to 400 m bsl). Day's temperatures in summer range from 35 C0 to 50C0 and in winter from a few degrees above zero to 30C0. The relative humidity ranges from 45% to 80% and rainfall from an average of 350mm/yr in the northern parts south of Lake Tiberias to 100mm/yr in the Dead Sea area, mostly committed in the period November to March (DOM open files).

4. Suitability of groundwater recharge programs

For aquifer recharge the following preconditions must be fulfilled:

- Availability of a suitable aquifer with potential storage capacity
- Availability of suitable recharge water
- Adequate ground surface conditions not contradicting existing land use
- Groundwater flow direction and artificial extraction sites are clear and defined
- Water /water and water/rock interactions are environmentally tolerable.

5. Geologic Formations

The JV area is covered by recent rock deposits originating from the surrounding mountains. They generally consist of unconsolidated sediments of gravel, sand, clay and marl. Partially these sediments interfinger with sediments deposited in the ancestor lakes of the present Dead Sea. Table 1 gives the formations building the upper part of the JV and their thicknesses. Of direct relevance to the recharge study are the alluvial fans and the recent deposits.
Locating the alluvial fans in the JV area faces a major problem due to the sinistral movement along the N-S striking strike slip fault, partly of which is found along the eastern side of the JV, between the valley itself and the mountain foothills. Accordingly wadis, which deposited the sediments as alluvial fans have shifted to the north leaving behind the alluvial deposits exposed to erosion and leveling. Locating these fossil alluvial deposits requires therefore, indirect geologic and geophysical methods. Locating them will allow their full utilization in groundwater artificial recharge potentials.

Formation Name	**Thickness m**
Alluvial fans and recent deposits	0 – 100
Lisan	40 – few hundred
Ubeidiya and Samra	35
Abu Habil Conglomerates	100
Shagur Conglomerates	~ 350 m

Table 1: Shallow geologic formations in the Jordan Valley area and their thicknesses

According to ongoing geologic and geophysical investigations, rechargeable aquifers which can serve as storage reservoirs are found in the JV (Fig. 1, Al-Amoush 2006). They can accommodate several tens of millions of cubic meters of recharge water. Figure 2 shows a geologic cross section in the northern part of the JV, illustrating the geologic and geomorphologic situation in the area.

The clastic alluvial deposits in the JV fine up towards the Jordan River which is an advantage to recharge schemes, due to decreasing permeability and hence limited groundwater flow into the Jordan River.

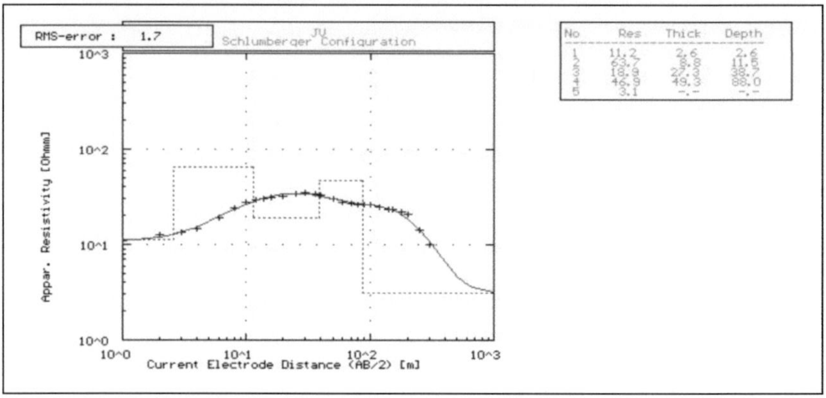

No	Res	Thick	Depth
1	11.2	2.6	2.6
2	63.7	8.8	11.5
3	18.9	27.3	38.7
4	46.9	49.3	88.0
5	3.1	-.-	-.-

Fig. 1: Example on geoelectric soundings in the northern Jordan Valley area

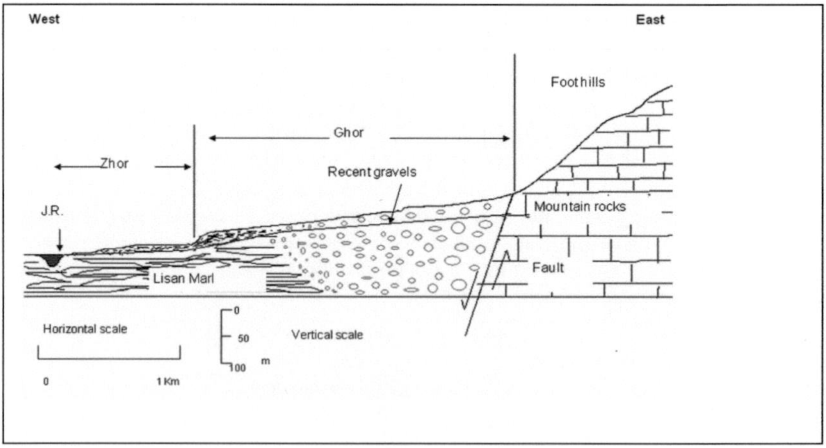

Figure 2: Geologic cross-section in the northern Jordan Valley (Suleikhat area)

6. Availability of recharge water

Flood waters of inter catchments and wadis downstream of dams in addition to treated waste water, especially during the rainy season still flow unutilized and can be used as recharge water (Table 2). Increasing water imports into the Jordan River eastern catchment area for domestic and industrial use is gradually resulting in increasing amounts of treated waste water which should be stored for times of irrigation needs.

	Flow Mio m3/a.		
Intercatchment Between:	**Flood**	**Base**	**Total**
Arab and Ziqlab (AB21)	5	1	6
Ziqlab and Jurm (AB22)	3	1.5	4.8
Jurm and Yabis (AB23)	1.5	0.0	1.5
Yabis and Kufranje (AB24)	0.1	1.3	1.4
Kufranje and Rajib (AB)	0.2	1.2	1.4
Catchment in the downstream of:			
Wadi Arab Dam	Negligible	0.0	0.0
Ziglab Dam	0.5	0.0	0.5

Table 2: Average discharge of intercatchments between each two neighboring side wadis of the Jordan River (Water Authority of Jordan).

7. Land suitability for recharge and extraction needs

Along the foothills of the eastern mountains of the JV land is generally not used and managed groundwater artificial projects can easily be implemented there either via recharge basins or through land application of recharge water.

Table 3 gives field test results on the infiltration potentials of the alluvial deposits in the JV.

Extraction of recharge water can be accomplished in areas in between the recharge areas and the Jordan River, before the groundwater enters the Lisan Formation regimes and becomes salinized.

Area	Testing time hours	Drop in water level (cm)	Drop in cm/day
Yabis	96	23	5.75
Waqqas	24	4.5	4.5
Manshiya	72	9.3	3.1
Tel Arabain	24	5.0	5.1
Zumayliya 1	24	3.2	3.2
Zumayliya 2	48	4.1	2.05
Arda	24	4.8	4.8
Gravels (6 tests)		10 cm/minute	14976
Gravel pool		35.167 cm/hour	844

Table 3: Infiltration tests results

8. Water /water and water /rock interactions

Mixing of groundwater possessing average composition in the fresh groundwater regimes of the northern JV area with water collected in the dams of that area, which is composed of a mixture of flood, base and treated waste water shows no major changes in the saturation indices of the resulting mixed water (Table 4). This indicates that mixing recharge water with the area's groundwater in different ratios will not produce major changes in the saturation indices of the mixed water and hence, it will not initiate any additional water rock interactions.

Sample ID	SI (Anhydrite)	SI (Calcite)	SI (Dolomite)	SI (Gypsum)	SI (Halite)
AVG GW to Max SW					
1 to 0	-4.4791	-1.7888	-3.1442	-4.2418	-9.0896
2 to 1	-4.6159	-2.5032	-4.5398	-4.3786	-9.2812
1 to 1	-4.6946	-2.7022	-4.9175	-4.4573	-9.396
1 to 2	-4.782	-2.8565	-5.2024	-4.5448	-9.5286
0 to 1	-4.9925	-3.144	-5.7161	-4.7552	-9.8779

Table 4: Saturation indices of average groundwater composition (AVG GW) and surface water of maximum composition (MAX SW) and of their mixtures in different ratios.

9. Conclusions

The JV area which is generally built of recent clastic sediments consisting mainly of course grained alluvial deposits possessing high porosities and permeabilities resulting in their high reservoir capacities. Groundwater levels in the area are dropping as a result of overexploitation and surface water in the form of flood waters of inter catchments and of wadis downstream of dams in addition to treated waste water, still flow unutilized and can be used as recharge water, especially during the rainy season.

Water /water and water /rock interactions show that the recharge water will not have any adverse impacts on the groundwater quality or on major dissolution precipitation processes in the aquifers.

Countries sharing the Jordan Valley area should utilize the valley's recharge potential to store by artificial recharge any excess water in the underground for times of need. That is especially valid for the rainy season and in times when there is no need for the excess surface water,. This seems a more appropriate way of storing water than surface reservoirs with all their problems of evaporation, eutrophication and pollution.

10. References

[1] Al-Amoush, H. 2006: Hydro-geophysical investigations for the purpose of groundwater artificial recharge in the Jordan Valley area. Unpublished Ph. D. thesis University of Jordan.

[2] MOWI: Ministry of Water and Irrigation 2010: Mega water projects in Jordan (Open files).

[3] Potash Company Jordan (2009) Open files

[4] RWC: Royal Committee on Water 2008: Water for life, Jordan's Water Strategy 2008-2022.

[5] Salameh, E. 2002: The potential of groundwater artificial recharge in the Jordan Valley area, Jordan Freiberger Forschungshefte, C494, Geowissenschaften 2002.

Development of water resources in Oman Mountains by capturing storm water runoff and underground water storage in artificial galleries

Günter HAHN, Heinz HÖTZL & Klaus-Dieter SCHLEGEL

Abstract

Extra water reserves are to be won in the Oman Mountains during the seasonal precipitation, when the surface water is captured and conserved for the March to December dry season. This sets the choice of preferred regions, which have the meteorology, hydrogeology, and geography as advance conditions for the collection of water, the underwater retention, conservation and economic distribution available. Important factors, because of the loss through seepage and evaporation, are the precipitation rate of over 15 mm per event and a corresponding drainage basin. Planning guidelines are made accordingly.

Gaining this surface water requires specific dimensioned overspill dams with individual measures for the debris collection in the overflow drainage areas. These serve in part as a preliminary screening and collection of water in the tributaries. The collected and purified water is stored in underground manmade reservoirs according to the grade of purification. These reservoirs are built up either as a single reservoir or as a combination system depending on the logistical and expected uses. Suitable is the same technology that is used in traffic tunnel building.

Dr. Günter Hahn, M.D.; M.Sc., Executive Scientific Consultant, D-76534 Baden-Baden, Sommerstrasse 12, Germany, E-Mail: guenterhahn@ yahoo.de
Prof. Dr. Heinz Hötzl, Institut für Angewandte Geowissenschaften, KIT; D-76131 Karlsruhe, Adenauerring 20b, Germany,E-Mail:heinz.hoetzl@kit.edu
Klaus- Dieter Schlegel, Dipl.-Ing (FH), Herrenknecht AG Tunnelling Systems; D-77963 Schwanau-Allmannsweier, Schlehenweg 2, Germany, E-Mail: schlegel.klaus-dieter@herrenknecht.de

1. Introduction

Water shortage was in arid and semiarid regions often a reason for restricted civil and economic development. Wide ranging possibilities of urbanisation arose first with the desalination of ocean water in the coastal regions, also in the Sultanate of Omen. Through this the population in the greater region Muscat developed from 800.000 to a total population of ca. 2.8 mil. Also along the coastal zone of the Gulf of Oman the population grew over proportionally, whereby the water supply largely comes from wells. The ingression of ocean water into aquifers because of the lowering of the ground water retention leads to the increased salinization of the useable agricultural areas.

The north of Oman is shaped through an up to 3000 m mountain range (Jabal Hajar), this reaches from the peninsula Musadam to its foothills in the Arabic See parallel to the Gulf of Oman. Geologically there are varied types of rock forms. In Jabal Al Akhdar the highest mountain range of Jabal Hajar there is an enlarged limestone area; other regions are dominated by ophiolithes.

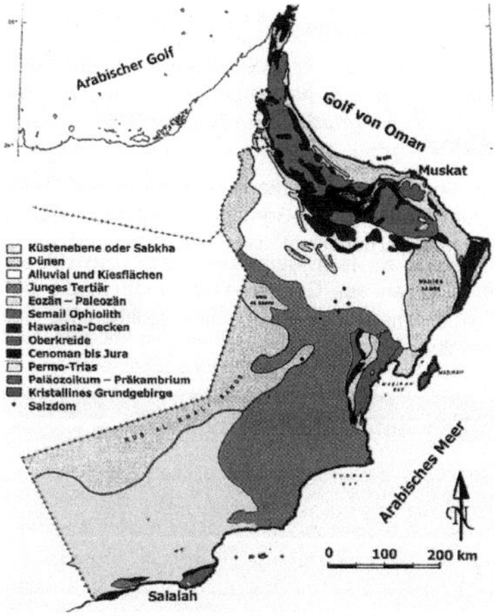

Fig. 1: Geological Map of the Sultanate of Oman

Together both rock forms clearly reduce the water infiltration into the underground; in region of the carbonate rocks by the partly steeply inclined smooth bedding planes on the surface, in regions with ophiolithes through the dense rock properties. Therefore the seepage from precipitation is largely reduced. This results in the quick run off from precipitation. Vegetative growth is seldom present. In the dry river beds and in the offshore plateaus we find enlarged areas of alluvial fans, which ramble up to 40 km from the Gulf of Oman respectively extending west towards the desert.

This situation already leads by precipitation quantities above 15 mm per event to short lived high water levels with a considerable degree of large area flooding. A large quantity of debris is carried with the flow; this becomes evident from several dams built parallel to the Gulf of Oman, which had been built to enhance the coastal ground water retention and were filled up with debris in few years.

Characteristic for the precipitation in the north of Oman is a period between December and March followed by a longer dry period. The main amount of precipitation usually results from a few events (about 1-3 per precipitation period).

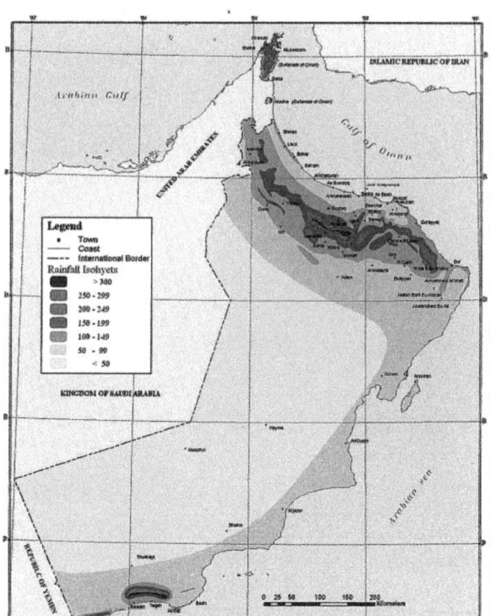

Fig. 2: Distribution of precipitations in the Sultanate of Oman

Despite seasonal precipitation with partially yearly top values of over 600 mm in Jabal Al Akhdar, in the rainless time (dry season) from March to December begin massive regional inland water shortages. Presently these shortages in the communities are compensated through a governmental tank truck supply system.

This is true for many regions of the Jabal Hajar and its foothills. Exactly in these regions the water shortage has lead to migration in the coastal regions.

In Jabal Hajar respectively Jabal Al Akhdar and its regions of influence from the coast of the Gulf of Oman to the foothills west of the Oman Mountains very different precipitation heights are noted, on the coast an average of a little over 100 mm, on the west flank of the Oman mountains a +/- 200 mm fluctuation and in the heights of Jabal Al Akhdar it reaches up to 600 mm (2). The distribution of the precipitation (3) is shown in fig. 2, locally in the yearly cycle in fig. 3. In as far as it is possible, to use this seasonal precipitation in the dry season, for which a prerequisite it is for an appropriate economical water collection and retention system (4).

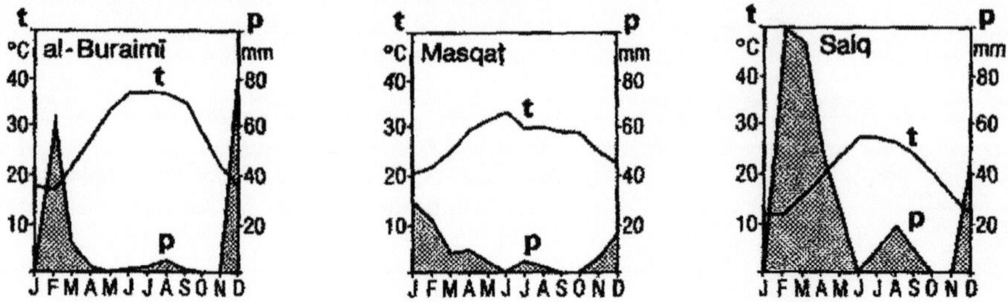

Figs. 3: Precipitation over a long time per year measured in different cities (3).

2. Prerequisites for the winning and storing of surface water

Among the first steps, for system optimization and choice of the preferred region, meteorological and hydrological evaluations are needed. These must be done in the desired regions as well as in regions where there are governmental guidelines for urbanization. Especially in the not yet habited regions of Jabal Hajar, specific water collection and retention would be an appropriate method for decentralizing the growing coastal population masses.

In the first step the following meteorological evaluations are needed:

- Statistics of the yearly precipitation over a longer time frame
- Recording of the precipitation occurrences
- Statistic of the intensity of the precipitation as well as the single total values
- Determination of losses through evaporation per precipitation occurrence

Through the currently available information in the Sultanate of Oman coarse assessment for regions of choice have become possible.

In addition gathering hydrogeological measurements are necessary to choose specific regions for the purpose of water collection. These are oriented on following facts:

- Determination of the run off amounts after strong rain fall for individually defined surfaces
- Determination of the necessary surface area corresponding to the needed amount of surface water
- Appropriate rock formations, in which water with help from overspill dams can possibly be collected into canyons and therefore stored

- Choice of appropriate underground diversion routes for underground retention
- Testing the technological possibilities of underground water storage with tunnelling technology

The effective retention quantity is arrived at through a possible retention amount per precipitation occurrence as follows:

(Total precipitation quantity – loss) x drainage basin = effective storage amount

Contingent on geology, the loss through evaporation on carbonate rocks of around 5 mm and by ophiolithes with a high solar thermal absorption of 5-10 mm per precipitation occurrence and hour can be presumed, the losses through seepage rest at by carbonate rocks 5-10 mm and by ophiolithes under 2 mm per occurrence and hour. In both cases a minimum loss of around 15 mm per occurrence and hour arise. These facts must be determined with exact studies at these locations.

If one sets a base need of daily 0.5 m³ water per person, exact plan guidelines for yearly minimum amounts per population quantity can be set. This per person need is made up of about 130 litres drinking water quality and three times this as usage water example: the watering of the wished for date palms and the small gardens in the Sultanate of Oman.

Fig. 4: New settlement nearby Nizwa in the Jabal Al Akhdar

Such a base need guideline can be demonstrated on an actual example from the Sultanate of Oman (fig. 4). This new settlement of 55 houses would, if as usual in the Sultanate of Oman be occupied by a 6-8 person households, needs 60000-80000 m³ water per year.

This water could easily be won from a near lying drainage basin of around 16 km² that flows from a valley just above the settlement. Out of the current information over this region an average yearly precipitation of 200 mm per year is expected, losses of 5 mm per hour each through evaporation and seepage and a minimum of 3 occurrences of at least 30 mm per hour, gives an effective run off amount of 800.000 m³ per year. With which up to 550 houses could be supplied.

3. Technical measures for the collecting and retention of surface water

The quick run off of surface water creates a large collection of rubble. Overspill dams are quickly made unusable, if this is not prevented in the upper reaches with a method of torrent lining. Beyond this there is the possibility of installing Tyrolean weirs in the individual channels (flow) that simultaneously serve to collect water and sort rubble (fig. 5).

Bestehende Fassung am Frilibach (Photo FMG SA)

Fig. 5: Example for a Tyrolean Weir for water collecting

Depending on the acceptable appropriate water drainage basin (ROI) to hold surface water, specified urbanisation planning guidelines, can be defined. For the collection of surface water either a single solution or a combination solution should be considered, in that two or more ROI are bound together. Thereby large amounts of water collected and retained regionally can go through a water distribution system to the end user.

The criteria for the single solution amount to:

- Solution for individual settlements
- Alternative to compensate for existing water deficits
- Defined governmental guidelines for a new grounding to water self-sufficiency
- Restricted geographic and geologic boundary conditions such as water basin region, urbanisation conditions, technical feasibility

The criteria for the combination solution amount to:

- Support solutions for more but regionally limited settlements
- Compensation for the water deficits in the urban regions of necessity
- Expansion of urbanisation within governmental designed guidelines
- At least one ROI with a high capacity or several ROI's on hand for the development of a network
- Geographic and geological boundary conditions

Fig. 6: Tunnelling Technology produced by Herrenknecht AG

Central core of the water collection system is the retention system that with tunnelling technology can be specifically, individually built and expanded (fig. 6). It concerns retention piping and connection systems with possible cross sections of up to 14 m. In opposition to surface reservoirs, that experiences especially in arid and semiarid regions large losses due to evaporation, produces these advantages:

- Individual accommodation to necessities
- Expansion of the system in reliance on the minimum available surface water
- The networking of the retention and supply systems with one another
- A self sufficient water supply system in identified deficit areas
- Separation of water into separate retention areas according to drink and usage water
- Substantial reduction of biological and /or chemical contamination
- The sustainability of the measures over a longer time frame, considering regular maintenance

The feasibility of a singular or a combination solution is connected to the geographic and hydrogeologic requirements, the dimensions of necessity and the estimated population development within the governmental planning guidelines. In all circumstances the "worst case" is based on the minimum available surface water.

For an optimization of the cost usage relationship, system development for regional deficit circumstances and the long range development of a total cost per m³ water must be fair, for the higher regions of Jabal Al Akhdar amount at present converted is € 8,-- per m³ water. Also to observe for the further sustainability of the method, its self-sufficiency e.g. versus other methods (e.g. Water desalination) as well as follow up costs. However the original investment cost, especially for the establishment of the storage tunnels is accordingly high.

A practical example for a singular as well as a combination solution can be drawn from the situation in the immediate area of the new settlement in fig. 4. Above the newly established settlement we find a water drainage basin of around 16 km² where the water flows off through a valley. At a distance of around 4 km is the boundary to a, in fig. 7, water drainage basin, whereby in similar manner the collection of surface water and the retention method can be used. Combining both systems would bring a minimum storage capacity of 1.6 Mio. m³ yearly.

Fig. 7: Water drainage basin demonstrated by satellite

To use retention tunnels, they must be carefully placed beneath the overflow dams in the valley. By a drill cross section of 10 m would be fig. 4 depicts that for the re-established tunnel length 1000 m would be needed, to cover the water needs for a year. To prevent losses through leakage in natural retention, it would have to be lined.

4. Conclusions

By the presented innovative methods for collection and underground retention of seasonal surface water in the Sultanate of Oman the methods manage to

- balance the considerable water deficits of the natural resources
- counteract conditional water shortage rural migration
- re-establish and re-urbanise through securing self-sufficient water supplies, also in uninhabited regions of Jabal Hajar e.g. Jabal Al Akhdar
- governmental security planning insures the allowance for decentralization of massive population centers

5. Literature

(1) HÖTZL, Heinz; HAHN, Günter; SABAH, Ali & WOLF, Leif 2009. Integrated Water Resources Management (IWRM), Concept for Arid and Semiarid Urban Regions. Proceedings of "International Conference on Capacity Building in Urban Water Management Under Scarcity Conditions", 13-15th December 2009, Muscat, Sultanate of Oman.

(2) HANNS, Christian; KÜRSCHNER, Harald. The Capital Area of Northern Oman, Teil 2; Dr. Ludwig Verlag, Wiesbaden 1998, pp. 44-45.

(3) Ministry of Regional Municipalities and Water Resources 2005, Sultanate of Oman. Water Resources in Oman, Cap. 2: Surface Water, p. 26.

(4) HAHN, Günter; HÖTZL, Heinz & BERSET, Joel Abd El Rahman 2010. Alternative Methods of Water Catchment in the Sultanate of Oman from Surface Water. "9th Gulf Water Conference: Water Sustainability in the GCC Countries", 22-24th March 2010, Muscat, Sultanate of Oman.

Hydrogeologic preconditions for a reasonable application of artificial aquifer recharge measures

J. Guttman[1)], H. Hötzl[2)], A. Marei[3)], A. Subha[4)] & L. Wolf[5)]

[1)] Mekorot Co., Tel Aviv, [2)] KIT, Karlsruhe, [3)] Al Quds University, Jerusalem, [4)] Jordan Ministry of Water and Irrigation, Amman, [5)] CSIRO, Brisbane, Australia

Abstract

Freshwater resources especially in arid and semiarid areas are under increasing pressure from partly extreme population growth, economic activities and intensifying competition for available water resources among users. To support regions which are suffering under seasonal water scarcity, artificial recharge of groundwater during the wet period of the year could be a solution to alleviate at least the lack of water during the dry season. The artificial recharge of groundwater preserves and stores the water in the aquifers. Compared with the surface storage of water in reservoirs the new technique has the advantage to avoid evaporation loses, which can range to more than 50 % of the surface reservoirs volumes.

Apart from temporarily excessive water availability the most important preconditions for water storage in the underground are the hydrogeologic properties and structural conditions of the underground in order to realize an economical reliable storage of the excess water. The main influence of the realisation of such a project depends on the hydraulic conductivity and the storage capacity of the rock formation. The structural conditions of the rock formations are responsible for the geometric dimension of the utilizable aquifer as well as of promotive spatial form, like synclines, troughs or traps with overflow conditions. For arid and semiarid areas the capture and underground storage of storm water run-off is of special importance. In this case extreme huge amount of water flow are available, however only over a short time. Therefore flood plains with specific infiltration capacity and aquifers with high storage capacity are requested, which can be frequently found in alluvial fans with only small gradients.

Catchwords

Freshwater, water scarcity, artificial recharge, groundwater, water storage, storm water run-off, arid and semiarid areas, hydraulic permeability, storage capacity of aquifers, retention time.

1 Introduction

The growing population and an increase of industrialization and agricultural production in numerous countries require more and more water of adequate quality. In many regions there is a lack of surface water and severe water contamination is to be found. Shallow groundwater resources are often of insufficient quality and over-exploited. Therefore, it is of high priority to take into consideration all the proved water techniques that could help to reduce the existing disaster.

Groundwater recharge is a hydrologic process where water moves downward from surface water through the unsaturated zone to groundwater. Groundwater is recharged naturally by precipitation and to a smaller extent by surface water from rivers or lakes. Groundwater recharge is an important process for sustainable groundwater management, since the volume-rate abstracted from an aquifer in the long term should be less than or equal to the volume-rate that is recharged.

Artificial recharge is the practice of increasing by artificial means the amount of water that enters a groundwater reservoir (Dillon 2005, Gale 2005). It is the planned human activity of augmenting the amount of groundwater available through works designed to increase the natural replenishment or percolation of surface waters into the groundwater aquifers, resulting in a corresponding increase in the amount of groundwater available for abstraction. Although the primary objective of this technology is to preserve or enhance groundwater resources, artificial recharge has been used for many other beneficial purposes. Some of these purposes include conservation or disposal of floodwaters, control of saltwater intrusion, storage of water to reduce pumping and piping costs, temporary regulation of groundwater abstraction, and water quality improvement

The concept of artificial recharge has been known for a long time. The practice began in Europe during the early nineteenth century. However, the practice has rarely been adopted on a large scale, with most large scale applications being found in countries such as the Netherlands, Germany, and USA. In the last fifty years the method has become a very important tool to improve the water scarcity situation particularly in semiarid to arid regions. Due to the growing population and the additional demand for water the municipal utilities as well as private suppliers are aimed to save each drop of the natural resources. One important part are the flood waters, which are generated sporadically and flow mainly as fast surface run-off without any utilization direct to the sea or in evaporation pans. For instance in the short wet periods Israel transports million cubic meters of water annually from the Lake Tiberius through the National Water Carrier System to the coastal plain and stores large amount of it underground. The water is used to meet high summer demands and offers a reliable source of supply during dry years. Of course, these flood water can be also stored in huge reservoirs. The disadvantage especially in the hot arid areas is the big loss by evaporation, which is demonstrated by the graph in Figure 1.

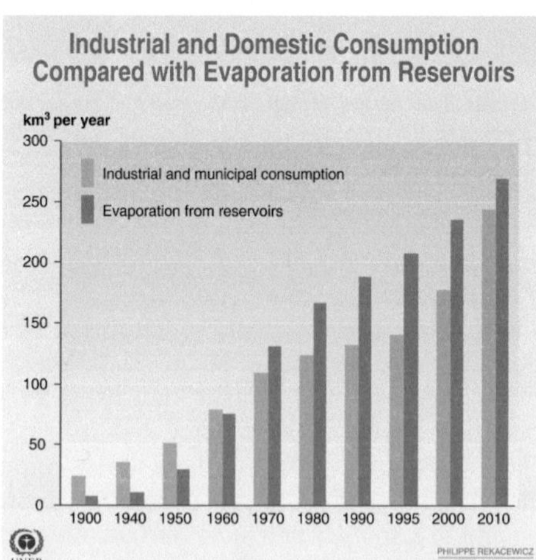

Source: Igor A. Shiklomanov, State Hydrological Institute (SHI, St. Petersburg) and United Nations Educational,Scientific and Cultural Organisation (UNESCO, Paris), 1999.

Figure 1: Global water losses in Reservoirs by evaporation compared with industrial and domestic Consumption (after UNESCO 1999).

2 Methods of Artificial Recharge

A variety of methods have been developed and applied to artificially recharge groundwater. The methods may be generally classified in the following four categories:

(1) Direct Surface Recharge Technique: Direct surface recharge techniques are among the simplest and most widely applied methods. In this method, water moves from the land surface to the aquifer by means of percolation through the soil (Fig. 2). In general, these methods have relatively low construction costs and are easy to operate and maintain. Direct subsurface recharge techniques convey water directly into an aquifer.

(2) Direct Subsurface Recharge Technique: Direct subsurface recharge methods access deeper aquifers and require less land than the direct surface recharge methods, but are more expensive to construct and maintain. Recharge wells, commonly called injection wells, are generally used to replenish groundwater when aquifers are deep and separated from the land surface by materials of low permeability. All the subsurface methods are susceptible to clogging by suspended solids, biological activity or chemical impurities. Recharge wells have been used to dispose of treated industrial wastewaters, to add freshwater to coastal aquifers experiencing saltwater intrusion

(3) Combination surface-subsurface methods, including subsurface drainage (collectors with wells), basins with pits, shafts, and wells. Combinations of several direct surface and subsurface techniques can be used in conjunction with one another to meet specific recharge needs.

(4) Indirect Recharge Techniques: these methods of artificial recharge include the installation of groundwater pumping facilities or infiltration galleries near hydraulically-connected surface water bodies (such as streams or lakes) to lower groundwater levels and induce infiltration elsewhere in the drainage basin, and modification of aquifers or construction of new aquifers to enhance or create groundwater reserves.

Figure 2: Infiltration basin of Menashe Artificial Recharge Plant, Israel, with intake building in the foreground (Gutman 2007)

3 Hydrogeologic Preconditions

Apart from temporarily excessive water availability the most important preconditions for water storage in the underground are the hydrogeologic properties and structural conditions of the underground in order to realize an economical reliable storage of the excess water. From the hydrogeologic properties of the rock the main influence of the realisation of such a project depend on the hydraulic conductivity and the storage capacity of the rock formation. Both properties determine what amount of water can be infiltrated and stored in a time unit and on the other hand how long the water will be available in the underground depending on the natural flow condition there. The structural conditions of the rock formations are responsible for the geometric dimension of the utilizable aquifer as well as of promotive spatial form, like synclines, troughs or traps with overflow conditions. The later can be extremely efficient by permitting the emptying of the whole trap during dry periods and their refilling after precipitation events before an overflow of the system occurs.

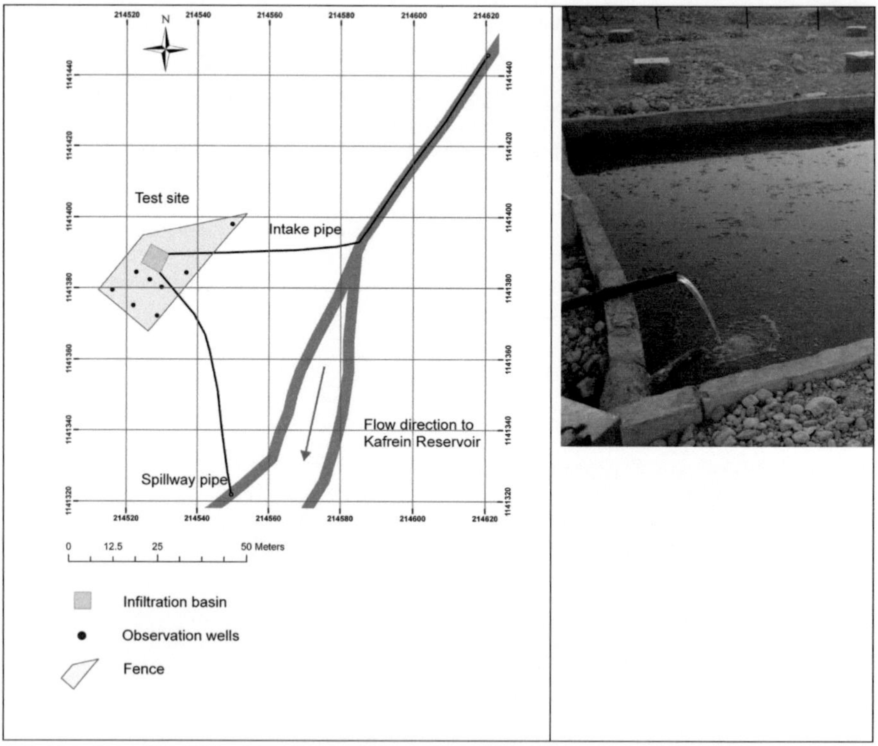

Figure 3: Artificial recharge test side at the Wadi Kafrein; left schematic map, right Foto of the infiltration basin (Wolf et al., 2009)

The realization and implementation of artificial groundwater recharge depends at first from the hydraulic properties of the rock material, in which the infiltration process will be carried out. A requirement for that is a certain vertical permeability of the soil and/or rock material. The infiltration capacity of the uppermost soil sequence is an important factor that governs the rate of soil moisture content in the unsaturated vadose zone and thereby the efficiency of the further percolation of the water down to the saturated zone. Infiltration capacities of different soil types are done by field-tests and are partly available by State Organizations. When the infiltration water reaches the groundwater table the horizontal permeability, expressed in the hydraulic conductivity is mainly relevant for the further dynamic of the water distribution. In general one can assume for sedimentary rocks, which are used in predominant cases for artificial recharge, a ratio between the horizontal and vertical hydraulic conductivity of 5:1. This means that the lateral flow of the artificially initiated percolating water should be in any case strong enough to deviate the arriving percolation water laterally. The desired effect of backwaters will occur in significant way, when the spatial extension of the aquifer cause backfilling or the thickness of the saturated aquifer is too small to discharge the whole percolating water. For the backfill process a high permeability is rather unfavorable, because high hydraulic conductivity would together with a significant gradient drain and discharge the infiltrated water in a short time.

The discharge behavior, however, doesn`t depend only from the high permeability, but in addition is a function of the hydraulic gradient, storage capacity and aquifer-geometry. Therefore the assessment of the appropriateness of an aquifer for artificial recharge should be done with respect to the hydrograph curve, indicating how long the stored water is delayed in the natural discharge process. This can be determined for instance by monitoring the decline of groundwater levels in observation wells or by measuring the discharge curve of springs. For the general application of surface direct recharge of seasonal floodwaters a retention time of at least one, better of 2 - 4 month are recommended. Flood waters occur in the rainy season, when partly still enough water is available in the soil for the plants. Therefore especially in Agriculture a longer retention time is favored, in order to extent the period of growing crops, vegetables and fruits. To have good recharge rate and to retain the recharged water for sufficient period for its use during longer period, moderate permeability is needed. Older alluvium, buried channels, alluvial fans, dune sands, glacial outwash etc. are the favourable places for recharge. In hard rock areas, fractured, weathered and cavernous rocks are capable of allowing high intake of water. The basaltic rocks i.e. those formed by lava flows, usually have large local pockets, which can take recharge water.

The second most important hydrogeological parameter for the assessment of the appropriateness of sediment and rock formations for artificial recharge is the storage capacity. This depends on one hand from the internal structure of the sediments and on the other hand from external factors like geometry, thickness and shape of the aquifer. The internal structure defines the porosity and the shape and type of the pores. The best preconditions for artificial recharge provide young loose sediment formations. While sand and gravels jut out with 20 and more percent of efficient porosity, one can find hardly more than 5% in hard rocks. Once again one has to consider all parameters to come to a final assessment. The low efficient porosity of hardrocks can be compensated by the size and volume of the hard rock aquifers

From the mentioned criteria follows a certain recommendation which type of soil and rocks are especially favorable for artificial groundwater recharge. High priorities have coarse unconsolidated sediments, like extended sand and gravel deposits, which are widely distributed in alluvial fans in the direct foreland of the mountains (Salameh 2002). But also other deposits like dune sediments and marine sand deposits are applicable so far they reach a certain thickness. From the consolidated sediments especially those are of interest, which are derivates from the original sand and gravel deposits. These are conglomerates and sandstones, which frequently still have a relative high percentage of residual porosity. Carbonate rocks are of interest, if they are highly karstified, so that due to this secondary solution porosity large amounts of water can be stored. Dense and tight rocks are of course unsuitable for artificial recharge. To these rocks belongs clay stones, siltstones, marls and most of the metamorphic and magmatic crystalline rocks. While the loose sediments are mainly horizontally stratified and occurring on the surface and covering other rock types, hard rock's can be deformed in wide troughs, synclines and down warped blocks. Some of these structure form large pockets or closed basin structure, which are specially proper for storage because of the can be treated like closed reservoirs, from where the whole water can be pumped out during the dry season and are filled up during the wet periods without requiring additional measures.

Figure 4: Bottom of the WALA Infiltration Reservoir, Jordan, covered with several meter thick silt and clays from four flood events. The sediments strongly reduce the original infiltration capacity of the reservoir.

For arid and semiarid areas the capture and underground storage of storm water run-off is of special importance. In this case extreme huge amount of water flow are available, however only over a short time. Therefore flood plains with specific infiltration capacity and aquifers with high storage capacity are requested, which can be frequently found in alluvial fans with only small gradients. One disadvantage of flood sediments, particularly under arid to semiarid conditions with missing protective vegetation cover, are the large amount of fine grained sediment loads. The deposition of this fine mud material is blocking the infiltration pores and therewith reduces dramatically the percolation of recharge water down to the groundwater (Fig. 4).

5 Conclusion

Freshwater resources in arid and semiarid areas are under strong pressure due to the natural climatic conditions. In many regions this pressure has increased dramatically due to partly extreme population growth and resulting additional water demand. To support regions, which are suffering under seasonal water scarcity, artificial recharge of groundwater during the wet period of the year could be a solution at least to reduce the lack of water. The options to install artificial recharge depend strongly from the hydrogeologic preconditions. In the report focus is drawn on the permeability and storage capacity of soil and rocks from which the efficiency and success of the recharge measures depend.

List of Illustrations, Abbildungsverzeichnis

List of Literature

Dillon, P. (2005). "Future management of aquifer recharge." Hydrogeology Journal(13): 313- 316.

Gale, I. (2005). Strategies for Managed Aquifer Recharge (MAR) in semi-arid areas. Paris, France, Unesco-IHP.

Guttman, J. (2007). Artificial Recharge in Mekorot - Examples from the Menashe and Shikma Plants, Internal presentation. Tel Aviv, Mekorot - Hydrology Department: 16.

Salameh, E. (2002): The potential of groundwater artificial recharge in the Jordan Valley area, in: Selected Contributions to Applied Geology in the Jordan Rift valley, Freiberger Forschungshefte, C 494, Geowissenschaften Freiberger Forschunsgheft C 494, ISBN 3- 86012-162-6.

Leif Wolf[1], Heike Werz[1], Heinz Hoetzl[1], Marwan Ghanem[2]: Exploring the potential of managed aquifer recharge to mitigate water scarcity in the Lower Jordan Basin ISMAR 2007, 30-46.

Supplying high purity recycled water for groundwater banking in rural areas – Approaches for the integrated management of common pool resources

Leif Wolf, Richard Creswell, Donald Begbie, Stewart Burn
CSIRO Land & Water, 120 Meiers Road, Indooroopilly 4068, QLD, Australia,
E-mail: leif.wo.f@csiro.au,

Abstract, Kurzfassung

The water supply security of South East Queensland has been increased recently by the provision of advanced wastewater treatment plants planned for indirect potable reuse of effluent from urban areas. With a maximum combined production capacity of 232 million litres of purified recycled water (PRW) a day, it is the third largest recycled water scheme in the world and the largest in the southern hemisphere. A significant amount (ca. 15-25 GL/a) of Purified Recycled Water (PRW) from urban areas is foreseen as augmentation of the depleted groundwater resources of the Lockyer Valley (approx. 80 km west of Brisbane). A current research project uses field investigations, lab trials and modelling techniques to address the associated management challenges: (i) how to quantify benefits for individual users from the augmentation of a natural common pool resource; (ii) how to minimise impacts of applying different quality water on the Lockyer soils, to creeks and on aquifer materials; (iii) how to minimise mobilisation of salts in the unsaturated and saturated zones as a result of increased deep drainage; and (iv) the potential for direct aquifer recharge using injection wells.

Keywords

Water recycling, water reuse, IWRM, groundwater,

1 Introduction

It is an increasingly accepted paradigm in semi-arid and arid contexts that future water demand can only be met in a sustainable fashion by employing multi-resource concepts, i.e. by integrating all available water resources (surface water, ground water, waste water, sea water) into a common management framework. The South East Queensland (SEQ) Water Grid is a highly dynamic example of such a system.

The water supply security of South East Queensland has been increased recently by the provision of advanced wastewater treatment plants planned for indirect potable reuse of effluent from urban areas. With a maximum combined production capacity of 232 million litres of purified recycled water (PRW) a day, it is the third largest recycled water scheme in the world and the largest in the southern hemisphere. The multiple barrier approach to ensure drinking water quality involves source control, tertiary wastewater treatment followed by micro- or ultra-filtration (MF), reverse osmosis (RO), and H2O2/UV advanced oxidation (Fig. 1).

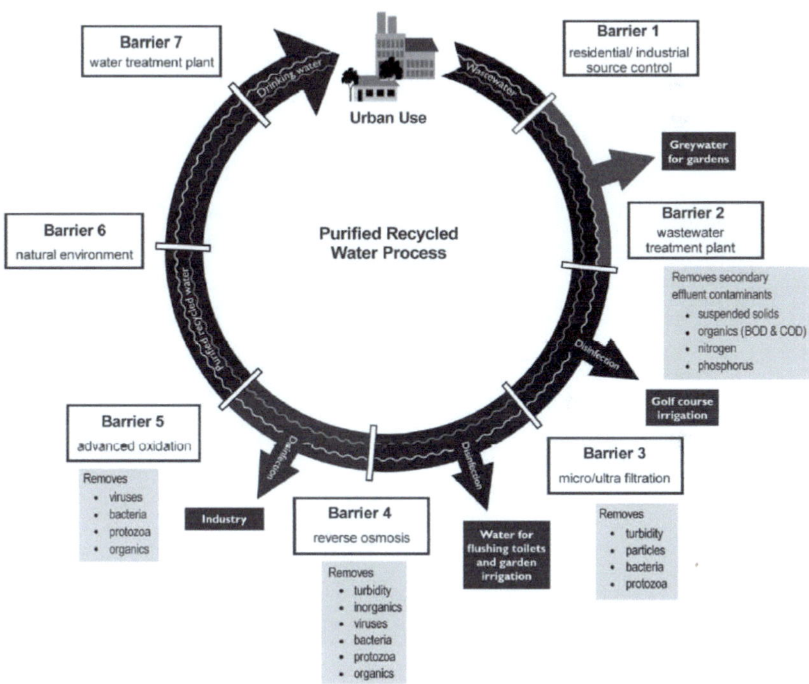

Fig.1: The multi-barrier concept for indirect potable reuse in Queensland [1]

2 The Lockyer Valley

A significant amount (ca. 25 GL/a) of the generated PRW is now foreseen for augmentation of the depleted groundwater resources of the Lockyer Valley (approx. 80 km west of Brisbane) (Fig.2). The Lockyer Valley is a heavily used agricultural catchment and a major contributor to the vegetable supply of Brisbane. The Lockyer Valley has undergone significant change over the past 100 years from a situation with perennial streamflow at the beginning of the 20th century to intermittent streamflow today. Irrigation bores have been drilled in increasing numbers since the early 1940s, with over 1,500 bores estimated to be currently servicing the Valley [2,3]. In selected areas, groundwater table drawdown manifested early and farmers have installed a total number of 17 weirs along the creeks of the Lockyer in order to temporary store runoff and to increase groundwater recharge since the 1950ies. Historically there has been an estimated annual groundwater withdrawal of up to 74,000 ML, although groundwater storage in the alluvial aquifers is estimated at a safe annual yield of only 27,000 ML/year [4]. The importation of additional water is thus required to sustain farming practices to their current extent. It is estimated that at least one-third of productive land already has been withdrawn temporarily from cultivation due to lack of water. Meanwhile, the over-appropriation has exacerbated salinity in some aquifers, as highly saline water seeps in from adjacent sandstone areas to replace the higher quality water taken from alluvial aquifers. In short, the Lockyer suffers from the difficulties of managing a common pool resource (compare with “the tragedy of the commons”, such as described in [5].

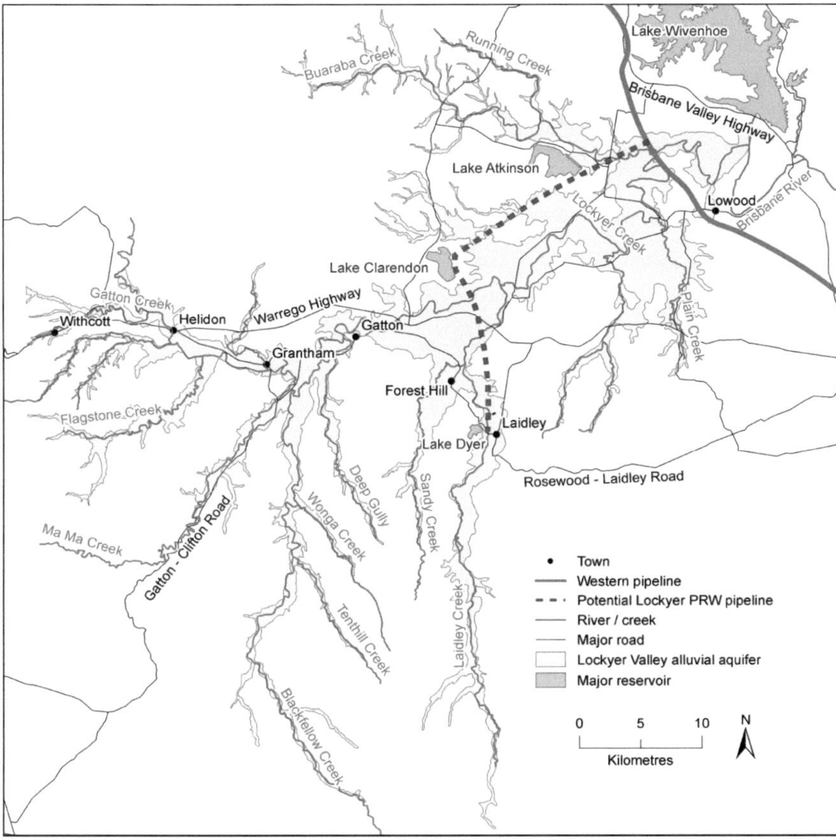

Fig.2: The Lockyer Valley alluvial aquifer and the planned diversion from the Western Pipeline

3 Outcomes

The suitable approach identified to address the common pool allocation problem is to define sustainable yields and volumetric water entitlements for the base case without PRW, then defining any additional water abstraction by individual users as being enabled through the artificial resource augmentation (PRW importation) [6]. A 3-D visualisation hydrogeological model was developed (Groundwater Visualisation System (GVS)) to integrate existing bore data and define the configuration of the alluvial aquifer system and its component volumes (Hawke et al, 2010)[7]. Available numerical groundwater models (MODLFOW SURFACT) were updated and critically assessed, and this highlighted the spatial heterogeneity of model sensitivities (RPS, 2010) [8]. The additional deep drainage from PRW application was modelled using a combination of the HOWLEAKY/APSIM model to predict likely deep drainage below the root zone, with these results fed into the HYDRUS model which was run for the main soil and crop types of the region. Within a sensitivity study, salt travel times across a 20 m unsaturated zone were modeled and found to range between 7 and 371 years [6]. The time lag between irrigation pulses and groundwater recharge pulses was found to vary between 150 days and 16 years, depending on subsoil hydraulic characteristics and existing moisture content. It can thus be concluded that no direct response of deep drainage rates to today's weather or farming conditions can be expected. The models are currently being verified via the extraction of a number of 20m deep soil cores, thus providing soil chloride-depth profiles for a chloride mass balance analysis. The response of clay minerals in soils and creek bed sediments to the changing electrolyte concentrations were also examined using column and batch tests.

4 Conclusion and Forecast

The supply of highly treated urban effluent to agricultural areas is potentially attractive despite the significant costs associated with treatment and transport which are mainly a function of the willingness and ability to pay by local farmers. Willingness to pay in the Lockyer currently ranges between 200 and 500 AU$/ML, but significantly prices up 20,000 AU$/ML have been reported for water rights in the wine growing areas of South Australia. In this context the release of recycled water to common pool resources such as aquifers or surface water systems can reduce the construction costs of water augmentation. To allow modeling of this augmentation, major model upgrades were found to be required in order to determine the spatial and temporal distribution of the potential benefits. Estimation of the change in diffuse recharge due to increased irrigation has been shown to be very sensitive to the crop types grown. Currently, an attempt is being made to utilize remote sensing information to generate landuse maps which are able to distinguish between different crop types. For the case study of the Lockyer Valley, the combination of Sodium Adsorption Ratios (SAR < 2) and Electrical Conductivity (E.C. <300 uS/cm) suggest no major problems will arise due to changes in soil structure upon the application of PRW water, but detailed tests are ongoing. Due to public acceptance issues, the indirect potable reuse scheme in South East Queensland remains an emergency top up for the urban drinking water supply, despite exceptional investments in water safety. Groundwater banking in an overexploited rural aquifer can constitute an elegant alternative method to respond to the seasonally varying urban demand for water from indirect potable reuse schemes and provide an alternative to PRW use for potable supplies.

Acknowlegements

The authors would like to acknowledge the contributions by David Rassam, Tim Ellis, Linda Foster, Jennifer Foley, Brett Robinson, Mark Silburn, Ashley Bleakley, Malcolm Cox, Amy Hawke, Allan James and last not least Jerome Arunakumaren and Peter Evans to the research project. The project was funded by the Urban Water Security Research Alliance (UWSRA), a joint undertaking between the Queensland Government, CSIRO and local universities.

.

List of Literature, Literaturverzeichnis ARIAL bold 13 pt

[1] W. H. Traves, E. A. Gardner, B. Dennien and D. Spiller (2008): Towards indirect potable reuse in South East Queensland, Water Science & Technology, vol **58.1**, p.153-161

[2] DPI (1994): Overview of the water resources of the Lockyer Valley.

[3] DNRM (Department of Natural Resources and Mines) (2005a). Lockyer Valley Discussion Paper: Declaration of the Whole Lockyer Valley as a Subartesian Area. Department of Natural Resources and Mines, Queensland.

[4] DNRM (Department of Natural Resources and Mines) (2003). Groundwater Trends in the Lockyer Valley, Brochure. Department of Natural Resources and Mines, Queensland

[5] Sarkera, A., Baldwinb, C. and Rossa, H. (2009): Managing groundwater as a common-pool resource: an Australian case study. Water Policy 11 (2009) 598–614

[6] Wolf, L.1, Cresswell, R.1, Rassam, D.1, Ellis, T.1, Toze, S.1, Foley, J.2, Robinson, B.2, Silburn, M.2, Bleakley, A.3, Arunakumaren, J.4, Evans, P.4, Hawke, A.5 and Cox, M.5 (2010): Urban-rural water exchange with Purified Recycled Water as an adjunct to

groundwater resources for irrigation in the Lockyer Valley, Proceedings Urban Water Security Research Alliance Science Forum, 27th/28th Sept.,Brisbane, Australia.

[7] RPS Consult (2010): Lockyer Valley Groundwater Modelling Sensitivity Analysis, 61p, Report to the UWSRA Alliance, Brisbane, Australia

[8] Hawke, A, Cox, M, Allen, J. (2010) Application of Groundwater Visualisation System (GVS) to the Lockyer Valley Alluvial Aquifer, 23p., Report to the UWSRA Alliance, Brisbane, Australia

Comparison of methods for isolation of natural organic matter (NOM) from loose deposits of drinking water distribution systems

Gruskevica K.,
Tihomirova K., Juhna T.

Department of Water Engineering and Technology,
Riga Technical University, 16 Azenes Street, Riga, LV-1048, Latvia,
phone/fax +371 67089085
(kamila.gr@inbox.lv; kristina.tihomirova@rtu.lv; talis.juhna@rtu.lv)

Abstract

Drinking water distribution systems are overloaded with loose deposits. Natural organic matter adsorbed on these sediments affect many chemical and microbiological processes, decreasing water microbiological stability. To estimate this effect it is necessary to know the amount of organic matter adsorbed on loose deposits. The special procedure for isolation of organic matter is needed for this purpose. Aim of this work was to compare commonly used methods for extraction and analyses of organic matter in loose deposits from drinking water distribution networks. For this extraction procedures (alkali extraction methods, sonication) were tested on 11 loose deposits samples. No correlations were found between extraction methods during this study. The natural organic matter extraction method using hot NaOH in combination with non-purgable organic carbon determination method is more accurate than other methods tested in this study for measure organic matter in loose deposits.

Catchwords

Loose deposits, Natural organic matter, extraction

Introduction

Drinking water distribution systems contain large amounts of loose deposits. Natural organic matter (NOM) attached at the interface between the water and the loose deposits plays significant role in many microbiological and physical processes and may affect water quality at the consumers' tap. Although the effect of dissolved organic matter entering from the treatment plant and leaching from the plastic pipes well studied (Van der Kooij,2003), the role of organic matter incorporated in the deposits has not been addressed sufficiently. Theses knowledge are especially important nowadays when the potential threat from deliberate terrorist attack is increasing whereas the loose deposits may serve as the temporary sink of the contaminants which are introduced in the networks.

There are two general approaches usually used: extraction of organic matter with the following determination of carbon and direct analyses of weight loss after combustion of the sample. Organic matter has a strong affinity to iron oxides (Gu et al., 1994) and thus an aggressive procedure of extraction is usually is needed. The aim of this study was to compare efficacy for determination of organic matter concentration in loose deposits from water distribution networks with conventionally used methods. Three methods which include extraction (strong base at different temperatures and sonication) with the following analyses of carbon and one combustion method was tested with loose deposits which were sampled from five networks in Latvia.

Materials and methods

Distribution system and sampling

Loose deposits samples were collected from several different drinking water distribution systems both from pipes and water reservoirs (WR). The samples of loose deposits were collected during a planned campaign of unidirectional flushing of distribution networks or from the bottom of WR in small towns of Latvia during the annual cleaning. All samples were collected in glass bottles or sterile plastic containers. After that the samples of loose deposits were delivered to the laboratory where they were concentrated on 0.45 μm pore-size cellulose filters, rinsed with 100 ml of sterile ultra pure water (Elga PureLab Ultra, Veolia Water Ltd., UK) and than dried at 57°C for 24 h before being analyzed.

Organic matter isolation methods

A. NOM isolation technique for freshwater sediments (Akkanen et al., 2005) was applied. Samples of loose deposits (0.5 g) were gently shaken with 1 ml of 0.5 M sodium hydroxide (NaOH) solution after which the mixture was heated for 30 min at 80°C. The suspension was centrifuged for 10 min at 6000g (g - gravitational acceleration). NOM extract was transferred into clean vial. The procedure was repeated. The first and the second NOM extracts were mixed (1:1).

B. A modified method developed by Thurman and Malcolm (1981) for the extraction of aquatic humic substances adsorbed on XAD-8 resin. 0.5 g of dried loose deposits sample was shaken with 1 ml of 0.1 M NaOH solution for 30 min at room temperature. NOM extract was transferred into clean vial.

C. The protocol was used by authors (Mermillod-Blondin et al., 2001) for detaching biofilms from sandy sediments. A sonication method using a narrow tip ultrasonic generator (Model: CPX130PB, 130W; Cole - Parmer USA) was used at: power = 40 W, for 180 s.

Volatile suspended solids determination method

Method of evaluation of volatile suspended solids (VSS) is described in Standard Methods for the Examination of Water and Wastewater (2540). Briefly loose deposits samples were dried at 105°C and weighted. The samples were then combusted at 550°C for 6h and weighted again. The mass difference after the combustion was attributed to volatile solids expressed as %.

Non-purgable organic carbon (NPOC) determination

The concentration of the organic carbon was determined as NPOC, because the samples contained higher concentration of the inorganic carbon compared to the organic carbon (European Standard EN 1484:1997, 1997). NPOC was determined of filtered and acidified sample using Shimadzu TOC 93 5000A total organic carbon analyzer (Shimadzu Corporation, Kyoto, Japan). Samples were spurged with CO_2 free oxygen for 6 min prior to analyses. The blank and control solution were analyzed with each series of NPOC sample in order to verify the accuracy of the results obtained by the method. All NPOC samples were tested in duplicate and the mean value calculated (CV≤2%).

Determination of total and organic carbon

Total organic carbon (TOC) and dissolved organic carbon (DOC) measurements were performed with a TOC-5000A Analyzer and auto sampler ASI-5000 (Shimadzu Corporation, Kyoto, Japan) based on high temperature and acidification of sample and by the difference of the total carbon and inorganic carbon measurement, according to the European Standard EN 1484:1997, 1997. For determination of DOC samples were filtered thought the 0.45 μm pore size membrane filters (Millipore Corporation, USA), which were carefully rinsed, first with sterile ultra pure water and then with the water sample. Each sample was tested in duplicate and the mean values were calculated (CV≤2 %). The blank and control solutions were analyzed with each series of sample in order to verify the accuracy of the results obtained by the method. The detection limit was 560 μg/l.

Statistical analysis

To compare all methods of NOM isolation and NOM quantity determination statistically significant assays of the differences (procedure for computing one way ANOVA) were developed, with paired samples when possible (Fower et al., 1998)

Results and discussion

Eleven loose deposits samples were obtained from five drinking water distribution systems and water reservoirs of Latvia and analysed for concentration of organic matter with three NOM extraction methods followed with analyses of carbon and direct determination method with combustion.

Firstly, we compared methods for determination of concentration of organic matter in loose deposits (TOC, NPOC). Concentrations of organic matter determined as organic carbon (OC) were in range from 0.7±0.3 to 279±68 mg per gram of sample. As the concentrations of inorganic carbon were high (70 - 80 % from the total carbon concentration in each sample) the conditions for determination of organic carbon using TOC can not be satisfied. Therefore TOC analysis can not give correct results and NPOC method must be used for determination of organic carbon. The highest concentrations of OC in samples were obtained by VSS method which was then excluded (as a significant error) by ANOVA analysis. It should be mentioned that VSS analysis may overestimate the organic matter fraction in loose deposits due to volatilization of metal oxides and hydroxides (Gauthier, 1999).

Secondly, we compared NOM extraction methods. To exclude significant error Z-test was applied. Z-test for NOM isolation methods B and C showed that the difference between TOC and NPOC results is not statistically significant: 0.02 and 0.2 respectively, which means that these methods of extraction probably are not suitable for loose deposits. Z-test and t-test for NOM isolation method A showed that difference between NPOC and TOC results is statistically significant: 22.5 and 38.8 respectively. This proves that the conditions for using TOC method are not reached; otherwise NPOC and TOC should have shown the similar results. Removing of the inorganic component during sample acidification and sparging, disturbing effect was reduced, and thus NPOC method potentially can give the most accurate results (repeatability=17 %).

It can be concluded that concentrations of OC (measured both by TOC and NPOC) obtained after isolation of NOM with methods B and C are understated. Method A for isolation of OC from loose deposits in combination with NPOC analysis gave the most accurate results (Figure 1).

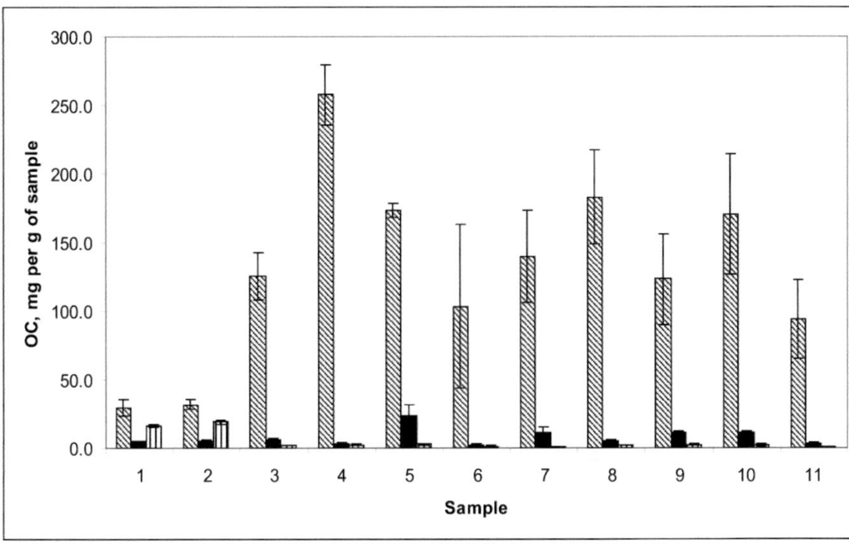

Figure 1. Organic carbon concentrations (measured as NPOC) obtained by different extraction methods from loose deposits samples

It should be mentioned that there is a great variations in results obtained with different NOM extraction methods (NPOC analysis was used for all the samples to determine OC concentrations). For extraction methods A and B the same reagent (NaOH) was used and differences were only in concentration of reagent and temperature), but results for A method for all of the samples were 82-98 % higher. Obviously with hot and more concentrated NaOH more (even 98% more) NOM can be extracted from loose deposits in comparison with room temperature NaOH. Also no correlations were found between extraction methods during this study.

Other studies (Stevenson, 1982) showed that using strong bases (like NaOH) it is possible to extract maximum 80% of humic substances from soil. To extract other types of organic matter other reagents (acids) should be used. On the other hand Chow et al. (2004) used 1 M NaOH (for DAX-8 and XAD-4 resins) and 10 M NaOH (for IRA-958 resin) to extract organic matter from resins. As a result very hydrophobic organic matter (VHA), slightly hydrophobic organic matter (SHA), charged hydrophilic substances (CHA) and neutrals (NEU) were the components of organic matter mixture, thus providing the full spectra of organic matter.

Conclusions

1. The most accurate results for determination of organic carbon in loose deposits showed NPOC determination method.
2. VSS method is not suitable for evaluation of OC in the loose deposits, because it may overestimate results even for 3 times.
3. The natural organic matter extraction method using hot NaOH (developed by Akkanen et al. (2005) for sandy sediments) in combination with NPOC determination method is more accurate than other methods tested in this study for measure organic matter in loose deposits.

List of Literature

[1] van der Kooij, D., Vrouwenvelder, J.S. and Veenendaal, H.R., 2003. Elucidation and control of biofilm formation processes in water treatment and distribution using the unified biofilm approach. Water Sci Technol. 47(5), 83-90.

[2] Gu, B., Schmitt, J., Chen, Z., Liang, L., McCarthy J.F., 1994. Adsorption and desorption of natural organic matter on iron oxide: mechanisms and models. Environ Sci Technol. 28(1), pp. 38-46.

[3] Akkanen, J., Lyytikäinen, M., Tuikka, A., Kukkonen J.V.K., 2005. Dissolved organic matter in pore water of freshwater sediments: effects of separation procedure on quantity, quality and functionality. Chemosphere. 60(11), pp.1608-1615.

[4] Thurman, E., Malcolm R., 1981. Preparative isolation of aquatic humic substances. Environ. Sci. Technol. 15 (4), pp 463–466.

[5] Mermillod-Blondin, F., Fauvet, G., Chalamet, A., des Chatelliers, M. 2001. A comparison of two ultrasonic methods for detaching biofilms from natural substrata. Internat. Rev. Hydrobiol. 86 (3), pp 349-360.

[6] Standard Methods for the Examination of Water and Wastewater, 1999 (20th Ed.) APHA, AWWA, WEF, Washington

[7] European comittee of standartization, 1997. EN 1484:1997 Water analysis – Guidelines for determination of total organic carbon (TOC) and dissolved organic carbon (DOC). Brussels: CEN.

[8] Fower J., Cohen L., Jarvis P. 1998. Practical Statistics for field biology. 1st ed. In: John Wiley&Sons Inc., USA, pp. 181 – 184.

[9] Gauthier, V., Gerard, B., Portal, J.M., Block, J.C., Gatel, D., 1999. Organic matter as loose deposits in a drinking water distribution system. Water Res. 33(4), pp.1014-1026.

[10] Stevenson, F.J., 1982. Humus chemistry: genesis, composition, reactions. Wiley, Ney York.

[11] Chow, K.C.W., Fabris, R., Drikas, M., 2004. A rapid fractionation technique to characterize natural organic matter for the optimization of water treatment processes. J. of Water Supply: Research and Technology – AQUA, 53(2), pp. 85-92.

Framework Master-Plan on Sustainable and Ecological Water Resources Management for the City of Changde/Hunan Province/PRC

Dr. Lothar Fuchs
Chiyan Peng, M.Sc.
Institut für technisch-wissenschaftliche Hydrologie GmbH, Engelbosteler Damm 22, 30167 Hannover, Deutschland, E-mail: l.fuchs@itwh.de, Tel: +49-511-971930

Abstract

Based on an EU funded project a framework master-plan for the city of Changde/Hunan province/PRC was developed. After an introduction of the recent international developments in the field of sustainable water management and water sensitive urban design, an analysis of the existing situation of the water bodies in the city of Changde was carried out. Based on this the general concept of the three "green-blue circles" was developed, describing the different use of the landscape and the water in and around the city of Changde. Seven key guidelines for the future development of the city and its water bodies in sustainable and ecological way were given. According to these guidelines, integrated measures for the improvement of the existing situation were developed and described. The effect of different measures was evaluated by simulation models. An action plan was set up, prioritizing, describing and evaluating the different measures including administrative and organizational aspects to ensure the implementation and long term success of the measures proposed.

Catchwords

Integrated water resources management, sustainable and ecological urban design, master-plan

1. Introduction

Within the project "Sustainable Problem Solutions for Asian Urban Settlements and Developments by Exemplary Analysis of Sewage and Waters of the Urban Settlement Changde and its Chuanzi River Basin" funded by the EU international know-how and best practice in the sustainable sanitation of polluted rivers was elaborated, discussed and partly applied. Based on this project the city of Changde started the development a "Framework Master-Plan on Sustainable and Ecological Water Resource Management for the city of Changde."

2. General aim of the framework master-plan

In this master-plan a joint vision by German and Chinese experts was created: How can Changde be developed if water in the city is not considered to be a problem but a starting point for an integrated urban development strategy? In this concept two design tasks of the (near) future are combined:

- The water management challenge (create more space and better quality for water)
- The urban development challenge (vital economy and high quality living environments).

The city of Changde follows the strategy to develop its character as the "city of three mountains and three rivers", a "garden and ecological city" as well as "one city, four districts, two riverbanks" and "modern fairyland city". In all these concepts water plays a major role and is seen as an important identity builder. But actually water within the city is hardly visible and of insufficient quality, whiles the landscape surrounding the city is dominated by streams, canals and lakes. The aim is to return the water to the experience of people's life also within the city.

This master-plan examined and gave guidelines how to create an urban water landscape by setting up an ecological watershed management within the city of Changde. This suggests that there should be clear connection between the underlying structures of topography, hydrology and soils and the major structuring elements of urban form. There is an obvious synergy between the need to create networks of open space to serve social and ecological needs and new approaches to open systems of urban water management.

This master-plan reintegrates different water-related urban issues towards comprehensive more sustainable urban design and improvement of environmental conditions. The interconnected systems of water, people and development on the land must first be seen and understood as an integrated whole before citizens, developers and their leaders can act intelligently to protect them. This master plan is intended to assist in building this understanding.

3. The Framework Master-plan

In the introduction the general aim was concretized and put into the context of several international case studies and best practice examples:

- In what ways are water quantity and quality affected by urbanization and growth?
- What does water sensitive urban design mean and why is it necessary?

- What are the general objectives to change the way the water can be managed in the urban environment?
- Where are water sensitive design principles already being applied and what kinds of guidelines already exist, that this project can refer to?

3.1. "Water city Changde" - Analysis of the actual situation in Changde

Changde is situated in the northwest of the province Hunan, with a population of 470.000 inhabitants downtown. The annual average rainfall is 1417 mm, but most of the rainfall concentrates in 4 months. The maximum daily rainfall measured was 251 mm. The Chuanzi River downtown is separated into several parts and strongly polluted by outflow of sewage or overflow from combined sewage systems. Changde is a region rich of water.

The analysis of the actual situation was carried out under the following aspects: Context (climatic and geo-hydrological, culture and urban development history, regional ecological connections etc.); water system (supra-regional watershed, regional water system, flood protection systems); utilization of water resources; urban development and urban drainage systems; water quality

The expected urbanization of the city of Changde will change the land-use of many areas from rural to suburban and city-like characteristics. Population growth and construction will inevitably impact the quality and quantity of water resources. As development decisions are made, it is important to understand how different patterns of development and applied techniques of water management will influence the water resources.

Any concept for a future water resource management must take into account the present use of the water resources. Some of the present uses must be accepted for future strategies, other might be questionable. In either case all uses must be considered for a well-balanced water management. An important reason for this is the acceptance of proposed measures by the population, industry and public authorities. For this the present uses of water resources and the constraints for future usage were identified and documented. This part of the master-plan also contains a review of existing studies, guidelines and plannings/ strategies related to water resources for the city of Changde. Their consequences for the water bodies are discussed critically as well as economic and legal issues. Mutual influences were detected and documented. Based on these data collected answers to the following questions were given:

- What are the effects of different land-uses on the existing hydrology and morphology of Changde´s water systems?
- How do different factors influence water quantity and quality?

- What are the most frequently occurring pollution impacts and their probable sources? How can they be identified, categorized and described?
- What are the effects of already applied techniques of water management on the existing hydrology and morphology of Changde´s water systems?

3.2. Objectives

Urbanization and the way in which we design and live our cities are directly responsible for the quality of water resources. Urbanization has the potential to alter the urban watershed hydrology by increasing peak water discharge and diminishing water quality. It is possible, though, for urban hydrology and water quality to improve as development proceeds. The goals and guiding principles for future water sensitive urban growth in the city of Changde were defined and discussed in the context of international legal directives and policies, in order to follow their standards and meet their demands:

- Sustainable land use planning: identify urban growth opportunities and boundaries, direct land-use activities and site-design strategies according to water resource requirements.
- Water flow management: optimize timing and quantity of water flow in streams, reduce surface runoff and safeguard flood-control.
- Water quality improvement: minimize wastewater, reduce wastewater and storm water pollutant loads, restore and protect water quality of watercourses.
- Improvement of ecology and biodiversity: protect and promote natural water systems and habitats.
- Aesthetic and community orientated design: integrate water management aspects into the urban design, create attractive environment for recreational use.
- Integrated management and administration: cooperation between different disciplines and sectors, requirements for management practice

Based on the analysis of the existing situation and the objectives seven key theses were defined for the city of Changde, which give a guideline for the future development of the water in the city:

1. Housing and living at the water improves the quality of life and new economic potential.
2. The identity of the "water city Changde" will be consolidated by the urban development at and with the water.

3. The urban development follows water-sensitive design principles.
4. The urban drainage system will be modified according to integrated, sustainable strategies
5. The ecological potential of the watercourses will be improved and used.
6. The flood protection will remain excellent.
7. The social aspect of the urban water management will be considered.

3.3. Measures and concepts

Based on these seven key thesis measures and concepts were developed for the urban development, the regional water management and the urban drainage system. Basic concept for the urban development is the concept of three circles as new basic network structure of the urban development. Together with other private and public green areas in the city, the three circles should improve the local climate and should be used as a network of bicycle roads and footpaths.

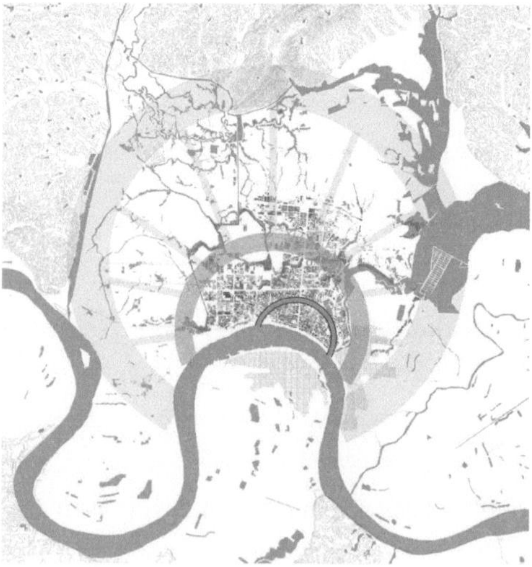

Fig. 1: The concept of the three blue-green circles

- Blue-green circle – the historic ditch – the circle of the protection river: The protection river has significant meaning for the identity of the city Changde and should be restored as an open ditch within the city as possible. It should be investigated, weather this circle could be connected with the golden circle.

- Blue-green circle – the golden circle – the circle of the Chuanzi River: The new modern city center is located at the banks of the Chuanzi River. It should be investigated to connect the Chuanzi River to a continuous river.
- Blue-green circle – the scenery circle – the circle surrounding the bottom of the mountains and the river banks: The scenery circle surrounds the city as a green blue water protection area, the transition between the city and the landscape. This area should be used for the regional water management, recreation and tourism.

On the regional scale concepts were developed to improve the water balance and the ecological capacity of the water bodies. Several elements were discussed and proposed.

Also for the urban drainage system and the integrated water sensitive urban development different elements were discussed and proposed like: green roofs, drainage of rainwater on the surface, green retention ditches, retention ponds, trough and trench systems, artificial wetlands, combination of ecological river banks and retention space. The main concept is that the city should work like a sponge to delay the run off process and to limit the discharge into river. For every different type of land use different concepts for urban drainage and urban design according to the situation and needs are recommended. All kinds of measure are not only used for one single target, the most measures have interactive relations with each other and should be integrated as an entire solution.

A simulation model was set up to simulate the outflow from the different sub-systems under quantitative and qualitative aspects. Different scenarios were investigated and the results show the different impacts and effects on water quantity and quality. Together with measures at the sewage system the emissions from the whole system could be reduced by more than 90%.

3.4. Action plan

Based on the different concepts and measures proposed an action plan was developed. In this action plan the different measures are prioritised, the reason for the measures are explained and the possible benefit quantified. The action plan is grouped into the three thematic fields

- Regional water balance and ecological cross linking of the blue green structures
- Urban development with a sustainable urban drainage system
- Sewage treatment

For each of these thematic fields relevant planning tasks, pilot- and initial projects and relevant construction works are arranged. The effect of a single measure in this context is not limited to the specific thematic field. That means that the proposed measures for the urban development also have a positive effect on the water balance and vice versa. Additionally some measures to safeguard supply of drinking water via alternative water resource are mentioned too.

Finally administrative and organisational measures are proposed, which are necessary for the implementation of the measures and the long term success. e.g. set up a watershed management organization. Also measures for the improvement of the basic data needed for advanced water management planning are mentioned.

4. Summary

The development of this framework master-plan was a challenge for the German as well as for the Chinese experts. Starting with the collection of basic data, knowledge exchange and capacity building a common understanding of a sustainable water system for the city of Changde could be developed.

The analysis of the existing situation showed that there is an urgent need for the improvement of the regional and local water cycle under quantitative and qualitative aspects and water sensitive urban design in Chinese cities. Different measures for the three main thematic fields were developed, prioritised and evaluated. Such an integrated interdisciplinarily concept as general strategy can guide a sustainable development of a city or a new development zone.

List of Illustrations

List of Literature

Sustainable Problem Solutions for Asian Urban Settlements and Developments by Exemplary Analysis of Sewage and Waters of the Urban Settlement Changde and its Chuanzi River Basin, Final report, EU Asia Pro Eco, 2008.

Wasser Hannover e.V. Wasserstadt Changde – Rahmenplan für eine wassersensitive Stadtentwicklung und nachhaltige Bewirtschaftung der Wasserressourcen, 2009.

The price to pay for treated wastewater:

an evaluation of water pricing scenarios in the Jordan Valley

Amani Alfarra[1], Ben Sonneveld[2], Heinz Hoetzl[3], Sameer Abed El Jabar[1]

[1]Ministry of water and Irrigation in Jordan. amani.alfarra@gmail.com, *[2], Centre for World Food Studies of the VU University Amsterdam[2],* bsonneveld@sow.vu.nl., *[3]Universität Karlsruhe, Institut für Formalerschließung und Regelwerkskunde* heinz.hoetzl@kit.edu.

Abstract Treated wastewater for irrigation could form a viable alternative to substitute fresh water resources in the Jordan Valley. Yet, investing in treatment infrastructure is costly and the subsidized water prices in the Jordan Valley results are low compared to the cost price for treated waste water. Therefore, this study examines the effect of additional water volumes on net profits of four farm archetypes and the possibilities for cost coverage of treated wastewater under various pricing scenarios. The study accounts, in a stylized manner, for positive (fertigation) and negative (salinization) effects of the treated wastewater quality. The results show that additional volumes are highly profitable for vegetable farms yet, salinity levels negatively affect banana farms. The scenario with the highest pricing variant provides some compensation, yet, most of the costs have to be covered by other public funds.
Keywords: Reuse, treated waste water, farmer, pricing

1 Introduction

The pressure on water resources in the Hashemite Kingdom of Jordan is most visible in the severe competition over water resources between the agricultural sector and a fast growing, and increasingly affluent, population who demands their guaranteed supply of clean water. This situation is likely to exacerbate in the near future as a mounting population continues to grow (2.6 %) to a 10 million people in 2050 and climate change is expected to affect the country negatively [5, 6]. The Jordan Valley (JV) is no exception; all water sources are fully committed in an advanced drip irrigation infrastructure that leaves little room for efficiency gains. Indeed a large part of the JV's agricultural potential remains idle and further intensification or expansion of the cultivated area remains fully dependent on the availability of additional water sources (Venot [1]). This study argues that treated wastewater (TWW) constitutes a viable alternative for irrigation in the JV as it allows further agricultural development and, simultaneously, substitutes the fresh water resources that are urgently needed for domestic use. Yet, more than 60 per cent of the wastewater is currently lost due to an inferior infrastructure and construction of sewage systems and treatment plants to collect and process the used water is costly.

In an ideal situation we could rely on water markets that efficiently reallocate available water sources of competing water suppliers between clients using prices that result from the supply and demand volumes of water. Under such conditions water markets could generate the required capital for investments in water infrastructure to tap from unconventional water resources as TWW. Yet, in the JV water markets are absent and water supply is under full control of the government, effectuated by the Jordan Valley Authority, who also determines a fixed and heavily subsidized price for water users.

The first water tariff in the Jordan Valley was implemented in 1961. Farmers paid 0.001 JD/m^3 independent of the amount of the water consumed. In 1966, this tariff was redefined to 0.001 JD /m^3 for the first 1,800 m^3 consumed, and 0.002 JD /m^3 for additional volume (JRIDI 2002). In 1995 agricultural water was again repriced by the Ministry of Water and Irrigation. Farmers pay now: 0.008 JD/m^{3}for the first 2 500 m^3, 0.015 JD for additional volumes exceeding the 2 500 up to 3 500 m^3, 0.02 JD for the next 1 000 m^3 up to 4 500 and 0.35 JD for higher volumes. The subsidy effect of the water price for farmers becomes visible when we compare these to households that pay a 4.5 JD/m^3 and higher prices when 20 m^3 is exceeded. Farmers water prices are also far from covering the costs for production of TWW that vary, according to the Ministry of water and Irrigation, from 0.026 JD per cubic meter without operation and maintenance (O& M) cost to, 0.63 JD when O&M is included up to 1.3 JD/m^3 [11], when capital costs are accounted. Even though the Jordan Water Strategy and Policies 2002, Article 43, declared that differential prices can be applied to irrigation water by quality, in practice the farmers in the Middle and

South JV where most of the TWW is used, are paying the same price as farmers in the North JV where the use of freshwater prevails.

Hence, the question is to what extent farmers can contribute to cover these costs of additional infrastructural investments that can generate the TWW. Earlier studies (Alfarra et al., 2009) confirm that a large majority (96%) of the farmers is willing to accept TWW and also willing to pay substantial higher prices. Yet, these studies did not include an assessment of the expected benefits for additional water volumes and the possibilities for cost coverage for water infrastructural works. Moreover, when evaluating the effect of TWW volumes famers will also consider the 'goods' and 'bads' of water quality as these differ remarkably from fresh water resources. At the positive side TWW might contain rich sources of nutrients with advantageous effects on crop growth if these nutrient concentrations are delivered in the correct proportions. Negative, is the increased salinity level that is caused by the dissolved nutrients, which might affect sensitive crops and lower production potential.

This paper is organized as follows. In section 2 we discuss the water quality aspects of the TWW and their impact on crop production. Accounting for these specificities of water quality we evaluate the impact of additional TWW volumes on farm income and cost coverage for four farming systems archetypes under 5 water pricing scenarios.

2 Water quality of TWW

Calculating Nutrients in Irrigation Water

Table 1 shows the nutrients expressed in their weight equivalent of commercial fertilizers. The nutrients in KTR and KAC-South are close to the ratio of commercial NPK fertilizers where we find 10 kg N, 20 Kg P2O5 and 30 K2O per 100 kg. The average commercial price for fertilizer in this composition in Jordan is 1500 JD per ton. Hence, as 1000 m3 water equals the amount of 100 kg of commercial fertilizers, it is equivalent to a value of JD 150, or 0.15 JD/M3. We conclude that when the price of water should consider the added value for nutrition in the TWW, as this can help to reduce fertilizer costs. A GTZ project has proved from that farmers can save about 50% to 60%of farm fertilizer in each season.

Impact of salinity

A major degradation factor of re-used waters can be its high salinity levels that are caused by high ion concentrations that have a negative effect on water intake of plants as it competes with the plants' osmotic potential. Moreover, high ion concentrations might reach toxic levels that impede proper plant growth. Finally, high concentration of alkaline damages the structure of the soil, with a dramatic loss of water holding capacity as a result.

Yet, the reaction of yield performance on higher salt concentrations is typically crop specific; crops might be highly sensitive or highly tolerant to salinity. Therefore, TWW with higher salt levels requires an appropriate selection of crops. Moreover, to prevent accumulation of salts in the root zone the water management should include a drainage system, regular leaching of the salts with fresh water, possibly with Calcium contents in case of high Alkaline concentrations.

Concerning the effects of salinity on crop yield, there is a wide range in plant species response to salinity. Sugar beet, sugar cane, dates, cotton and barley are among the most salt tolerant; whereas beans, carrots, onions, strawberries and almonds are considered sensitive [9]. In general, salinity decreases both yield and quality in crops and previous research has led to the development of large data bases on the salt tolerances of many crop species and varieties. Salt tolerance can be represented most simply based on two parameters: the threshold salinity (t) which is expected to cause an initial significant reduction in the maximum expected yield (Y); and the slope (s) of the yield decline. Slope is simply the rate that yield is expected to be reduced by for each unit of added salinity beyond the threshold value. The formula to calculate relative yields is [10]: $YR = Y - s\,(ECe - t)$ where $ECe > t$ [1]

Salts are added to the soil during each time of irrigation and accumulate in the root zone. In case that appropriate drainage systems are absent and insufficient freshwater is available for leaching soil salt levels might reach damaging concentrations. The crop removes much of the applied water from the soil to meet its evapotranspiration demand (ET) but leaves salts behind in the shrinking volume of soil water. The following table shows crop tolerance rating and their equivalent soil salinity.

Figure 1 stylizes the yield reducing effects for crops with different sensitivity levels for salinity. We will use this relationship in the next section when we evaluate the introduction of additional TWW in the JV.

The average salinity for treated wastewater at King Talal Reservoir (KTR) used in the Jordan Valley is 2.7 whereas the average salinity for freshwater resources from King Abdalah Canal (KAC) is 1.1. So, significant yield loss can be expected for sensitive crops that are cultivated on treated wastewater.

3 Pricing scenarios

In this section we will evaluate various water pricing scenarios and their impact on 1) farmers' income and 2) cost coverage for new TWW plants. We evaluate the situation for four prevailing farm archetypes, which are considered representative for the majority of farm households in the JV. Table 2 shows an agronomic-economic profile of the four archetypes. Water quota and net profits figures were derived from Venot (2007). Figures on fertilizer savings were obtained from [12]. Current water tariffs were provided by the JVA. Yield losses due to the sensitivity of crops and prevailing salt levels were estimated using the relationships explained in section 2. For the citrus and banana farms were assume that additional TWW volumes are still blended with fresh water and that the final ECe level is around 1.5 ds/m.

Currently the average total water volume that is supplied to the JV is 250 MCM, 87 MCM of which is TWW. The average demand/supply ratio in the JV is 64 per cent, which means that 90 MCM of additional volume is required to let the JV occupy its total water requirements. Using the above mentions cost estimates the total costs for: Running; Running and O&M; Running and O&M and capital costs are, respectively, 15 million, 57 million and 117 million JD.

The selection of water price scenarios is based on three criteria. First, we abstain from abrupt changes as a sudden raise in prices would be met with disapproval and cause social unrest. Second, we capitalize on findings of Alfarra et al. (20009) that show farmers willingness to pay four to five times the current water price. Third, and finally, from Venot, 2007 we calculated the marginal contribution of water to the net farm profit and used this price as an upper bound for our assessment.

We are now ready to run various water pricing scenarios and evaluate their impact on farmers' income and on cost coverage of new TWW infrastructure. We evaluate the scenarios over a period of twenty years. For each year an additional amount of TWW (4.5 MCM) volume is generated resulting in the 90 MCM after twenty years. The amount of money that is used to cover the cost of the TWW infrastructure is the difference between the total amount generated with the new and the old water tariff. The effect of farmers' income accounts for the effect of rising salinity levels on crop yields, savings made on fertilizer and costs incurred by water tariffs. We designed a simple model that varies the water tariffs as fixed amounts or with gradual annual increases. Of all the various possibilities we will run five water pricing scenarios:

Scenario I. BUA, business as usual, the same water tariff that currently prevails.
Scenario II. FLAT. A flat water tariff that covers the Running costs of the TWW plants.
Scenario III. GRADUAL/LOW. A gradual increase of the water tariff with 1 per cent per year
Scenario III. GRADUAL/MODERATE. A gradual increase of the water tariff with 5 per cent per year.
Scenario III. GRADUAL/HIGH. A gradual increase of the water tariff with 10 per cent per year.
Their results are discussed below:

4 Results

BUA. The results of the first scenario are depicted in Figure 2. Especially vegetable farms benefit from the additional water volume; vegetable crops are less sensitive to salt water and save substantially on fertilizer costs. Also Citrus and mixed farm increase their income with almost 70 per cent. Banana farms remain more or less the same, basically because they were already close to the maximum water level requirement (87 per cent) and the salt levels affect crop yields negatively. Yet, the coverage of cost for additional TWW infrastructure is extremely low. Under this scenario the entire implementation of TWW plants will be dependent on subsidy from the government or foreign donors.

FLAT. Figure 3 shows the results of the FLAT scenario. The income of the farmers is hardly affected as water only makes up a small amount of the total farm costs and benefits from the additional water volumes are substantial, except for the earlier discussed banana farms. Cost coverage is high initially but decreases rapidly to lower levels especially when O&M and capital costs are included. Hence, also in this scenario the subsidies will have to cover substantial amounts.

GRADUAL/LOW. Figure 4 shows the results of the GRADUAL/LOW scenario. We can conclude that the trends on farmer income and cost coverage remain more or less the same as compared to the BUA alternative.

GRADUATE/MODERATE. The results of the GRADUATE/MODERATE scenario are presented in Figure 5. The effects on farm income are noticeable. Banana farms are reducing their farm income while the increase in income for the citrus and the mixed farms is reduced. Cost coverage for the Running costs rise up to 30 per cent. We conclude that the annual increase of five per cent has on the long run some negative effects on income growth and only slightly compensate the TWW costs.

GRADUATE/HIGH. Finally, Figure 6 presents the outcomes of the GRADUATE/HIGH scenario. Here we see that farm income is affected negatively after some 10 years or so. Especially the Banana farms have a relatively substantial decrease, but also the lower income farms with citrus and mixed cultivation have negative net profits as compared to their starting year. Coverage of costs for running operations is almost a 100 per cent but coverage of the costs including O&M and capital is still small, despite the high increase in water tariff.

5 Discussion

In this paper we evaluated the effect of nutrients in TWW for its cost saving effects on fertilizers and quantified the crop specific effect of salinity levels on yields. This information was used to evaluate impact of additional TWW volumes on farmers' income under various water tariff scenarios. Moreover, we also considered the costs that had to be covered for additional water volumes.

We found that a considerable amount of nutrients can be saved as the nutrient composition in the KAC has a remarkable coincidence with the NPK ratios of commercial fertilizer. Meerbach and Böning-Zilkens 2006 [12] also found that up to 50 per cent of fertilizer costs can be saved at least when the TWW is used to fertigate the crops. Yet, the negative side of the TWW water for irrigation is the sensitivity of the main crops banana and citrus for its moderate salinity levels. Future water distribution schemes that supply TWW to these farms should be supplied with sufficient fresh water so as to mitigate the effect of salinity.

We found that farmers' income in general grows with additional TWW, except for banana which is already supplied for almost 87 per cent and is also affected by the TWW salinity level. Only when water tariff does increase at a high pace farmer incomes become lower as the total price for water starts to become a high share of the total costs. The coverage of cost for running costs, O&M and capital costs will be difficult to cover by farmer contributions alone.

Yet, treatment of waste water has also environmental and health benefits that are for the benefit of the society as a whole. Contributions to TWW infrastructure from other sectors in the society is, therefore, natural. We conclude that there are good prospects for agricultural expansion in the JV when the use of TWW in Jordan becomes more efficient through an increase in WRI. Farmer contributions through higher water seem justified as the benefits of an additional M3 TWW outweigh its costs by far.

The objective of introducing a new pricing mechanism that includes different factors not only for cost recovery and benefit, but also to enforce farmers changes their attitudes. Such as changing their crops which is sensitive to salinity and required high amount of water, such as banana and citrus, to crops less water demand and more tolerance to salinity. Water scarcity in the region required a more responsible behaviour from users to value water that they can receive.

In addition pricing can help farmers' to understand the true value of receiving treated wastewater in the region especially the coming era will bring more drought to the Jordan valley were fresh water will be more valuable for domestic uses. We recommend the introduction of a gradual tariff rise to let farmers get accustomed to a new water tariff situation. From field experience we know that appropriate extension programs that explain the changes in water tariffs will be indispensable. Finally, we suggest that water tariffs are differentiated with lower tariffs for the poorer farmers and their families in the JV.

References

[1] Venot, J.-P., F. Molle, and Y. Hassan, Irrigated Agriculture, Water Pricing and Water Savings in the Lower Jordan River Basin (in Jordan), in Comprehensive Assessment of Water Management in Agriculture Research Report 18. 2007, International Water Management Institute (IWMI): Colombo, Sri Lanka. p. Pp.66.

[2] Howitt, R. and K. Hansen, The Evolving Western Water Markets Choices , The magazine of Food , Farm and resources issues, 2005(1st Quarter 2005): p. Pp. 59-64. .

[3] Chong, H. and D. Sunding, Water Markets and Trading. Annual Review of Environment and Resources, 2006. 31(1): p. 239-264. WMO. THE DUBLIN STATEMENT ON WATER AND SUSTAINABLE DEVELOPMENT. in

[4] The Dublin International Conference on water and Environment, Economic value of water, women, poverty, resolving conflicts, natural disasters, awareness. 1992: The World Meteorological Organization (WMO). http://www.wmo.ch/pages/prog/hwrp/documents/english/icwedece.html.

[5] World Bank, Water Resources Management in Policy paper. 1993, International Bank for Reconstruction and Development / The World Bank Washington, D. C. p. P.140.

[6] Fraiture, C.d. and C. Perry, Why Is Agricultural Water Demand Unresponsive at Low Price Ranges?, in International Water Management Institute (IWMI). 2007, CAB International: www.iwmi.cgiar.org. p. Chapter 3, P.14.

[7] Merrill, S., et al., Driving Economic Growth - chapter 2, in Foreign Aid in the National Interest - Promoting Freedom, Security, and Opportunity, B. Ross-Larson, et al., Editors. 2002, U.S. Agency For International Development: Washington, D.C. p. Pp.169.

[8] GTZ, Guidelines for Reclaimed Water Irrigation in the Jordan Valley - Practical Recommendations for Farmers and Extension Workers 2006, Ministry of water and Irrigation (MWI), Jordan Valley Authority (JVA) ,Commissioned by German Technical Cooperation (GTZ). p. P. 81.

[9] Dajic, Z., SALT STRESS, in Physiology and Molecular Biology of Stress Tolerance in Plants, K.M. Rao, A. Raghavendra, and K.J. Reddy, Editors. 2006, Springer Netherlands: Netherlands. p. 41-99.

[10] Allen, R.G., et al., Crop evapotranspiration - Guidelines for computing crop water requirements - Chapter 8 - ETc under soil water stress conditions, in FAO Irrigation and drainage paper 56 1998, Food and Agriculture Organization of the United Nations (FAO): Rome - Itali. p. 326.

[11] MWI, The operation report for the sewer system, S. department, Editor. 2009, The Hashemite Kingdom of Jordan, Ministry of Water & Irrigation. p. P.22.

[12] Meerbach, D. and M. Böning-Zilkens, Analysis of Irrigation and Fertigation Practices Using Reclaimed Water in the Jordan Valley. 2006, German Technical Cooperation (GTZ) and Ministry of Water and Irrigation (MWI): Amman, Jordan. p. Pp. 112.

[13] WB, Jordan at Glance (2001 -2008)/ database- statistic. 2009, World Bank: Amman, Jordan.

[14] Harald, K., et al., Regional climate change in the Middle East and impact on hydrology in the Upper Jordan catchment. IAHS-AISH publication 2007: p. P.9.

[15] Bates, B., et al., Climate Change and Water: Technical Paper. 2008, Intergovernmental Panel on Climate Change (IPCC): Geneva. p. Pp.210.
[16] FAO., Agricultural Drainage Water Management in Arid and Semi-Arid Areas, in FAO Irrigation and Drainage Paper 61. 2002, FOOD AND AGRICULTURE ORGANIZATION OF THE UNITED NATIONS (FAO): Rome, Italy. p. Pp.120.
[17] Ayers, R.S. and D.W. Westcot, Water quality for agriculture, in FAO Irrigation and Drainage Papers - 29 1994, Food and Agriculture Organization of the United Nations (FAO) - Agriculture and Consumer Protection Rome -Italy. p. Pp.234.

Table 1: Amounts of nutrients in the irrigation water sources in the Jordan Valley

Water source	N (kg/1000 m^3)	P_2O_5 (kg/1000 m^3)	K_2O (kg/1000 m^3)
KTR	18.6	8.9	31.4
KAC-south	18.4	7.05	31.3
KAC-north	1.4	0.52	12.7

Table 2: Four archetypes of farming systems in the JV: an agronomic-economic profile.

	Commercial vegetable farm	Citrus family farm	Commercial banana farm	Family farm, mixed
Before TWW				
water quota (m3/ha/yr)[1]	5050	10100	15000	5050
Fertilizer (JD/ha/year)[2]	695	496	993	298
net profit (JD/ha/year)[1]	5319	1550	8865	745
area ha[1]	8	4	4	7
Total water	40400	40400	60000	35350
Water costs (JD/farm)	323	323	480	283
Fertilizer costs (JD/farm)[2]	5560	2234	2979	596
Farm income JD/yr	42553	6200	35461	5213
After TWW				
Saving fertilizer (JD/farm)	2224	894	1191	238
Yield reduction: salinity	5	10	15	10
salinity losses (JD/farm)	2128	620	5319	521
Supply/demand ratio	64	62	87	64
Nett profit (JD/farm)	57969	8830	35943	6806
Nett increase	15416	2630	482	1594
%increase per farm	1,36	1,42	1,01	1,31

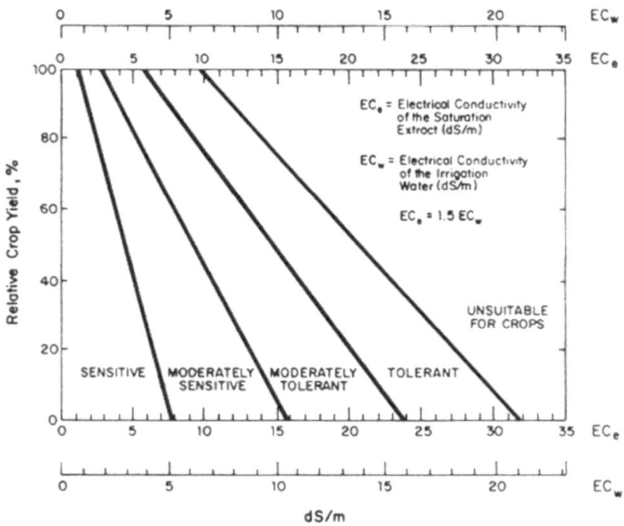

Figure 1: Effect of salinity levels for crops with different degrees of salt sensitivity. Source: [17]

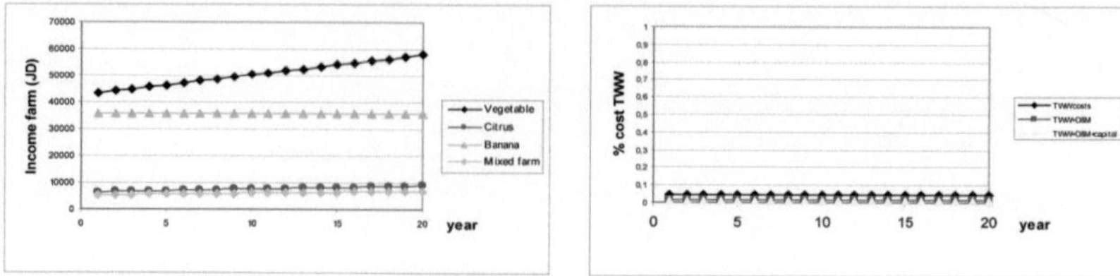

Figure 2. Effect on farm income (a) and cost coverage (b): scenario BUA

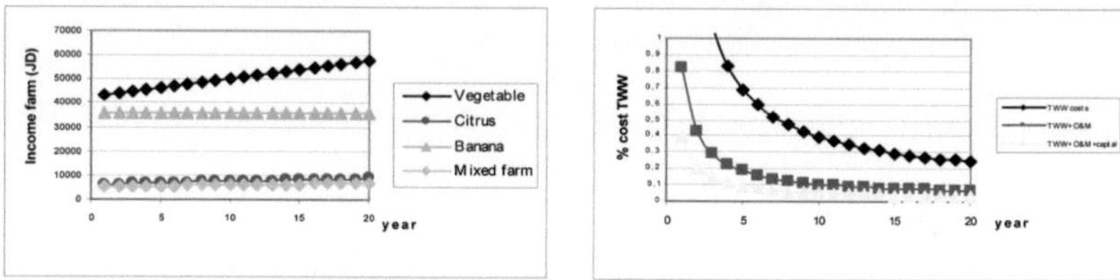

Figure 3 Effect on farm income (a) and cost coverage (b): scenario FLAT

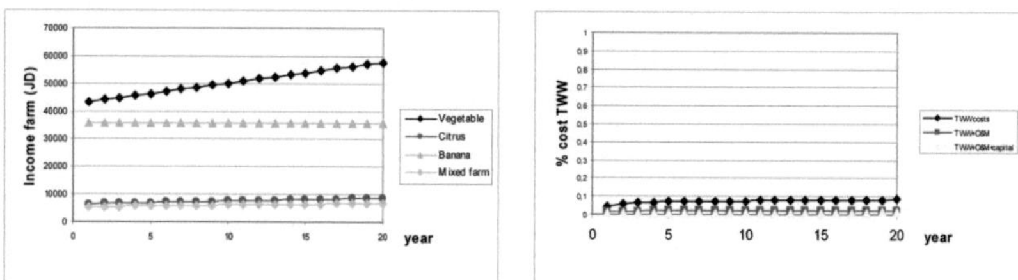

Figure 4. Effect on farm income (a) and cost coverage (b): scenario GRADUAL/LOW

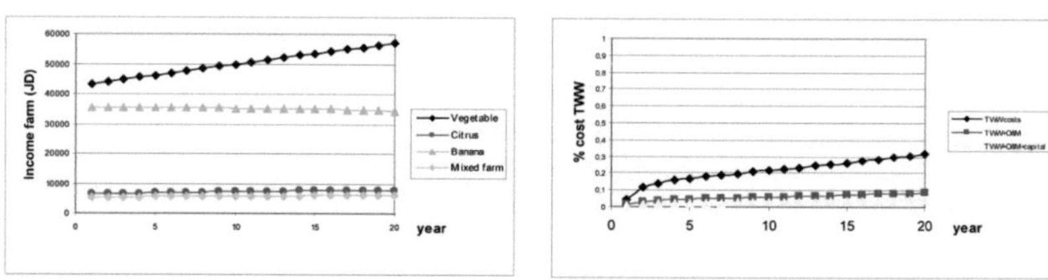

Figure 5. Effect on farm income (a) and cost coverage (b): scenario GRADUAL/MODERATE

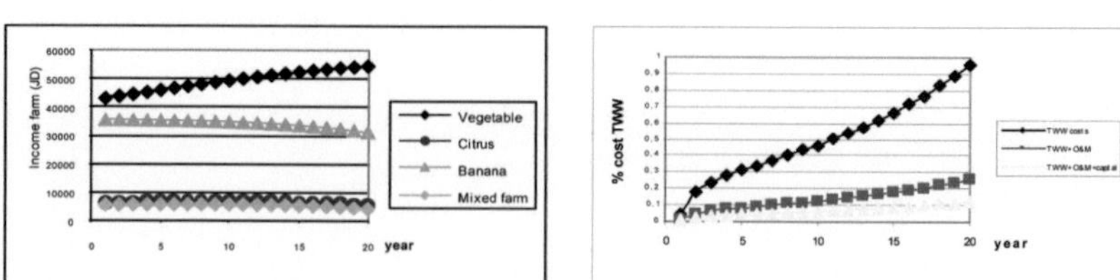

Figure 6-1: Effect on farm income (a) and cost coverage (b): scenario GRADUAL/HIGH

A simulation based integrated Water Management System for sustainable Arid Zone Water Resources Management

Jens Grundmann
Niels Schütze, Gerd H. Schmitz
Institute of Hydrology and Meteorology, TU Dresden, 01062 Dresden, Germany,
email: jens.grundmann@tu-dresden.de, phone: +49/351/46335556

Abstract

For ensuring an optimal sustainable water resources management in arid environments, we develop a new simulation based integrated water management system. It aims at achieving best possible solutions for groundwater withdrawal including saline water management together with a substantial increase of the water use efficiency employing novel optimisation strategies for irrigation control and scheduling. To achieve a robust and fast operation of the management system, it unites process modelling with artificial intelligence tools and evolutionary optimisation techniques for managing both, water quality and water quantity. We demonstrate our methodology by an exemplary application of the south Batinah region in the Sultanate of Oman which is affected by saltwater intrusion into the coastal aquifer due to excessive groundwater withdrawal for irrigated agriculture. Based upon of a generated database of crop water production functions for several predominate crops in the Batinah region we show the effectiveness of our methodology for the evaluation and optimisation of different irrigation practices, crop pattern and resulting abstraction scenarios. Due to contradicting objectives like profit-oriented agriculture or aquifer sustainability a multi-objective optimisation is performed.

Catchwords

integrated water resources management, process modelling, irrigated agriculture, artificial intelligence, saltwater intrusion, multi-objective optimisation

1 Introduction

Due to rapid population growth and increasing industrialisation, the water demand in arid regions is continuously on the increase. More than 80% of the limited water resources are used for irrigated agriculture in this regions but the water use efficiency is quite low. Groundwater aquifers, which represent the predominant natural water resources, are overused. Thereby continuously sinking groundwater levels are often associated with water quality problems occurring e.g. due to marine saltwater intrusion into coastal aquifers or a changing groundwater chemistry in the case of pumping in fossil resources. Especially saline irrigation water causes a destruction of agricultural resources and menaces the economical basis of farmers. Considering the uncertainty of climate and global change development a sustainable solution to the tight water situation can be only achieved by a comprehensive, integrated and highly efficient water management based on innovative approaches. In doing so physically based process models are necessary for a reliably portraying of the nonlinear phenomena and their interdependency in a changing environment as well as a prognosis of future developments. However it is often the case, that these highly accurate yet complex and intricate components fully contradict the general needs of potential users e.g. farmers who require a simple, easy to use and robust operation without numerical inconvenience. For this reason, recently developed new approaches serve to overcome this problem: synthesizing process models and artificial intelligence techniques [5, 7] will provide a convenient, easy to use but nevertheless highly reliable, rigorous management tool. Within the paper we present some key issues of a new simulation based integrated water management system for sustainable arid zone water management. It demonstrates how a highly coupled and dynamic agriculture-aquifer system – a land use-water interaction – can be integrated in an optimisation environment for finding optimal management strategies under multiple but mostly contradicting viewpoints and trade offs. By means of an exemplary application of the south Batinah region in the Sultanate of Oman which is affected by saltwater intrusion into the coastal aquifer due to excessive groundwater withdrawal for irrigated agriculture, we illustrate the functionality of the management system for multi-objective decision making.

2 Methodology

The optimisation framework of the simulation based management system, illustrated in Figure 1c, consists of three major modules, (i) an agricultural module for simulating the behaviour of farms, (ii) a groundwater module for calculating the aquifer response and (iii) an optimisation module.

The agricultural module

For simulating the behaviour of high productive agricultural farms a database of crop water production functions (CWPF) is generated. It describes the relationship between the amount of irrigated water and the potential crop yield for each crop assuming an optimal water application. For constructing CWPF's we use the soil-vegetation-atmosphere-transport (SVAT)-model APSIM [3], adapted to the regional climate conditions, together with the GET-OPTIS evolutionary optimisation algorithm for optimal irrigation scheduling and control [6]. Both together are used for maximizing the crop yield for given amounts of water and salinity levels which results in this case in a 2D-CWPF (Fig. 1a).

The groundwater module

The aquifer behaviour, including the seawater interface, is modelled by an artificial neural network (ANN) namely by a MLP-ANN [4]. It was trained using a scenario database containing the responses of the numerical density depended groundwater flow model OpenGeoSys [2] for all realistically feasible abstraction scenarios (Fig. 1b).

The optimisation framework

The optimisation framework connects the agricultural and groundwater module and aims for managing both – water quality and quantity – according to the formulated objective function (Eq. 1). For this purpose we use the CMAES global evolutionary optimisation technique [1] in order to estimate the optimal water demands $Q(t_i)$ and cropping pattern $A(t_i)$ for m crops over a certain period T (Eq. 2, Fig. 1c). The water demands are evaluated by the groundwater module which calculates water $h(t_i)$ and salinity $S(t_i)$ levels at the beginning and end of an extraction period i. Fixed and variable costs for pumping $CP(t_i)$ are considered in the objective function (Eq. 1). Furthermore, the salinity of the abstracted irrigation water is provided and directed to the agricultural module (Fig. 1c). Together with the cropping pattern and the water demands the salinity values serve as input for estimating the crop yields Y_j by means of the 2D-CWPF database for each crop. Based on the yields revenues P_j*Y_j are estimated which are part of the objective function as well as the irrigation costs $CI(t_i)$ resulting from the irrigation schedule. Finally we calculate a sustainability index SI which evaluates the change of the aquifer state between the end and the initial state of the simulation period. This index is included in the objective function (Eq. 1) which aims for maximising the net profit and the sustainability considering varying weights w between the two contradicting objectives.

$$max(OF) = w_1 \sum_{i=1}^{n} \left[\left(\sum_{j=1}^{m} P_j Y_j(t_i) - CI_j(t_i) \right) - CP(t_i) \right] + w_2 SI(t_n) \quad (1)$$

$$\left(A_1(t_1) \ldots A_1(t_n), A_m(t_1) \ldots A_m(t_n), Q(t_1) \ldots Q(t_n) \right) = arg\,max\,(OF) \quad (2)$$

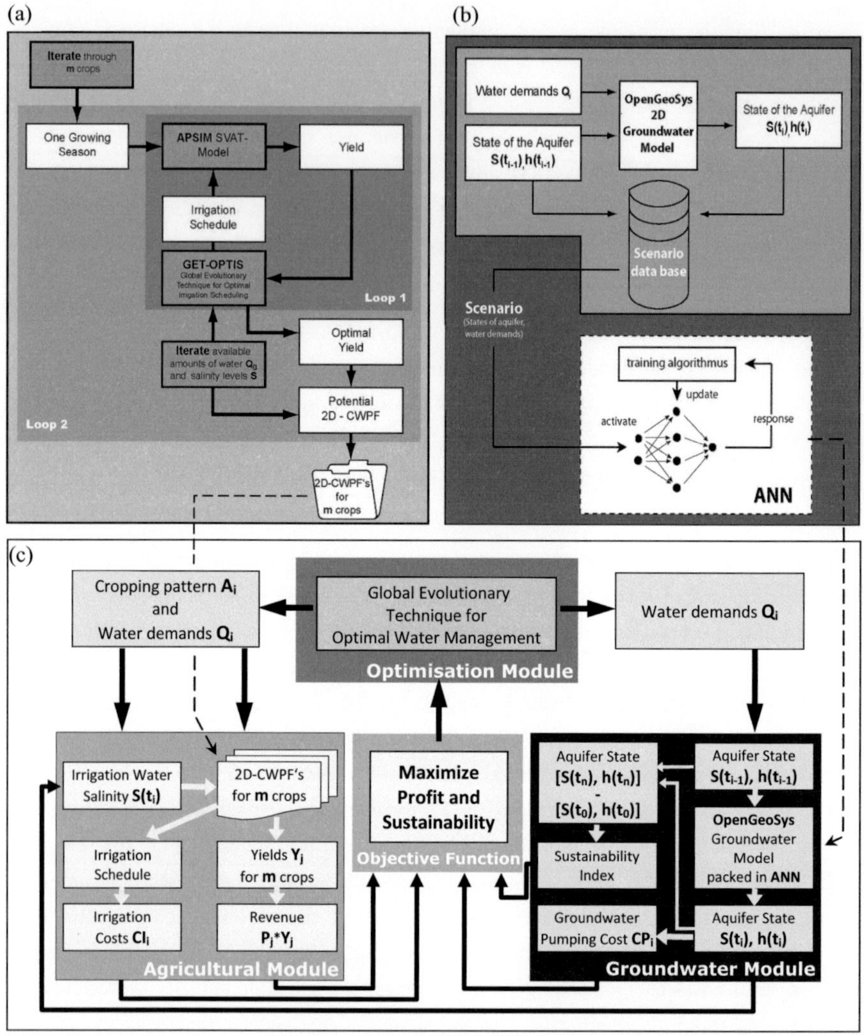

Fig. 1: Estimation procedure for the 2D crop water production functions (a) and the ANN groundwater model (b) and their linking within the optimisation framework (c)

3 Exemplary Application

We set up the exemplary application according to the characteristics of the wadi Ma'awil catchment located on the northeast coast of the Sultanate of Oman. Long-time overpumping of the coastal aquifer for irrigated agriculture leads there to an inversion of groundwater's natural flow direction resulting in marine saltwater intrusion and finally to a continuously increasing number of abandoned farms. To demonstrate the optimisation framework functionality we build a 2D slice in x-z-direction of the coastal aquifer to model the groundwater flow inclusive saline transport. For sake of simplicity 2D-CWPF's are constructed for two different crops, maize – a salt sensitive cash crop – and sorghum – a lower priced but more salt resistant crop – which are grown in two seasons per year [8]. We incorporate fixed and variable costs for installation and operation of pumping and irrigation equipment and assumed that the farmers decisions are profit based. So we performed three long term optimisation runs over the next 20 years to maximise (1) the short term profit (only for the next 4 years) and (2) the long term profit. In the third optimisation run we carry out (3) a multi-objective analysis of the profit together with the aquifer sustainability by using equal weights within the objective function. The decision variables are the yearly cultivated acreage, the yearly cropping pattern and the yearly irrigation water demands which results in a number of 60 decision variables for the 20 years period. Exemplarily for run 2, the optimisation algorithm needs 1200000 evaluations of the objective function and a computing time of only 1,5 days on a standard PC for finding an optimal variable set. This relatively short computing time will be possible by the use of the MLP-ANN groundwater model which operates approximately 1000 times faster than the physically based OpenGeoSys model.

4 Results

The results for the different weighted objectives are shown in Figure 2. The lower part illustrates the temporal course of the profit which rises up very quickly in the first two years for the short term profit optimisation run (Fig. 2c). After then the profit is decreasing quite fast due to increasing salinity in the irrigation water, like shown in Figure 2b, which also leads to a change in the cropping pattern from maize to sorghum (Fig. 2a). The results for the long term profit optimisation run shows also a rising and falling course of the profit (Fig. 2f). Nevertheless, the peak comes later with a lower peak value in relation to the short term profit optimisation run. The crop pattern changes also later (Fig. 2d). The salinity of irrigation water (Fig. 2e) increases in the last 5 years very quickly and achieves a level at the end of the 20 year period on which the farmer does not get a profit from the agricultural production. The results of the multi-objective optimisation run (run 3) shows very smooth graphs (Fig. 2g-i). Maize is grown all the time but on a lower acreage than on the beginning stages of run 1 and 2 (Fig. 2g). Therefore the salinity of irrigation water does not raise (Fig. 2h) and a profit is made the whole time (Fig. 2i). Comparing the total return between the three runs estimated by the line integral of the profit graphs, then the best return is made in run 2. However, at the end of the 20 years the farmer probably has to give up farming like the farmer in run 1 due to the high salinity levels of the irrigation water. In contrast the farmer in run 3 can work additional 20 years and get then a higher return than the farmer in run 2 because of a more sustainable aquifer management.

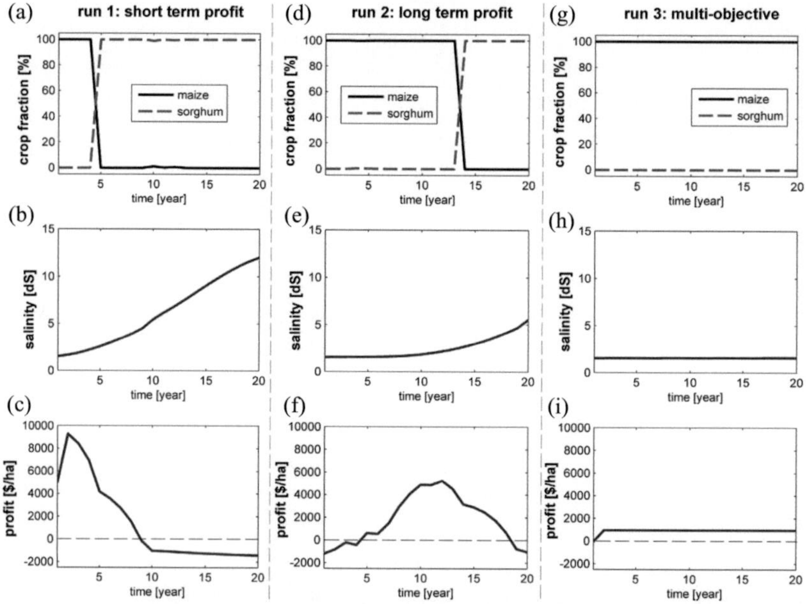

Fig. 2: Results of the three optimisation runs illustrated by the percentage of acreage, the salinity and the profit for the short term (2 a-c) and the long term (2 d-f) profit based optimisation and the multi-objective optimisation run (2 g-i)

5 Conclusion and Outlook

The presented prototype of a simulation based water management model allows for the management of both water quality and quantity of a coupled, dynamic agriculture-groundwater system. Thereby the modelling of the density driven groundwater flow is mandatory for calculating crop yield damages due to aquifer overpumping and irrigation with salty water. The exemplary application shows that farm operation focusing exclusively on profit maximization leads sooner or later to further progress of the saltwater front. Therefore the sustainability must be considered as a second optimisation objective. In this context a multi-objective optimisation can provide sustainable solutions both in an environmental and social sense. To achieve this task in a reasonable time a rigorous application of artificial intelligence methods is necessary in order to speed the process models up. The presented prototype forms the basis for the development of a large-scale management and planning system focusing on a sustainable arid zone water resources management.

List of Literature

[1] Hansen, N., Ostermeier, A.: Adapting Arbitrary Normal Mutation Distributions in Evolution Strategies: The Covariance Matrix Adaption. Proceedings of the 1996 IEEE International Conference on Evolutionary Computation (ICEC '96), S. 312-317, 1996.

[2] Kalbacher, T., Wang, W., McDermott, Ch., Kolditz, O. Taniguchi, T.: Development and application of a CAD interface for fractured rock. J Environ. Geology (47), 1017-1027, 2005.

[3] Keating, B.A. et. al.: An overview of APSIM, a model designed for farming systems simulation. European J. Agron. (18), 267-288, 2003.

[4] Nabney, I.T.: Netlab. Algorithms for Pattern Recognition. Springer Verlag, London, 2002.

[5] Rogers L.L., Dowla, F.U.: Optimization of groundwater remediation using artificial neural networks with parallel solute transport modeling, Water Res. Research, 30(2), 457-481, 1994.

[6] Schmitz, G.H., Schütze, N., Wöhling, T: Irrigation control: towards a new solution of an old problem. IHP/HWRP-Reports (5). Dt. Nationalkomitee of the Int. Hydrol. Programme of Unesco and the Hydrology and Water Res. Programme of WMO, Koblenz, Germany, 2007

[7] Schmitz, G.H., Puhlmann, H., Droge, W. & Lennartz, F.: Artificial neural networks for estimating soil hydraulic parameters from dynamic flow experiments. European Journal of Soil Science, 56, 19-30, 2005.

[8] Schütze, N., Kloss, S., Schmitz, G.H.: Evaluating Potentials and Corresponding Risks of Optimal Deficit Irrigation Strategies in Oman Under Climate Change. In: W. K. Zubari, editor, Proceedings of the 9th Gulf Water conference, Muscat, Oman, 2010.

Upcoming Water Quantity and Quality Problems in Arid and Semi Arid Climatic Zones

Elias Salameh[1] and Fritz Frimmel[2]

[1] *Faculty of Science, Department of Geology, University of Jordan, Amman-Jordan*
[2]*Engler-Bunte-Institut, Department of Water Chemistry, University of Karlsruhe, 76128 Karlsruhe, Germany*

Abstract:

In the MENA countries extensive overexploitation of fossil groundwater for irrigational use is gradually leading to aquifer depletion and salinization. Of the use sectors which will be most suffering of that depletion is the irrigation sector with negative impacts on societies and countries. The catastrophes will affect the socio-economic situation of agricultural depending on these groundwater resources. Industries will be less affected and domestic uses have to resort to other sources, especially desalination which nowadays can be afforded by most users.
In addition, increasing use of detergents and discharges of treated and untreated municipal effluents containing phosphate and nitrate to diminishing surface water sources, with high sun illumination and other conditions are gradually leading to increasing eutrophication problems of major surface water bodies. This is expected to render these water sources, presently used for drinking purposes, unsuitable for the same purpose, requiring advanced and expensive additional treatment and control procedures.
Unless the MENA countries take the necessary proactive measures to alleviate the problems of resources overexploitation and eutrophication they will in the near future face immense water problems with all the socio-economic and social comfort implications of these problems.

Catchwords/key words: Upcoming water problems, MENA, overexploitation, fossil water, eutrophication

1 Introduction and Problem Statement

The water sector in arid and semi-arid climatic zones has since decades been exposed to increasing stresses, such as diminishing surface water resources, depleting aquifers, quality deterioration, inadequate water supplies and infrastructure, and salinization processes.
Countries in the arid and semi-arid climatic zones are, since a few decades, witnessing:

- Increases in population
- Expansion in industrialization, especially mining industries

- Improving living standards
- Development of agriculture and increasing use of chemicals especially agrochemicals and detergents.

To satisfy the increasing needs of development non-renewable groundwater resources were developed, not only for municipal use, which can somehow be justified, but also for irrigation purposes without sound economic analyses or provision for substituting these resources or putting investments to enable coming generation to generate similar resources. The results are depleting groundwater resources and salinized aquifers.

In arid and semi-arid regions the degradation in the environmental quality is on the increase negatively affecting surface and ground water resources used for human consumption.

In these zones provision of waste water treatment and reuse, and recycling of solid waste are not adequate and are expensive to implement. Wastes, especially liquid waste end up in surface water bodies causing the major quality problems facing the use of surface water for drinking purposes, which are eutrophication and formation of chlorinated hydrocarbons. These problems are a direct result of human activities and are, with time, intensifying in the MENA countries.

Nutrients, such as nitrogen, phosphorus and silicate in lakes, reservoirs and some streams, rivers, and near shore marine water are prerequisites for life, and in natural systems do not pose environmental problems. They become a problem when too large inputs affect the original character, properties or functions of the ecosystem.

Increasing use of detergents and discharges of treated and untreated municipal effluents containing phosphate and nitrate to diminishing surface water sources, in addition to the high sun illumination and other conditions are gradually leading to increasing eutrophication problems. This is expected to render major water sources, presently used for drinking purposes, unsuitable for the same purpose, requiring advanced and expensive additional treatment.

Other conditions found in arid and semi arid areas, such as high UV- and other waves seem to hinder eutrophication processes. The interplay between eutrophication enhancing and hindering factors seems to play a major role in the eutrophication conditions of surface water bodies.

The diminishing water resources and their quality deterioration, especially eutrophication processes are illustrated on a case study from Jordan as an example for the MENA countries.

The measures and the additional treatment processes applied to alleviate these problems in Jordan are explained to serve as case studies for other areas, where eutrophication problems are expected to increasingly face the water resources of these areas.

The counter actions to reduce the negative impacts of such water quality deteriorations are: Setting strategies, introduction of plans and relevant programs in addition to implementation of remediation measures. These include waste water treatment and reuse schemes, avoidance of surface storage of treated water during the dry season and reducing the use of chemicals.

Unless the necessary measures are undertaken to alleviate the above mentioned problems countries in the MENA region using surface water for drinking purposes will, in the near future, face very intensive challenges in their drinking water supplies.
The solution of these problems although technically possible, the cost of treatment remains prohibitive for the majority of these, generally developing countries with weak economies.

2 Extensive aquifers

In the MENA region extensive fresh groundwater bodies are found to underlie vast desert areas such as **Disi –Saq** between Jordan and Saudi Arabia, **Erg** between Algeria and Tunisia, **Fazzan** in Algeria and Libya, **Kufra** and **Sirte** in Libya, **Rub'Al Khali** in Saudi Arabia among others (Table 1). Naturally, these groundwater bodies are under equilibrium conditions where recharge equals discharge. Artificial extractions of water from the up-gradient areas of these water bodies take place on the account of the stored amounts.

Area	Age (K years)	Salinity (mg/l)
Kufra and Sirte	10 – 30	250 – 350
Disi /Saq	2 – 36	220 – 350
Erg	0 – 15	400 – 800
Saudi Arabia (Rub' Al Khali)	5 – 25	300 – 450

Tab.1: Major fresh groundwater bodies in the MENA region with their salinities and average ages

3 Fossil Gradient and Rising land surface

The suggested existence of fossil gradients (Lloyd 1980) as a cause of the continuous groundwater movement in these aquifers can not hold scientific argumentation, because groundwater attenuates very fast and in a few hundred years.
Rising land surface or uplift along the up-gradient areas of groundwater bodies as a cause of groundwater gradient conservation can not be substantiated by scientific argumentation, because uplifts are generally up to 1mm/yr while drops in water levels in aquifers without recharge are more than 10mm/yr.
In extensive aquifers, gradients range form around **3* 10 -3 to 10* -3**, and permeability values from **10 -6 to 5*10 -5 m/s.**
To maximize the available groundwater quantities for calculation purposes a high porosity of 10% is assumed. The result is, without recharge the water level will naturally drop by 15 cm/year which is equivalent to 150m/ 1000 years as a minimum. But since water levels are in most aquifers at a depth of less than 200m below ground surface and the ground surface itself is 150 to 400masl (Table 2), the groundwater can not possess fossil gradients otherwise, its level before around 1500 years must had been above ground surface for which no evidence exists.

In the Gulf of Aqaba the total rise in land surface during the last 120 thousand years amounted to 80m as proven by the age of the raised reefs (Al Rifaiy 1988). Using the same Ref argumentation about the fossil gradient the total drop of water level would equal in 120 000 years * 150m/ 1000yr = 18 000 m, which contradicts all ground facts.

Therefore, in extensive aquifers the suggested hypotheses of fossil gradients and uplift as causes of the continuous groundwater flow do not hold scientific argumentation.

The conclusion is that all these types of groundwater bodies have a type of recharge; otherwise they must have emptied during the last thousand years at most. Natural discharges from all these groundwater bodies will continue irrespective of all our extractions along the head waters and as long as we are logically up-gradient of the discharge sites. Therefore, artificial extractions go on the account of the stored groundwater amounts. This means that even if these huge groundwater bodies receive recharge, that recharge compensates only the natural discharge from the aquifers, and not the artificial extraction.

Therefore, extractions from groundwater bodies are mining processes.

In that concern two types of thinking can be distinguished:

- Exploiting these resources along their discharge sites which will not affect the stored groundwater and will safe the resources for coming generations.
- Mining of these waters at pleasure and next generations will take care of themselves, by supplying water using other technologies especially, desalination.

The decision about what way to follow is a national concern and should be addressed by each country on a national level.

Some two decades ago desalination cost was $US 1.5 to 3/M3 and although oil prices went up by 2 to 3 fold during these two decades, desalination cost at present is one half to one third of the former cost. This means that countries which decided to choose desalination instead of overexploitation made losses whereas those overexploiting their resources have made gains.

But, at present desalination cost seems to be at its limits and countries having the choice between overexploitation and desalination have to rethink their policies.

There remains one issue for proper consideration and that is, some of these aquifers lie at elevations of hundreds of meters above sea level, which means, in addition to their value as fresh water, their position above sea level have potential energy value (Table 2). In other words; in addition to the cost of desalination of sea water to substitute the fresh groundwater, the pumping of the desalinated water to the locations of the fresh overexploited groundwater has to be added.

Country/	Ground water level (masl)	Distance to discharge site at sea level (km)
Libya	200 – 300	700
Algeria	300 – 400	1000
Algeria /Tunisia	150	250
Jordan/ Saudi Arabia	700 - 800	300 - 600

Tab. 2: Groundwater levels of the main extensive aquifers and the distance to discharge sites

Overexploitation resulting in groundwater depletion and quality degradation is indeed a real problem with negative impacts on societies and countries. But the major and direct problems and catastrophes will affect the socio-economic situation of agricultural depending on these groundwater resources. Industries will be less affected and domestic uses have to resort to other sources, especially desalination which nowadays can be afforded by most users. As an example on water resources and uses in the MENA area Table 3 lists the case of Saudi Arabia. Around 85% of the used water is fossil water.
Average drop in the water levels in the different aquifer parts 2.3 to 10m/yr.
Desalinated water is not expected to become that cheap for use in irrigation and to compete with available fresh surface and groundwater resources in the same areas where overexploited groundwater resources are found.
Therefore, it is expected that irrigation projects depending on fossil groundwater will gradually, (with the depletion and degradation of the groundwater resources) become a burden to farmers and investors and will be closed in the future.

The total water uses in Saudi Arabia equal	$24*10^9$ m^3/yr
Desalinated	$1 *10^9$ m^3/yr
Renewable resources	$2.4*10^9$ m^3/yr
Fossil water	$20.6*10^9$ m^3/yr
Industrial	3%
Municipal uses	9%
Irrigational uses	88%

Tab. 3: Water sources and uses in Saudi Arabia

Unless proper programs are advanced to solve the problem, nature is expected to put its solution in the form of depletion and salinization of resources. But, this is expected to result in unemployment, poverty and social discomfort and eventually social and economic unrest.

4 Eutrophication Processes and their Consequences

A combination of conditions, such as high sun illumination and high temperature with increasing additions of nutrients of NO3, SO4, K, Zn, Fe and others into surface water bodies due to the different human activities is gradually leading to increasing eutrophication processes (Carpenter 1998). This is especially true in arid areas, including the MENA countries.
Eutrophication processes create immense problems to water works purifying raw water to drinking quality. Clogging filters in water purification plants, formation of carcinogenic THMs upon water chlorination and water supply with bad odor and taste are some of the problems accompanying eutrophication processes (Rook 1974 and Barney et. al.1983). They lead also to difficulties in water use in industry and in irrigation due to clogging of irrigation and filtration systems, bad taste and odor, suffocation of plants, growth of algae in industrial facilities etc.

Human activities can accelerate the rate at which nutrients enter ecosystems. Runoff from agriculture and urban development areas, pollution from septic systems and sewers and other human related activities increase the flux of both inorganic nutrients and organic substances into terrestrial, aquatic, and coastal marine ecosystems.

In the MENA region extensive use of detergents and fertilizers started to be reflected in the NO3 and PO4 concentrations in surface water bodies, which presently are used for drinking purposes; i.e. the rivers; Nile, Euphrates, Tigris and Jordan and reservoir dams established in their catchments.

The problems in the future are expected to be increasing eutrophication, complicated and expensive water purification procedures, health risks of system failures, formation of trihalomethanes and taste and odor problems with all their socio-economic ramifications.

4.1 Case study King Abdullah Canal (KAC)

King Abdullah Canal conveys water along the eastern side of the Jordan Valley for agricultural, domestic, and small industrial use. The canal was constructed in 1961 and was planned to serve irrigation purposes only. Then, it was extended three times between 1969 and 1987, and now has a total length of 110 km with a head discharge capacity of 20m3/s. The KAC water is presently used for irrigation and municipal water supply with minor amounts used in industry.

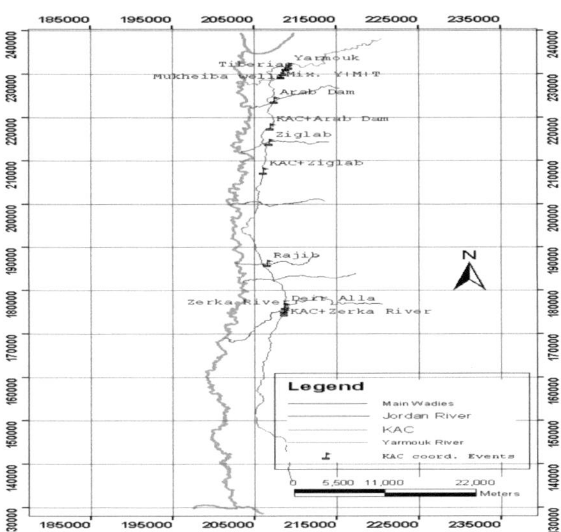

Fig. 1: King Abdullah Canal

Two main problems face the use of the KAC water for domestic purposes. These are eutrophication and formation of chlorinated hydrocarbons during water treatment (Salameh1987).
All these factors (high nutrient concentration, high sun illuminations, suitable pH, high temperature, and the low velocity of water) are available in KAC to cause eutrophication processes (Al-Khoury 2005). All these conditions provide ideal environment for big variations of algal species to grow and increase in numbers forming algal blooms (Al-Harahsheh 2008). Nevertheless, the occurrence of eutrophication blooms is limited in KAC. The UV radiation which may restrict eutrophication is less pronounced in the Jordan Valley area, lying at 220 to 250 m below sea level because of the thicker atmospheric column compared to other places in the world.
In 1987 and in 1998 the water originating from KAC had experienced several incidents of severe taste and odors problems. Zai WTP failed to produce drinkable water for the inhabitants and supplied them with odorous and with algae contaminated water. In addition, very high concentrations of trihalomethanes were found in the supplied water.
The quality of the raw water sources and the warmer summer temperatures in early July 1998 initiated the taste and odor problems. The taste and odor problems were far greater than experienced before and lead to the catastrophe.
Almost without any exception all major surface water bodies in the MENA countries such as the Nile, the Euphrates, the Tigris the Jordan , the Orontes and others are showing increasing concentrations of PO4, NO3, BOD, COD, SS, biocides, chlorophyll and algal masses. In some of these surface water bodies eutrophication processes have started with all their subsequent implications to water treatment for drinking purposes.

5 Conclusion

Overexploitation resulting in groundwater depletion and quality degradation is indeed a real problem with negative impacts on societies and countries. But the major and direct problems and catastrophes will affect the socio-economic situation of agricultural depending on these groundwater resources. Industries will be less affected and domestic uses have to resort to other sources, especially desalination which nowadays can be afforded by most users.
Unless proper programs are advanced to solve the problem, nature is expected to put its solution in the form of depletion and salinization of resources. But, this is expected to result in unemployment, poverty and social discomfort and eventually social and economic unrest.
Continuing along the same developmental lines in urbanization, industrialization and irrigation activities without the proper provision for wastewater treatment and reuse schemes, protection of surface water resources, reduction in the use of fertilizers, detergents and biocides severe problems will face the MENA countries concerning their municipal drinking water supplies, especially eutrophication and trihalomethane formation problems.
This will not only mean increasing purification cost, but also advanced technologies with sophisticated control and care forced by higher risks of failure and risks to human health.

List of Illustrations

List of Tables

6 References

[1] Al-Harahsheh, S.T. (2008): Eutrophication processes along King Abdullah Canal-Chemistry, organism, and resulting compounds. Unpubl. Ph.D. thesis, University of Jordan, Amman-Jordan.

[2] Al-Khoury, W.E. (2005): The water constituents of King Abdullah Canal and their role in the eutrophication processes. Unpubl. MsC. thesis, University of Jordan, Amman-Jordan.

[3] Al-Rifaiy, I.A., Cherif O.H. (1988): The fossil coral reefs of Al Aqaba, Jordan. Facies Vol. 18, no.1, pp 219-230.

[4] Barney, G., Oliver, Earl M. Thurman, (1983): Influence of aquatic humic substance properties on Trihalomethanes potential, Water Chlorination Environmental Impact and Health Effects, Vol.4, Book 1, Chemistry and Water Treatment by Robert L. Jolley, Ann Arbor Science Publishers, 1983.

[5] Carpenter, S.R., Caraco, N.F., Correll, D.L., Howarth, R.W., Sharple, A. N., Smith, N. H., (1998): Non-point pollution of surface Waters with Phosphours and Nitrogen, Ecological Application, Vol. 8, No. 3, pp.559-568.

[6] Lloyd, J.J. (1980): An examination of recharge mound decay and fossil gradients in arid regions sedimentary basins. In: Arid-zone hydrology: Investigations with isotope techniques. IAEA Publications, STI/PUB/547.

[7] Rook, J.J., (1974): Formation of Haloforms during chlorination of natural waters, Proc. Soc. Water Treat. Exam. 23: 234-243.

[8] Salameh, E., (1987): The potential of surface water utilization for domestic purposes in Jordan, Intern. J. Environmental Studies, 1987, Vol.28, p.291-300.

Appropriate Water Supply Techniques for Semi-Arid Regions

The Cuvelai-Etosha-Basin in Namibia

M. Zimmermann
M. Brenda, A. Jokisch, W. Urban
Technische Universität Darmstadt, Department of Civil Engineering and Geodesy, Institute IWAR, Chair of Water Supply and Groundwater Protection, Petersenstr. 13, 64287 Darmstadt, Germany, E-Mail: m.zimmermann@iwar.tu-darmstadt.de, Tel.: +49 (0) 6151 16 29 48

Abstract

The Cuvelai-Etosha-Basin is located in central northern Namibia. The precipitation in this region is extremely variable, evaporation rates are high, perennial rivers do not exist, and groundwater aquifers are saline. The water supply is mainly based upon a pipeline grid fed by the distant Kunene River at the Namibian-Angolan border. Approximately 1 million people or 50 % of the Namibian population live in this region. Population growth, migration, and urbanization will probably enhance the regional demand for water. Hence, the research project "CuveWaters – Integrated Water Resources Management (IWRM) in the Cuvelai Basin (Northern-Namibia)" has been constituted to develop endogenous water resources and alternative techniques as well as to examine their social and technological feasibility. Regarding the water supply side, several adaptable decentral techniques for rural areas are investigated, namely rainwater harvesting (RWH), groundwater desalination, and subsurface water storage (SWS).

Catchwords

Central northern Namibia, groundwater desalination, Integrated Water Resources Management, rainwater harvesting, Subsurface Water Storage, water supply

1 Introduction

The Namibian part of the Cuvelai basin is located in central-northern Namibia (Fig. 1). Approximately 1 million people or 50 % of the Namibian population live in this area (Kluge et al. 2008), which comprises only about 7 % of the country's area.

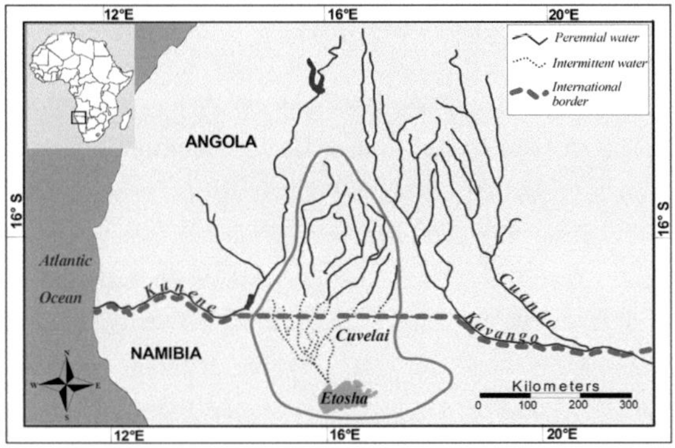

Fig. 1: The Cuvelai catchment area (map created by Steffen Niemann)

The Cuvelai basin is named after a system of ephemeral and intermittent rivers which seasonally drain the rainwater runoff of southern Angola into the region. Central northern Namibia is characterized by high precipitation variability (50-990 mm per year, including consecutive droughts) and seasonal alternations of a dry period during winter and heavy rainfall during summer (including floods).[1] Up to 96 % of the precipitation occurs in the rainy season. Especially the lack of perennial rivers and the salinity of groundwater aquifers are a challenge for the regional water supply which is fed by the Namibian-Angolan border river Kunene. The system consists of a 150 km long open canal and a pipeline grid with an overall length of about 2,000 km (Zimmermann, Urban 2009). The water is withdrawn at Calueque dam on Angolan territory.

Social and economical factors make the situation even more difficult (Kluge et al. 2008). A population growth of approximately up to 2.8 % and migration into the urban centres of the region will probably enhance the demand for water. Further problems can be seen in the high livestock density, overgrazing, soil degradation, and deforestation. At the same time, it is expected that the withdrawal of water from the Kunene River on Angolan territory will grow due to socio-economic developments (e. g. irrigation, hydropower). On the whole, the regional water demand exceeds the local natural resources. This is why an integrated management approach is necessary. In the following, the research project CuveWaters and the water supply techniques proposed will be introduced.

[1] These data were obtained from the Namibia Meteorological Service.

2 The research project CuveWaters

The main objectives of the research project "CuveWaters – Integrated Water Resources Management (IWRM) in the Cuvelai-Etosha Basin (Central Northern Namibia)" are to reduce the dependency on the water of the Kunene River, to develop endogenous water resources, to introduce innovative and adapted techniques, and to examine their social as well as technological feasibility.[2] The central idea is to test and establish a multi-resources-mix of diverse water supply and sanitation techniques. This means that different water qualities are supposed to be used for different purposes. These tasks are accompanied by capacity building and participation of all relevant stakeholders, including the water users. In terms of the water supply side, several adaptable decentral techniques for rural areas are investigated, namely rainwater harvesting (RWH), groundwater desalination, and subsurface water storage (SWS).

2.1 Rainwater harvesting (RWH)

Rainwater is hardly collected in the region by technical means despite an annual mean of 472 mm (Sturm et al. 2009). Precipitation can be harvested from roofs or other relatively impermeable surfaces with high runoff coefficients. Regarding the roof catchments, corrugated iron roofs with runoff coefficients of 0.8-0.85 (Gould, Nissen-Petersen 2003) are more effective than thatched roofs. Compacted soil, plastic sheeting, or concrete can be used for the construction of ground catchments. Concrete lined surfaces feature relatively high runoff coefficients of 0.73-0.76 (Gould & Nissen-Petersen 2003) and are thus preferred over other options for the pilot phase.

The most cost-intensive part of a RWH system is the reservoir. The storages have to be constructed on-site and adapted to local conditions. Hence, attention has to be paid to the availability of construction materials and workforce. Furthermore, selected designs have to be cost-effective and robust. Four pilot plants for RHW were constructed in the jointly selected village of Epyeshona (Okatana Constituency, Oshana Region) from Oct. 2009

[2] Project partners are the Institute for Social-Ecological Research (ISOE) in Frankfurt/Main and the Chair of Water Supply and Groundwater Protection (Institute IWAR) of the Darmstadt University of Technology. CuveWaters is funded by the German Ministry of Education and Research (BMBF). Namibian cooperation partners are the Desert Research Foundation of Namibia (DRFN), the Namibian Ministry of Agriculture, Water and Forestry (MAWF), and branch offices of the GTZ (German Society for Technical Cooperation) and the BGR (German Federal Institute for Geosciences and Natural Resources) in Namibia. Visit http://www.cuvewaters.net for details.

until Feb. 2010. The village community chose households during several workshops as sites for three roof catchment facilities with catchment areas between 90 and 110 m^2. Differing materials were applied for the tanks of these systems: polyethylene, ferro-cement, and concrete bricks. The reservoirs have a capacity of 30 m^3 each. The harvested rainwater can be used for small-scale irrigation but also domestic purposes and is furthermore of good quality, especially compared to traditional sources.

A 120 m^3 cistern was constructed for another pilot plant with a concrete-lined ground catchment which supplies approximately six households. The institutional setting of this facility was developed by the users themselves in participatory workshops. The collected water is mainly supposed to be used for horticulture. Hence, gardening plots were created and users were trained in basic gardening techniques (Fig. 2). All in all, up to 24 technicians from the village received capacity building measures to build, operate, and maintain the RWH systems. At the moment, actual system performances in terms of water quality and quantity, technology, utilisation, and institutionalisation are monitored.

Fig. 2: Workshop at the rainwater harvesting pilot plant with ground catchment in the village of Epyeshona (Okatana Constituency, Oshana Region) in 2010

2.2 Groundwater desalination

People in remote areas without access to the piped scheme use water from hand-dug wells for domestic purposes and livestock watering. This water becomes brackish due to hydrogeological conditions and evaporation during the dry season. Moreover, many wells are contaminated with algae, faeces, and parasites since they are unprotected. Hence, decentralised and robust techniques for groundwater desalination are proposed. Pilot plants were installed in two jointly selected villages of the Omusati Region. The facilities are completely solar-driven due to the lack of conventional sources of energy such as electricity, oil, or gas. Furthermore, the region features a solar radiation of more than 6 kWh/(m^2*d) and a mean sunshine duration of 8-9 h/d (Mendelsohn et al. 2003).

Two pilot plants (Fig. 3) were constructed in the village of Amarika (Otamanzi Constituency): a membrane distillation plant of the Fraunhofer Institute for Solar Energy Systems (ISE, Freiburg) and a reverse osmosis plant of the proaqua company (Mainz). Each system produces 3 m^3/d of drinking water. The brine is mainly reinjected into deeper groundwater layers but partly also evaporated in a pond. Furthermore, a multi-effect-humidification plant of the Terrawater company (Kiel) with a capacity of 3.5 m^3/d was installed in the village of Akutsima (Okahao Constituency). Additionally, a multi-stage-flash plant of the Solarinstitute Jülich with a capacity of 600 l/d is planned. The distillate of both systems is mixed with raw water in order to achieve drinking water quality by remineralisation. The brine is completely evaporated in ponds.

The pilot plants are operated by trained local caretakers since Jul. 2010. They are supported by a service provider (Aqua Services & Engineering, Windhoek) who regularly checks and maintains the facilities. In addition, the systems are monitored by the German manufacturers. Institutional arrangements are again developed in participatory workshops with the users to raise their understanding and responsibility.

Fig. 3: Desalination plant in Amarika (Otamanzi Constituency, Omusati Region)

2.3 Subsurface water storage (SWS)

The technique of Subsurface Water Storage is intended to store the natural runoff from the above mentioned Oshanas of the Cuvelai-Etosha-Basin. While in the rainy season (from Nov. until Apr.) the region is experiencing floods, it can happen that the whole area suffers from droughts in the dry season. Especially in the 1960s and 1970s, it has been tried to store Oshana water for utilisation in the dry winter season in pump storage dams and excavation dams (Stengel 1963). Major problems of these techniques are the high evaporation rate in northern Namibia (2,700 mm/a) and the declining quality of the water stored. Instead of using open reservoirs, covered subsurface storages are proposed to avoid these disadvantages (Fig. 4). However, it is assumed that the provided water will be of medium quality and thus is intended to be used for small-scale irrigation and livestock watering. The capacity of the SWS will be at approximately 1,000 m³ whereby it is able to supply roughly five households for a period of two years.

SWS is mainly considered for the storage of flood water in areas without suitable aquifers. The technique does not interfere with saline or fresh water aquifers since it is independent from local groundwater. Furthermore, the components of SWS are supposed to be constructed from local or at least easily available materials. Within the project, it is planned to clarify methodical and technical questions as well as the conditions of operation and maintenance since a general state of the art for SWS is not documented yet.

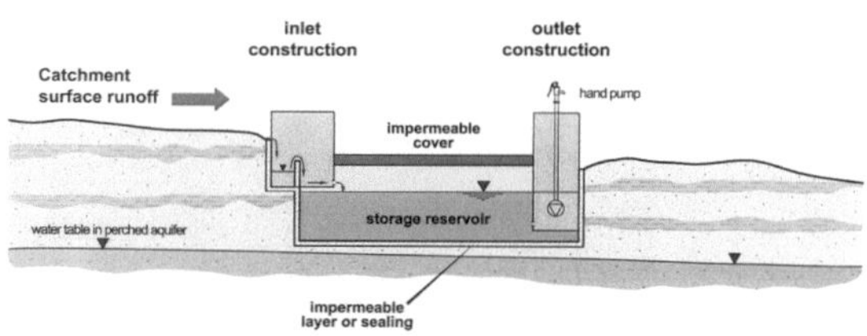

Fig. 4: General construction of a Subsurface Water Storage: Catchment surface runoff is diverted into an inlet construction and into a closed artificial subsurface reservoir.

3 Conclusions and Forecast

The discourse on IWRM leaves many questions regarding operational aspects unanswered. However, there is a broad consensus that sustainable development is a core element of IWRM, and poverty alleviation as well as ensuring basic services are two essential tasks of the social dimension of sustainability. A major goal of CuveWaters is to improve the livelihoods of the rural population in central northern Namibia by increasing the security of water supply and thus fostering regional economic activities. Beside the political aspects of IWRM, demand management and technical solutions play an important role. The proposed techniques are able to contribute to these goals by developing endogenous water resources (as in the case of RWH), by strengthening the population's resilience potential (SWS), and by appropriate and adapted treatment processes (groundwater desalination). Eventually, a constant water availability, quantity, and quality for several purposes such as human consumption and irrigation or livestock watering can be provided throughout the year.

The monitoring and evaluation of the pilot plants in terms of their performance and acceptance will be continued during the second phase of CuveWaters until 2011. An innovative and adapted tool for technology assessment was developed to enable policy-makers and stakeholders to take decisions. In the third phase of the project, the regional diffusion of successful and viable techniques is planned.

List of Illustrations

List of Literature

[1] Gould, J.; Nissen-Petersen, E. (2003): Rainwater Catchment Systems for Domestic Supply – Design, construction and implementation. ITDG Publishing, London.

[2] Kluge, T.; Liehr, S.; Lux, A.; Moser, P.; Niemann, S.; Umlauf, N.; Urban, W. (2008): IWRM Concept for the Cuvelai Basin in Northern Namibia. Physics and Chemistry of the Earth, 33, 48-55.

[3] Mendelsohn, J.; Jarvis, A.; Roberts, C.; Robertson, T. (2003): Atlas of Namibia – A Portrait of the Land and its People. Ministry of Environment and Tourism of Namibia.

[4] Stengel, H.W. (1963): Wasserwirtschaft in Südwestafrika. Afrika-Verlag Der Kreis, Windhoek.

[5] Sturm, M.; Zimmermann, M.; Schütz, K.; Urban, W.; Hartung, H (2009): Rainwater harvesting as an alternative water resource in rural sites in central northern Namibia. In: Physics and Chemistry of the Earth, Vol. 34 (2009), S. 776–785, Elsevier.

[6] Zimmermann, M.; Urban, W. (2009): Wasserversorgungstechnische Lösungsansätze innerhalb des IWRM im Norden Namibias. In: Linke, H. J. (Edt.): 1. Darmstädter Ingenieurkongress – Bau und Umwelt, 14. und 15. September 2009, Tagungsband, Technische Universität Darmstadt.

Estimating resource costs of meeting EU Water Framework Directive ecological status requirements at the river basin scale

Niels Riegels
Valerie Sturm, Roar Jensen, Lisa Benssason, Haris Doulgeris, Flemming Moeller, Peter Bauer-Gottwein
Department of Environmental Engineering, Technical University of Denmark, 2800 Kgs. Lyngby, Denmark, E-mail: ndr@env.dtu.dk, Telephone: +45 45251452

Abstract

Resource costs of meeting EU Water Framework Directive (WFD) water status requirements at the river basin scale are estimated. Resource costs are interpreted as opportunity costs of water use arising from water scarcity. An optimization approach is used to identify economically efficient ways to meet WFD water status requirements. The approach is implemented using a river basin simulation model coupled to an economic post-processor; the simulation model and post-processor are run from a central controller that iterates until an allocation is found that maximizes net benefits given WFD requirements. Water use values are estimated for urban/domestic, agricultural, industrial, livestock, and tourism water users. Water status is estimated using metrics that relate average monthly river flow volumes to the natural hydrological regime. The decision variable in the optimization is the price of water, which is used to vary demands according to microeconomic theory. Results from a real-world application in northern Greece show the suitability of the approach for used in complex, water-stressed basins.

Keywords

Hydro-economic models, Integrated water resources management, Water value, Environmental flows, Positive mathematical programming

1 Introduction

The EU Water Framework Directive (WFD), introduced in 2000, outlines requirements for economic analysis in water resources planning. The central purpose of the WFD is the protection of water resources within the EU, with the goal of achieving "good surface water and groundwater status". (EU Commission, 2000).

The assessment of environmental and resource costs has emerged as a challenging aspect of the WFD (e.g., Heinz et al., 2007). Brouwer (2004) provides a useful overview of what appears to be the current interpretation. Resource costs are interpreted as opportunity costs arising from inefficient allocation of water by existing institutions as well as opportunity costs of depletion of water stocks beyond natural states of recharge (e.g., groundwater mining). Environmental costs are interpreted as lost environmental benefits, where these benefits are measured using environmental valuation methods.

This analysis interprets resource costs as opportunity costs. However, no attempt is made to value environmental costs of water use. WFD water status objectives are implemented as constraints. Resource or opportunity costs of complying with those objectives are then estimated relative to an existing baseline. The resulting estimate is proposed as the resource cost of implementing WFD water status objectives. The resource cost should not be interpreted as a marginal opportunity cost. It is assumed to be the difference between average annual net benefits to all water users at an existing baseline and average annual net benefits after WFD ecological status objectives have been achieved.

The WFD defines "good surface water and groundwater status" in terms of parameters that are used to assess the ecological and chemical state of water bodies. In this analysis, only one parameter is used to assess water status: hydrological regime. The term "hydrological regime" refers to the pattern of a river's flow quantity, timing, and variability (Poff et al., 1997). An analysis limited to hydrological regime has value because other parameters used to assess ecological status are impacted by the hydrological regime; the hydrological regime may be a "master variable" affecting the distribution and abundance of species in the river ecosystem (e.g., Power et al., 1995).

An economic optimization approach is used to estimate resource costs. In this approach, economic values are estimated for water uses including urban/domestic use, irrigated agriculture, industry, livestock production, and tourism. Water demands are varied until an optimal set of demands are found that allow for a hydrological regime that meets the "good status" objectives of the WFD while maximizing the economic value of water to other users.

2 Methods

The river basin decision support system MIKE BASIN is used in this application (DHI, 2009). The model is run on a daily timestep for a 20-year period using 20 years of historical hydrology (1981-2000) intended to capture a reasonable range of hydrological conditions. The hydrological inputs to the model are river inflow and groundwater recharge timeseries values. Crop water requirements are calculated dynamically using an approach based on FAO-56 (FAO, 1998). Crop yields are modeled using the FAO-33 methodology (FAO, 1979).

The Aggitis River basin in northern Greece is used as the case study area. The Aggitis River is a major tributary to the Strymonas River. The river basin model of the Aggitis basin is adapted from a MIKE BASIN representation developed as part of an effort to comply with WFD requirements (ENM Ltd., 2008). 59 anthropogenic water use locations are included in the representation. A small reservoir in the upper basin is operated for flood control and to supply water to downstream irrigation locations.

This analysis uses a method based on Arthington (2006) to estimate the status of the hydrological regime. In this method, hydrological regime status is estimated as a function of the extent to which the distribution of flow volumes for each month matches the unmodified distribution. Figure 1 compares the unmodified cumulative density function (CDF) of August flow volumes for two river reaches in the study area to the CDF for a baseline simulation including all current water uses.

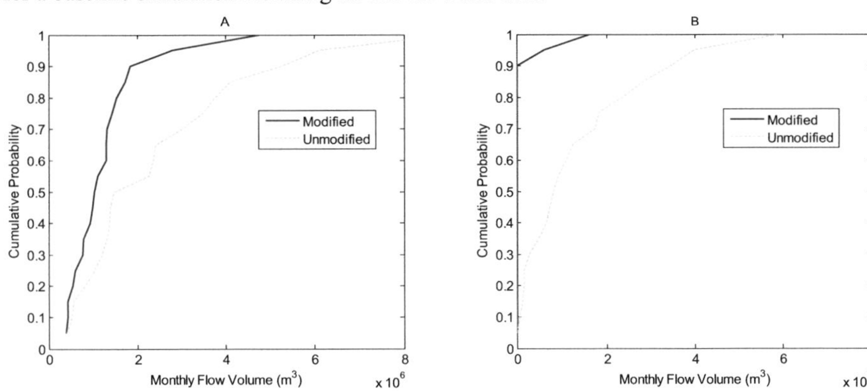

Fig.1, Cumulative density functions of August flow volumes

The maximum difference along the cumulative probability axis between modified and modified CDFs was used to assess the status of the hydrological regime. It is assumed that if the maximum difference between CDFs is less than 0.25 for all reaches in the basin during all months, then the status of the hydrological regime is “good”.

Urban/domestic water users are modeled as utility-maximizing consumers, while irrigation, industry, livestock, and tourism water users are modeled as profit-maximizing producers. The value of urban/domestic water use is estimated using consumers' surplus, with demand function parameters are estimated using the constant elasticity point expansion method. Irrigation, industry, livestock, and tourism water use values are estimated using residual imputation (Young, 2005).

The optimization proceeds by running a hydrological simulation model iteratively until an allocation plan is found that maximizes net benefits of water use while meeting the water status requirement. During each iteration, water price is used to identify demands at each node. These demands are given to the hydrological model, which runs for a 20-year simulation period and attempts to meet demands given hydrological, mass balance, and reservoir operations constraints. After each iteration, water use timeseries values at each demand location are given to an economic post-processor that computes average annual economic values. The river basin simulation model and economic post-processor are treated as a single function: the input is a water price, and the output is the average annual net benefit of water use. The objective function is formulated as a continuous function and a non-linear gradient-search optimizer from the Matlab optimization toolbox (The Math Works, 2010) is used to find the maximum.

$$\max \frac{\sum_{t=1}^{20}\left(\sum_{i=1}^{N} NB_{it}(\overline{w_{it}}) - \sum_{i=1}^{N} c_{it}\right)}{20} \cdot erf\left(s \cdot \left(status_t - status_a\left(\mathbf{Q}_{mj}\right)\right)\right)$$

where

$NB =$ annual net benefit of water use at location i during year t

$c_{it} =$ annual cost of water supply at location i during year t

$t =$ year of simulation

$i =$ water use location

$N =$ number of water use locations

$\overline{w_{it}} =$ vector of daily use volumes at water use location i during year t

$s =$ error function scale parameter

$status_t =$ target value for ecological status

$status_a =$ actual value for ecological status

$\mathbf{Q}_{mi} =$ matrix of monthly volumes m at each river reach j [1]

The optimization function is subject to the following constraints:

$$\overrightarrow{w_{it}} \leq \overrightarrow{d_{it}} \text{ (for each entry of } \overrightarrow{w_{it}}, \overrightarrow{d_{it}})$$
$$\overrightarrow{d_{it}} = f(p_w) \text{ (for urban/domestic, industry, livestock, and tourism water users)}$$
$$\overrightarrow{d_{it}} = f(\overrightarrow{a_i}) \text{ (for irrigation water users)}$$
$$\overrightarrow{a_{ik}} = f(p_w) \qquad [2].$$

where

$\overrightarrow{d_{it}}$ = vector of daily water demands for location i and year t

p_w = water price

$\overrightarrow{a_i}$ = vector of land allocations to various crops grown at location i

3 Results and Conclusions

Summary results comparing water use values by sector for the baseline and "good" ecological status scenarios are presented in figure 2. The overwhelming majority of opportunity costs appear to be borne by the agriculture sector, which could lose as much as 90% of the value attributed to agricultural water use in the baseline scenario.

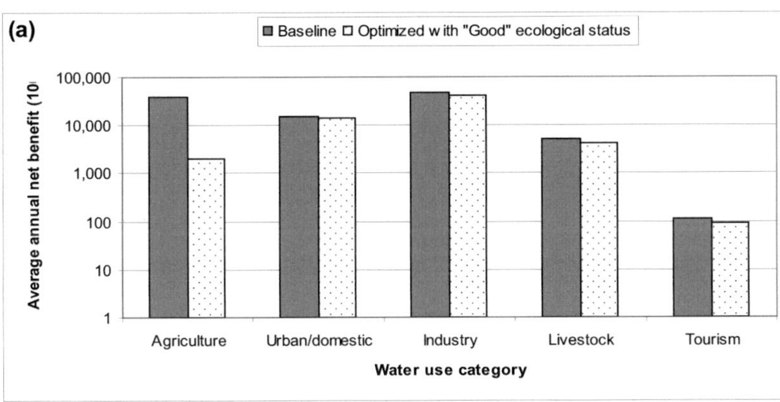

Fig. 2: Annual average net benefits by water use category

The optimal water price is 1.06 €/m^3. Figure 3 gives an indication of the impact of the water price increase on the flow regime. The figure shows cumulative density functions the month of August for a representative river reach. The figure indicates that the optimized flow distribution matches the unmodified distribution more closely than the modified distribution.

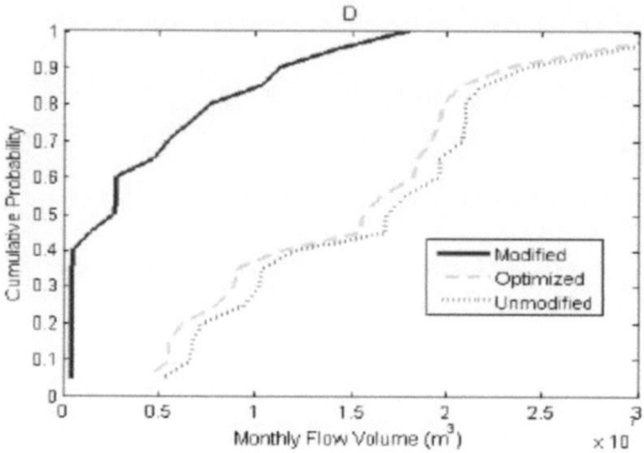

Fig.2, Comparison of August CDFs

The price-based optimization approach presented here implements the ecological status objectives of the WFD by eliminating water uses with low marginal values. Given the assumptions used in this analysis, most of these are agricultural water uses. The overwhelming majority of resource costs are therefore borne by the agriculture sector.

Viewed from another perspective, it could be argued that the agriculture sector causes an environmental loss to society. This highlights the importance of including environmental values in this kind of study. If an ecological status target can only be achieved with significant opportunity costs, then the environmental benefits to be gained from reaching the target should be quantified using a common metric.

The approach could be improved by the addition of continuous, multi-input production functions, particularly for agriculture. Marginal willingness to pay for water is currently determined by dividing the residual profit for each production process by the water requirement for that process; if the price of water exceeds marginal willingness to pay, then the analysis assumes that the process is abandoned. In the agriculture sector, the consequence is that significant amounts of farmland are assumed to go out of production.

In reality, it seems likely that farmers would shift land to crops that are more profitable per unit water use, or invest in other factors of production, such as water-saving technologies.

References

[1] Arthington,A.H., Bunn,S.E., Poff,N.L., and Naiman,R.J., 2006. The challenge of providing environmental flow rules to sustain river ecosystems. Ecological Applications, 16 (4): 1311-1318.

[2] Brouwer, R., 2004. The concept of environmental and resource cost: Lessons learned from ECO2. In: Brouwer, R. (Ed.), Environmental and Resource Cost and the Water Framework Directive. An Overview of European Practices. Amsterdam, Holland, RIZA Working Paper 2004.112x.

[3] DHI, 2009. MIKE BASIN user manual Hørsholm, Denmark.

[4] ENM Ltd.,2008. Development of systems and tools for the management of water resources in the water districts of Western Macedonia, Central Macedonia, Eastern Macedonia, and Thrace, Hellenic Republic, Ministry of Developement (in Greek).

[5] EU Commission, 2000. Directive 2000/60/EC of the European Parliament and of the Council of 23 October 2000 establishing a framework for Community action in the field of water policy. Official Journal L 327 , 22/12/2000 P. 0001 - 0073.

[6] FAO, 1979. Yield response to water. Doorenbos, J and Kassam, A. H. FAO irrigation and drainage paper 33. Rome.

[7] FAO,1998. Crop evapotranspiration: Guidelines for computing crop water requirements. R.G.Allen, L.S.Pereira, D.Raes, and M.Smith. FAO irrigation and drainage paper 56. Rome.

[8] Heinz,I., Pulido-Velazquez,M., Lund,J.R., and Andreu,J., 2007. Hydro-economic modeling in river basin management: Implications and applications for the European Water Framework Directive. Water Resources Management, 21 (7): 1103-1125.

[9] Poff,N.L., Allan,J.D., Bain,M.B., Karr,J.R., Prestegaard,K.L., Richter,B.D., Sparks,R.E., and Stromberg,J.C., 1997. The natural flow regime. Bioscience, 47 (11): 769-784.

[10] Power,M.E., Sun,A., Parker,G., Dietrich,W.E., and Wootton,J.T., 1995. Hydraulic Food-Chain Models. Bioscience, 45 (3): 159-167.

[11] The Math Works, Inc., 2010. Matlab optimization toolbox user's guide. Natick, MA.

[12] Young,R.A., 2005. Determining the Economic Value of Water: Concepts and Methods. Resources for the Future, Washington, DC.

Adaptive Water Resources Management under Extreme Climatic and Hydrogeological Conditions

Interdisciplinary Research Activities in Karst Regions of South East Asia

Franz Nestmann
Peter Oberle, Muhammad Ikhwan, Philipp Klingel
Karlsruhe Institute for Technology
Institute for Water and River Basin Management, 76131 Karlsruhe
Germany, E-Mail: Franz.Nestmann@kit.edu, telephone +49(0)721/6082194

Abstract

In many developing countries, despite of much national and international effort to improve the water supply situation, the technical and ecological as well as economical solutions are still insufficient. The situation in karst regions is even more severe. For the implementation of a water resources management in karst regions adapted and innovative technologies for exploration, distribution, usage as well as protection are required in order to assure sustainability. This paper describes the results and experiences of water resources management in the karst region Gunung Kidul, Indonesia. They evolved from a long-term cooperation (project duration: 2002-2007 and 2008-2013) between Germany and Indonesia, supported by Federal Ministry of Education and Research (BMBF).

Keywords

Water availability, water usage, vulnerability, appropriate technology, sustainability, karst water

1 Introduction

According to WHO/UNESCO (2002) more than 1.1 billion people throughout the world are affected by water shortages and the vast majority of these people are living in developing countries. This situation of water shortage is even more severe in karst regions. Karst is related to the occurrence of carbonate and gypsum rock. Approx. 20 % of the world's population lives on carbonate rock. More than a quarter of mankind depends on karst aquifers as their source of water. However, due to the absence of surface water storage possibilities, people living in karst regions are often suffering from acute water shortage, especially during dry seasons.

On the other hand in many karst regions exist large networks of underground rivers which lead water continuously, also during dry season. Nevertheless, very often there is limited accessibility to these rivers due to their location deep underground (in some cases the groundwater level is more than 100 m deep). This high difference of altitude causes significant operational costs for water extraction using conventional techniques such as diesel or electric driven pumps. In addition, due to high infiltration rates, the situation is even more complex because of the vulnerability to contamination e.g. from agriculture and urban waste water.

As a consequence, for an exploitation of the underground rivers, adapted and innovative solutions for management, distribution, usage as well as protection of the water are required in order to assure the sustainability of the regional development. Based on this situation an interdisciplinary research group from Karlsruhe Institute for Technology (KIT) was established. In co-operation with industrial partners they intensively worked on the development and implementation of concepts and technologies for adaptive water resource management in karst regions in Southeast Asia (e.g. Indonesia, Thailand and Vietnam). The aim was to establish a long-term integrated water resource management (IWRM) concept which can be used as a basis for further research and development work with global relevance.

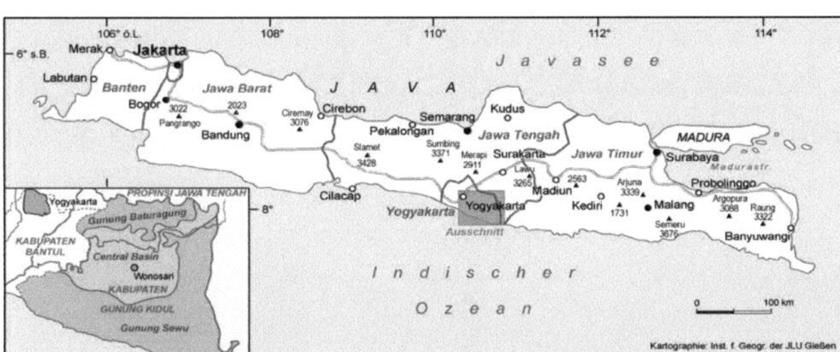

Fig. 1 Java Island and location of district Gunung Kidul

The present paper will focus on the past and current activities in the karst region of Gunung Sewu in the district of Gunung Kidul, Java, Indonesia (see Fig. 1). The district is situated within the Yogyakarta Special Province along the southern coast of the

Indonesian island Java (area approx. 3.000 km^2). The south eastern region of Gunung Kidul is called Gunung Sewu ("the 1000 hills"), defined by cone formations of karst formed by tropical erosion. Because of the natural hydrogeological conditions as well as the lack of lasting technologies for water supply, the Gunung Sewu region is faced with acute water shortage especially during dry season (see. Fig. 2). Therefore it is categorized as 'poorhouse Java'.

Fig. 2: Gunung Sewu during rainy season and dry season [Source: IfG – Giessen]

2 The Climate Conditions

The study area shows a tropical climate with high rain intensity. In average the area has a total precipitation of 2,000 mm/year with an evapotranspiration rate of approximately 1,500 mm/year. However, the total annual precipitation is distributed intermittently between rainy season and dry season. During rainy season (October until April) the average rain frequency reaches 20 days per month, during dry season this value decreases to 3 days per month.

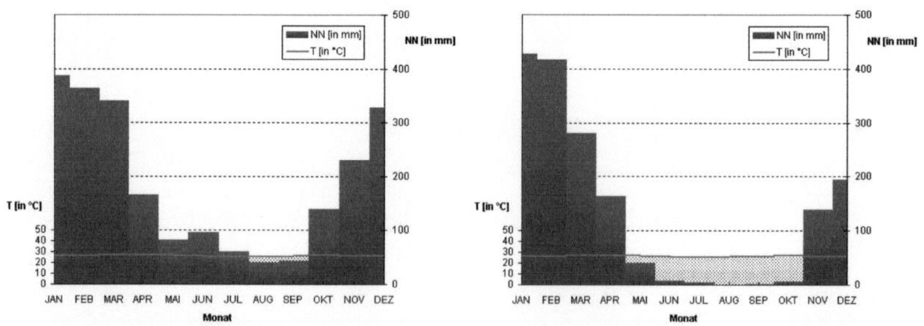

Fig. 3 Annual distribution of average rain intensity in Gunung Kidul during period 1971 – 2000 (left) and during El-Niño phenomena in 1991, 1993, 1994, 1997 (right). [Source: IfG – Giessen]

The inconstancy is getting worse during years of El-Niño phenomenon (ENSO – El Niño Southern Oscillation), which in total has occurred eight times during the last 30 years [2]. While rainfall is influenced by El- Niño phenomenon only marginally during rainy season, the effects during dry season are tremendously. The El-Niño events are associated with an increased risk of dry conditions that are more pronounced and longer lasting than in 'normal years' [1]. The dry season can extend up to eight months; extreme water shortages (precipitation rates decline by more than 50%), drought, forest fires and reduced rice harvests are common results [6]. Fig. 3 shows the extremely reduced average rain intensity in Gunung Kidul during these El-Niño phenomena.

3 Innovative and Sustainable Technology for Water Supply

In the year 2002, KIT and University of Giessen initiated a German-Indonesian joint project to exploit underground rivers. Under the coordination of the Institute for Water and River Basin Management (IWG/KIT) and in co-operation with German companies a pilot hydropower plant was successfully implemented in the cave "Gua Bribin". The intention was to integrate innovative technologies as well as renewable energy sources into the concept of the plant. Due to their high availability, low investment costs and good maintainability reverse driven centrifugal pumps were used instead of complex turbines. The produced mechanical energy is transmitted via a gearing to the feed pumps for water transport. Thus economical as well as ecological aspects were met in an ideal way. Nowadays "Gua Bribin" serves as water supply facility as well as "cave laboratory" (see Fig. 4). At full capacity 65 l/s of water can be supplied to a reservoir at 220 m height. From this reservoir the water will be distributed to the communities mainly by the impact of the gravitation field. Due to the fact that cost intensive external energy will not be needed this facility can be operated 24 hours a day to supply 80.000 inhabitants with 70 lpcd, see [3] for more details.

Fig. 4 Successful test operation in 10/2008 (left), Outline of hydro power plant (right)

4 Integrated Water Resources Management in Gunung Kidul

In 2008 both Indonesian and German sides agreed to extend the cooperation with the follow-on project Integrated Water Resources Management (IWRM) in Gunung Kidul. The IWRM should contain all aspects of research and development of water resources covering infrastructure (civil works), the water distribution system, water quality regulations and wastewater treatment and disposal. Operational and economical aspects must be taken into consideration independently from the hydrological, hygienic, ecological, social and cultural boundary conditions. Fig. 5 illustrates the basic conception of the implemented IWRM. For the implementation of this basic conception, the project is subdivided into seven work packages (WP) which are described below. From German side scientific institutions from KIT, University of Giessen and the Technologienzentrum Wasser (TZW), as well as six industry partners are involved in the implementation of this project. From Indonesian side governmental institutions, research centers and universities as well as non-governmental organizations participate.

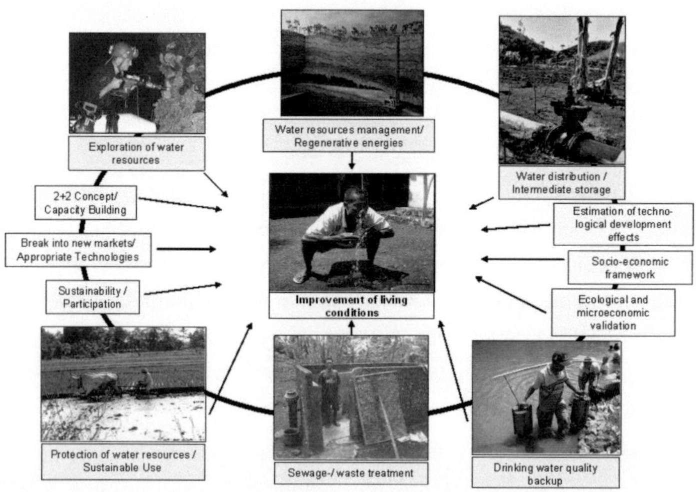

Fig. 5 IWRM basic conception

WP1: Water Resources Exploration

Due to an intermittently rainfall distribution throughout the year typically the main concern in karst regions is an unstable groundwater regime [5]. Therefore the basis for an adapted IWRM concept is the evaluation of the hydrological, hydraulic and hydrogeological conditions in the target area. Geophysical methods, investigation of the network of underground rivers through speleological survey and tracer methods, geodetic detection of the surrounding topography are used to assess the water availability. The data and information which was already collected during the first project in Gua Bribin shall be extended e.g. by the development of additional monitoring systems and further speleological explorations.

WP2: Water Supply Methods / Renewable Energy

Objective of work package 2 is to make the underground water resources exploitable. This includes the development of geotechnical concepts for water storage in karst rock (e.g. use of existing karst cavities, underground barrages, application of injection technologies), working out of concepts for sustainable reservoir management by using renewable energies and adapted construction techniques (e.g. advancement of "pump as turbine" [4] together with the industry partner KSB AG, wooden pressure pipeline). Furthermore the implementation of a monitoring system for long-term observation of barrages (deformations, side underflow; in cooperation with GIF GmbH) is a content of this division. A new concept for producing the needed water pressure contains a long wooden pressure pipeline instead of a barrage. This concept shall be implemented in a second cave named "Gua Seropan".

WP3: Water Distribution, -Conditioning and -Quality Assurance

In the rural areas of Gunung Sewu the main focus lies on the redevelopment and optimization of structure and operation of existing water distribution systems. The operation will involve a combination of real time optimization models and adapted controlling technologies which are provided by IDS GmbH. Decentralized energy recovery in the distribution network and the development of adapted technologies for conditioning, distribution and use of drinking water are further aspects that shall be elaborated within this work package. As an example for chemical and microbiological quality control together with CIP GmbH a pilot plant (container system) for water treatment will be installed in the hospital of Wonosari which can be regarded as a laboratory model for further decentralized plants in the region.

WP4: Wastewater and Refuse Treatment

The aim is the development of adapted technologies for separation, treatment, usage and recirculation of wastewater and refuse flows. Wastewater and organic refuse shall be treated in a way that accomplishes a recycling concept for the nutrients while at the same time preserves and ensures the scarce water resources. This also includes a microbiological-hygienic evaluation of the applied technologies. The basis for the implementation of a sustainable disposal concept is the development of a material flow model of the region, in which all relevant water-bound nutrient flows are represented. The model should reveal existing deficits and help to determine potential starting points for improvement. Due to the grave differences between rural and urban areas in the model region, the requirements for differentiated approaches and working priorities for the de- or semi-decentralized wastewater and refuse treatment are given. Currently the implementation of co-fermentation and co-composting of septic sludge and bio-waste in semi-decentralized pilot plants in cooperation with Huber AG is in progress.

WP5: Socio-Economic Conditions, Ecological Analysis and Technology Assessment

For creating a basic requirement profile for a water supply system the analysis of the economic and social conditions, as well as the current supply situation for drinking water in the project area is indispensable. Generalizable modelings of material flow and energy requirements of the water supply- and water disposal-systems will help to improve the current approaches. Through Life Cycle Assessment (LCA) and Life Cycle Cost Analysis (LCCA) both environmental potentials and economical consequences are determined. The assessment concerning the compatibility of the applied technologies regarding ecological and economical aspects is essential for a before-after analysis of the project area.

WP6: Capacity Development and Technology Transfer

The sustainability of the technical concepts (incl. awareness for water resources protection) can only be achieved with the acceptance of the target groups as well as their participation in all project phases. Design and implementation of the water management concepts are therefore supported by accompanying measures, such as workshops, awareness-raising campaigns and an intensive knowledge transfer. This transfer coined as 'capacity development' should specifically include the local communities (end-users) as well as the concerned administrative authorities, scientific institutions and the chambers of industry and commerce.

WP7: Interdisciplinary Data Management System

In order to assure the sustainability of the IWRM achievements, an integrated management instrument in the water sector is highly required. For this purpose in cooperation with COS Systemhaus OHG a web-based GIS system for water resources management will be developed and implemented. With this instrument interconnection, processing and evaluation of various data collections concerning water management will be enhanced. Furthermore, simulation and optimization tools for water distribution systems (WP3) as well as material flow models (WP4) will also be coupled.

5 Outlook

The utilization of karst aquifers for water supply in a sustainable way to improve the living conditions of the population has a global relevance. In many regions of the world (e.g. Vietnam, Thailand, south China, Japan, Philippines, Laos and South America) a multitude of unexploited cavernous water flows exist, while the people living in the concerned region suffer from severe water shortage. Based on the experiences from Indonesia, currently an intensive study to disseminate the IWRM concept to Geopark Dong Van Karst Plateau at the Chinese border in North Vietnam is carried out. The first intensive exploration will be done in October 2010 and the results will be presented during the international IWRM Conference in Karlsruhe.

Literature

[1] D. Kirono, N. Khakhim, Rainfall and El Nino Southern Oscillation: Links and its Impact on Crop Production. The Indonesian Journal of Geography, Vol. 30, No. 76, S. 21-34. 1998.

[2] National Oceanic and Atmospheric Administration (NOAA). Climate Prediction Center - Changes to the Oceanic Niño Index (ONI). URL: http://www.cpc.ncep.noaa.gov. 12.08.2009.

[3] F. Nestmann, P. Oberle, M. Ikhwan, T. Lux, U. Scholz. Bewirtschaftung unterirdischer Fließgewässer in Karstgebieten. WasserWirtschaft 99, Heft 7-8.2009

[4] P. Oberle, M. Ikhwan, F. Nestmann, J. Fritz, T. Ravisundar. Angepasste Wasserfördertechnologie unter Einsatz von Pumpen im Turbinenbetrieb. WasserWirtschaft 99. Heft 7-8. 2009

[5] Z. Stevanovic, R. Eftimi. Karstic sources of water supply for large consumers in Southeastern Europe - sustainability, disputes and advantages. Geologia Croatica. Ausgabe 63/2, S. 181. 2010.

[6] T. Whitten, R. Soeriaatmadja, S. Afiff. The Ecology of Java and Bali - The Ecology of Indonesian Series, Volume II. Singapore. 2000.

Name: Ms. Nagwa Elnwishy
Corresponding: Biotechnology Research Center, Suez Canal University, Ismaillia, Egypt. nwishy@yahoo.com
Abstract type: Oral presentation
Topic: *irrigation and salination*
Title: influencing smallholder farmers to water reuse in irrigation

Abstract:

Agriculture is the largest water use sector of all water based sectors; Irrigation of agricultural lands accounted for 70% of the water used worldwide (Ximing *et al.,* 2003). In Egypt, irrigation represents up to 95% of all water uses (NAP, 2007), and plays a major role in the national food security. Thus, water reuse approaches were immerged.

Farmer indigenously practiced water recycling for years by recycling agriculture wastewater in/with fish farming, but it lead to a huge number of environmental risks and human health problems (Tawfic *et al.,*2003). Currently, it has been developed by biotechnology and biochemistry involvement.

However, for a smallholder like farmers, such technologies are unaffordable, and the concept of environment protection and climate change mitigations becomes insignificant as it affects the farmer's financial stability.

Therefore, the objective of paper was to emphasize an affordable methodology for waste water recycling in integrative agriculture – aquaculture to encourage such farmers to its application. The method works backwards the traditional recycling; as waste water of aquaculture is recycled number of times in aquaculture and agriculture. Meanwhile, this can increases the irrigation water quality for agriculture use; it was proved to increase waters content of major fertilizers needed to plants and improve the soil for irrigation, and reduces soil salinity with massive leaching (Elnwishy, 2008). This can be economically applicable mainly by smallholders, thus enforce its wider application and ensure economical and environmental benefits.

The innovation in the method is the possibility to utilize a certain quantity of fresh and/ or brackish water effectively up to 5 times recycling, and the recycling process does not require expensive additives. It is applicable, effective, affordable for all stakeholders levels, and it has no environmental or health risks. if applied by farmers, it can be a valuable economical and productive water management methodology which also produce organic products, save water and improve soil properties for agriculture purposes.

The paper reviews the risks derived from the current and previous mistaken indigenous recycling processes in the Nile Basin as study cases analysis, and strived to develop a simple methodology for water recycling which can be applied by small farmers, and to encourage greater experience and knowledge. Application of the methodology may be an opportunity to shrink the gab between science and farmers, resulting in encouraging the wise use of water and mitigate the impact of climate variability and protect the environment especially in such arid areas.

Assessment of Phreatic evaporation in irrigated lands of Fergana Valley by using stable isotopes

Gulomjon Umirzakov

Institute Landshaft ecology and recources management, Justus-Liebig-University Giesse35392 Giessen, Germany, E-mail: umirzakovgulomjon@yahoo.com,

Phone +49 (0)641 99 37394/-12780 Fax +49 (0)641 99 19872

Abstract

From ancient times agriculture in Central Asia has been developed concerning with water recourses. This territory situated in arid region always relied on irrigated agriculture. In recent years rapidly increase of irrigated zones has caused the increase of water demand. The Fergana Valley is the most important irrigated agricultural zone not only in Uzbekistan, but in Central Asia. The affect of groundwater water on agricultural crops is one of the vital issues in water and land use. Especially, this problem has been urgent for irrigated lands of the Fergana Valley. The main reason of that is the Valley's irrigated lands situated in cavity are covered with mountain ranges. However, the Valley borders on the Hunger-Steppe in the south-west with a narrow gully. Through this territory the main water resource of irrigated lands of the Valley the Syrdarya flows, at the same time the river takes over as a "drainage" which takes out returned water. The river channel cannot take out returned water totally for the reason that motion of groundwater is slow in the centre of Fergana Valley. (Krilov 1977). In recent years rise of groundwater level has been observed as a result of inefficient use of water resources in irrigation in the territory (Ikramov 2000). Firstly, groundwater deteriorates soil condition of irrigated lands, moreover, it affects to plant physiology and productivity of crops. Consequently phereatic evaporation or evaporation from underground occurs. In this case it is important to study the impact of groundwater on soil condition in root zone and plant physiology.

Catchwords

Irrigated lands, groundwater level, stable isotopes, phreatic evaporation.

Introduction

One of the most important parameters related to soil salinization is the direct evaporation from the groundwater. If the groundwater table is sufficiently close to the surface, groundwater will evaporate through capillary rise. Classical soil physical and meteorological methods can be used to estimate evaporative fluxes over short time periods, but are difficult to use over longer times which are appropriate for most water balance studies ((Brunner and et al. 2008)). The advantage of phreatic estimation using stable isotopes is that it provides an integrated measure over relatively long periods of time. In this paper method of estimation of phreatic evaporation in intensively irrigation land in the Fergana Valley is shown. The study in that territory started by Starke M. in 2009 and it has been continued on developing estimation and mapping of phreatic evaporation by using GIS. Some results of last year's research and work continued this year are presented.

Text

Research method: Water balance researches have been developed in irrigated territory of Central Asia by many native and Russian scientists (Alimov A., Rachinsky A., Khodjiboev N., Dukhovny V., Kharchenko S., Rubinova F., Rafikov A., Ikromov R. et al.). They have studied various elements of water balance: water intake, returned water, evaporation, transpiration, soil moisture, groundwater dynamics etc. Furthermore several scientists have investigated evaporation from groundwater which is the essential part of irrigated lands hydrology. However, this approach has not been investigated in details, phearatic evaporation values in irrigated lands have not identified yet. If we look at world water balance investigation essential part of water balance-phreatic evaporation has been developed by using stable isotopes since 1960.

Nowadays, analyses based on stable isotopes are applied in various fields of study. Especially in hydrology, isotopes play an important role since they provide information about origin, formation and flow path of water (Gat, 1996). According to that, it is possible to trace local and global water movement.

Stable isotopes are nuclides of the same element which possess the same number of protons but differ in the amount of neutrons, and thus showing a mass difference (Mazor, 2004). In research, advantage is taken from the process of isotope fractionation, i. e. the partial separation of light and heavy isotopes due to physical processes and chemical reactions (resulting in a changed composition of isotopes) (Mazor, 2004).

This study uses stable isotopes of hydrogen and oxygen, more precisely ^{1}H and ^{2}H (deuterium – D) as well as ^{16}O and ^{18}O. In order to measure differences in isotope ratios of water samples, the ocean, as biggest and relative homogeneous water reservoir (Gat, 1996), has been used as reference standard for the delta-scale of oxygen and hydrogen isotopes (Craig, 1961a and 1961b). The ∂-(per mill) value is defined as

$$\delta\text{‰} = \frac{R_x}{R_{std} - 1} \cdot 10^3 \qquad (1)$$

where R can be the ratio D/H or $^{18}O/^{16}O$ (Craig, 1961a and 1961b). Positive ∂-values compared to the standard (Vienna Standard Mean Ocean Water – V-SMOW) mean an enrichment of heavy isotopes, whereas negative values stand for their depletion (Craig, 1961 b; Gat, 1996).

Craig and Gordon (1965) developed a model that describes the movement of water isotopes undergoing evaporation from a free water body into the atmosphere. In 1967, Zimmermann et al. found that, in contrast to the previous study on evaporation from open waters, the presence of soil inhibits turbulent vertical mixing and a typical isotopic signature develops in a saturated soil column owing to evaporation.

According to Zimmermann et al. (1967) the isotopic profile of a saturated soil can be expressed as

$$R(z) = R_{\infty} + (R_{\text{eff}} - R_{\infty}) \cdot e^{-\frac{z}{z_l}} \qquad (2)$$

where R(z) is the isotopic signature as a function of the depth (z), R_{∞} the isotope ratio of the water entering the column from below and R_{eff} the isotope ratio at z=0. z_l is defined as

$$z_l = \frac{\theta \cdot f \cdot D}{E_P} \qquad (3)$$

where θ is the volumetric water content, f the tortuosity factor of the soil, D the self-diffusion coefficient for liquid water [m^2 s^{-1}] which is approximately equal to isotopic diffusivity in liquid phase (Mills and Harris, 1976) and E_P the phreatic evaporation rate [m s^{-1}].

In 1988 Barnes and Allison found to calculate phreatic evaporation rate from unsaturated soil:

$$R(z) = R_{\infty} + (R_{\text{eff}} - R_{\infty}) \cdot e^{-\frac{z - z_{\text{ef}}}{z_v + z_l}} \qquad (4)$$

where z_{ef} is the position of the depth of the evaporating front (maximum in the profile). It corresponds approximately to the position in the profile above which vapour movement prevails and below which liquid movement dominates (Barnes and Allison, 1983). z_v is defined as

$$z_v = \frac{D_v \cdot N^{\text{sat}}}{\rho \cdot E_P} = \frac{(n - \theta) \cdot f \cdot D_v \cdot N^{\text{sat}}}{\rho \cdot E_P} \qquad (5)$$

where D_v is the diffusion coefficient for water vapour in air [m^2 s^{-1}], N^{sat} the density of saturated water vapour [kg m^{-3}], ρ the density of water [kg m^{-3}] and n the porosity of the soil.
In 2008 Brunner et al. developed to calculate phreatic evaporation. The decay length λ can be obtained by fitting an exponential curve to the isotope data. λ can be used instead of zv +zl:

$$\lambda = z_v + z_l = \frac{1}{E_P}\left(\theta \cdot f \cdot D + f \cdot D_v (n-\theta)\frac{N^{sat}}{\rho}\right) \tag{6}$$

According to Brunner et al. (2008) the obtained evaporation rates can be used to establish a relationship between depth to groundwater and evaporation rate which in turn can be included in a groundwater model to calculate phreatic evaporation.

The research area is irrigated lands of the Fergana Valley in Central Asia. The Fergana Valley is an intermountain depression in Central Asia, between the mountain systems of the Tien-Shan in the north and the Alai in the south. The valley is approximately 300 km long and up to 70 km wide, forming an area of 22,000 sq.km (Goudie, 1996). Its position makes it a separate geographic zone. It is drained by the Syr Darya River and numerous mountain streams that are fed by glaciers in the mountains (Horst et al., 2005).

Fig. 1: Topographic and hydrographic map of Fergana Valley (UNEP/GRID-Arendal, 2005).

The prevalent soil types of the region are meadow soils and Sierozems, while they consist mainly of loam and clay loam (Stulina, 2002). In general, these grey soils are heterogeneous layered, light textured, gypsum-bearing and contain high contents of carbonates (Stulina, 2002). The landscape is marked by an intensive agricultural land use, where mainly cotton and wheat are grown.

Soil samples were collected from chosen fields and were taken simultaneously for different analyses. Bulk Density and was determined for different points per genetic horizon using soil sample rings with a known volume of 100 cm^3. The first collected samples were extracted by cryogenic vacuum distillation. Afterwards analysis of $\partial^{18}O$ and ∂D value was done by using a LGR DLT-100 (liquid-water isotope analyzer, Los Gatos Research, Inc., Mountain View, California).

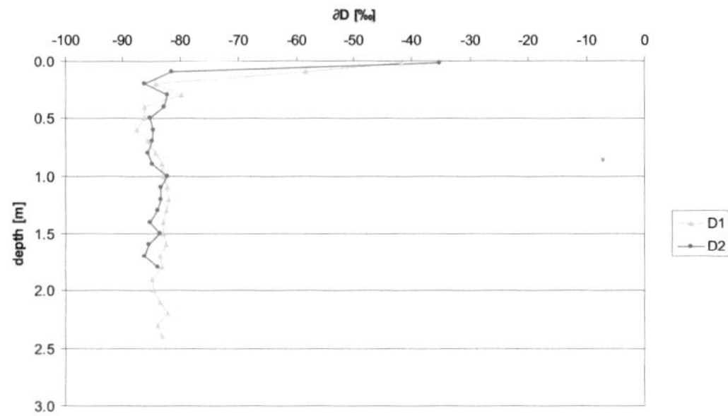

Fig 2. ∂D-profiles of the irrigated sites in WUA Akbarabad.

The graphs of determined $\partial^{18}O$ and ∂D value are drawn on changing by depth (Fig 2). The graphs of boreholes D1 and D2 have a peak at 0.2 m, where the soil water is depleted in contrast to the soil surface at which the water shows a value of -41.85‰ for D1 and -35.31‰ for D2, respectively. However, the peaks are hardly to differ from the normal staggered run of the curves. From 0.3 m on, the ∂-values have almost reached the corresponding groundwater value of about 83‰. It is important to take γ value from this curve to calculate phreatic evaporation rate. Phreatic evaporation rate is calculated on other sites but some of them are not reliable. There are 3 sites whose R^2 is more than 0.78. These sites have dependence on groundwater level (Table1).

		A1	A2	B1	B2	C1	C2	D1	D2
∂D	B	-4.42	0.35	-1.04	-8.47	-0.41	0.11	-8.33	-29.42
	R^2	0.26	0.02	0.35	0.78	0.10	0.00	0.82	0.96
	Λ	0.23	-2.82	0.96	0.12	2.45	-9.43	0.12	0.03
	E_P [m a^{-1}]	**0.07**	**-0.01**	**0.02**	**0.13**	**0.01**	**0.00**	**0.14**	**0.57**
$\partial^{18}O$	B	-1.84	-0.95	-7.68	-9.03	-0.23	-29.36	-8.71	-28.92
	R^2	0.72	0.07	0.58	0.63	0.05	0.85	0.95	0.97
	Λ	0.54	1.05	0.12	0.11	4.37	0.03	0.11	0.03
	E_P [m a^{-1}]	**0.03**	**0.02**	**0.11**	**0.14**	**0.00**	**0.57**	**0.15**	**0.57**
	Depth to gw [m]	2.60	2.10	2.40	2.30	2.20	1.30	2.40	1.80

Tab. 1: Calculated evaporation rates on the basis of isotope profiles and corresponding parameters (results of Starke, 2009)

Conclusion

From the first results a relation between phreatic evaporation and the depth to groundwater was identified in some irrigated lands even there were some unreliable results. It is planned to investigate this approach wider in this area. The irrigated lands were chosen on various types of soil and different groundwater level in two Water Users Associations (WUA) in central Fergana Valley. The soil was chosen due to soil texture on Kachinsky classification since the map of soil types in the territory was created. Soil classification in the figure belongs to sandy, sandy loam, silt-loam types. Groundwater level which is up to 3 metres in this area was taken into consideration while phreaic evaporation does not almost occur in groundwater more than 3 metres. A correlation value of phreatic evaporation and groundwater depth in various types of soil enables to make phreatic evaporation mapping.

List of literature

[1] Times New Roman 11,5 Y. Yoshida, K. Dowaki, Y. Matsumura, R. Matsuhashi, D. Li, H. Ishitani, and H. Komiyama. Comprehensive comparison of efficiency and CO2 emissions between biomass energy conversion technologies-position of supercritical water gasification in biomass technologies, Biomass and Bioenergy. 2003.

[2] Barnes C., Allison G. The distribution of deuterium and 18O in dry soils. 1. Theory. Journal of Hydrolohy 60, pp.141-142. 1983

[3] Brunner P, Li H T, Kinzelbach W, Li W P, Dong X G: Extracting phreatic evaporation from remotely sensed maps of evaporation. Water resources researches 44. W08428. 2008.

[4] Goudie D. (1996) An Overview of the Fergana Valley. Perspectives on Central Asia,Volume 1, Number 1 January 1996, Eisenhower Institute's Center for Political and Strategic Studies.

[5] Horst M G, Shamutalov S S, Pereira L S, Gonçalves J M. Field assessment of the water saving potential with furrow irrigation in Fergana, Aral Sea basin. *Agricultural Water Management* 77. 2005

[6] Ikramov R. Principles of water-salt regime management of irrigated lands in Central Asia under condition of water resources shortage. "GydroInGeo", Tashkent. 2001.

[7] Kovda V, Rozanov B. Soil and soil formation. "High School", Moscow. 1988.

[8] Krilov M. Basis of meliorative hydrology in Uzbekistan. "Fan", Tashkent. 1977.

[9] Mills R, Harris K. The affect of isotopic substitution on diffusion in liquids Chemical Society Reviews 5.1978.

[10] Stulina G (2002): Water and Salt Transfers. Experimentation and modelling with RZWQM, ISAREG and KCISA. Final scientific report. Department of Agricultural Engineering, Institute of Agronomy, Technical University of Lisbon, Lisbon. 2002.

[11] UNEP/GRID-Arendal (2005): UNEP/GRID-Arendal Maps and Graphics Library, http://maps.grida.no/go/graphic/topography-and-hydrography-of-the-ferghana-valley (accessed December 13, 2009).

[12] Zimmerman U, Ehhalt D. Munnich K. Soil-water movement and evapotransipiration: changes in the isotopic composition of the water. In: International Atomic Energy Agency (IAEA) (Editor), Proceedings of the Symposium on Isotopes in Hydrology, Vienna.1967.

Prediction of Erosion and Sedimentation by Runoff in Kurukavak Creek Basin by the WEPP Model

Mustafa TOMBUL[1]

Hasan TAŞ[2]

[1]Anadolu University Engineering Faculty of Civil Engineering Department Eskişehir/TÜRKİYE, e-mail: mtombul@anadolu.edu.tr, Tel: 90-222-3213550 ext:6608 Fax:90-222-3239501
[2]Dumlupınar University Engineering Faculty of Civil Engineering Department Kütahya/TÜRKİYE, e-mail: hasantas@mail.dumlupinar.edu.tr

ABSTRACT

Soil sediment production and runoff are common occurrences in harvested landscapes that have the capacity to lower soil and site productivity and contribute to loss of water quality. Landscape features such as soil type, slope, slope length, rainfall quantity, and rainfall intensity, ground cover and management practices influence the amount of runoff and soil loss. More in depth knowledge of the amount of runoff and soil loss that could be expected under conventional harvesting and site preparation practices in an intensively managed pine plantation will provide important data for the control and prediction of soil erosion.

Estimating of soil loss is one of the greatest challenges in natural resources and environmental planning. As known water erosion and sedimentation have a complex structure related to a lot of parameters as soil type, climate, slope, management, land use and vegetation. However erosion and sedimentation have great importance for dam lakes. Economical life of a dam is closely concerned with sedimentation in dam basin. Because of this before construction of a dam erosion and sedimentation characteristics have to be known well. There are a lot of types of estimating erosion and sedimentation yield and some computer simulation models have become increasingly popular in predicting erosion and sedimentation for various characteristics.

In this study sediment yield of Kurukavak Creek Basin was tried to estimate by runoff using WEPP model.

Keywords: soil erosion, sedimentation, WEEP model, runoff

1 Introduction

Soil erosion from productive farmlands decreases soil quality and crop production, diminishes on-site land value, and causes off-site environmental damage. Soil erosion is one of the most serious environmental problems in the world today because it threatens the natural environment. Increased erosion can create conditions which are hazardous, transported sediment often has negative impact on the environment and costly repairs may be necessary.
The sediment transported in Creeks originates either from the Creek channel or from the soil surface in the watershed. The transport depends on the shear stress induced by flowing water and the transportability of a particle depends on its fall velocity. The fundamental erosion processes on land surfaces are detachment and transport by raindrop impact and flow, and deposition by flow. The amount of sediment eroded from a soil surface equals the amount of detachment or the transport capacity of rain and runoff, whichever is less. Soil erosion from soil surfaces increases with rainfall intensity, slope, and surface runoff.
Soil erosion models can be divided into empirical and physically based models. Empirical models usually establish relationships between runoff, sediment yield and precipitation, plants, soil types, land use types, tillage styles, water conservation measures and so on. In the last decades, several studies have been carried out to build models suitable for quantifying soil erosion. Among these models, the Water Erosion Prediction Project (WEPP, Flanagan, D.C., Nearing, M.A., 1995. USDA-Water Erosion Prediction Project: Hillslope profile and watershed model documentation.
The Water Erosion Prediction Project (WEPP) model is one of the most utilized tools for simulating water erosion and sediment yield. WEPP has been tested and applied in different geographic locations across the United States (Savabi, 1993; Savabi et al., 1995; Huang et al., 1996; Laflen et al., 2004), in Australia (Rosewell, 2001) and in the United Kingdom (Brazier et al., 2000).
In this study sediment yield of Kurukavak Creek Basin has tried to estimate by the WEPP model and the applicability of this model to basins like Kurukavak Creek Basin has researched.

2 Model Description

We selected the Water Erosion Prediction Project (WEPP) model (Flanagan and Nearing, 1995) for estimating soil erosion and water runoff. WEPP has been a good predictor of soil erosion at time scales ranging from individual events to annual averages (Laflen et al., 2004).
Modeling soil erosion is the process of mathematically describing soil particle detachment, transport, and deposition on land surfaces. There are at least three reasons for modeling erosion: (a) erosion models can be used as predictive tools for assessing soil loss for conservation planning, project planning, soil erosion inventories, and for regulation; (b) physically-based mathematical models can predict where and when erosion is occurring, thus helping the conservation planner target efforts to reduce erosion; (c) models can be used as tools for understanding erosion processes and their interactions and for setting research priorities (Nearing et al. 1990).

The Water Erosion Prediction Project (WEPP) developed by the US Department of Agriculture (Nearing *et al.*, 1989) is a new generation, process-based, soil erosion prediction model based on fundamentals of infiltration theory, hydrology, soil physics, plant science, hydraulics, and erosion mechanics. It is a continuous simulation model for predicting daily soil loss and deposition from rainfall, snowmelt, and irrigation. It consists of nine components: climate generation; winter process; irrigation; hydrology; soils; plant growth; residue decomposition; hydraulics of overland flow; and erosion and deposition. Data input is divided into four categories: management (vegetation and management practices), slope soil and climate. The model has two run models, depending on whether the area under study is defined as cropland or rangeland. There are also different run models for continuous and single-storm simulation. The surface hydrology component of WEPP computes the surface runoff and peak discharge using the kinematical wave equation. The WEPP erosion model computes soil loss along a slope and sediment yield at the end of a hill slope. Interrill and rill erosion processes are considered, and it uses a steady-state sediment continuity equation as a basis for the erosion computations.

The sensitivity analysis of the WEPP Hill slope model was performed (Nearing *et al.*, 1990). The dominant factors related to model response were precipitation, rill erodibility, rill residue cover, and rill hydraulic friction factors. Saturated hydraulic conductivity and interrill erodibility were moderately sensitive parameters. Other factors that had less influence on output were canopy height, interrill cover, soil bulk density, antecedent moisture, peak rainfall intensity, time to peak rainfall intensity, rill width and spacing, and sediment characteristics. The performance of the WEPP Hill slope model in predicting runoff and soil loss under cropped conditions were evaluated (Zhang *et al.*, 1996). They found that the runoff and soil losses were slightly over predicted for small storms and for years with low runoff and soil loss rates. However, average runoff and soil losses for different cropping and management systems were predicted adequately. The overall runoff and soil loss predictions of the model compared well to measured values.

3 The Study Area

Kurukavak Creek, which is a subbasin of middle Sakarya basin, is selected as the study area. It is located in the northwest part of Turkey and has a drainage area of 4.25 km^2 The elevation variation in the basin is between 830 m and 1070 m. The basin is equipped with three rain gages (R1, R2, and R3) and one runoff recording station (H1) (Figure 1). The mean annual rainfall varies from 420 mm to 950 mm. It has a lot of sub rill and it flows from west to east. Average annual runoff ranged from a measured low of about 47.41 mm/yr to a maximum of over 252 mm/yr. Sediment yield ranged from about 0.260 t/yr to over 1.079 t/yr (Karaş, 2006). Soils ranged from a sandy clay loam to a sandy loam and management included conventional and tillage, meadow.

Several field measurements and mapping of soil properties (unsaturated hydraulic conductivity, saturated hydraulic conductivity, soil moisture values, and soil texture) and topographic elevations were carried out on this basin. The digital elevation model (DEM) with 30 m square grid was derived from the triangular irregular network modeling of the elevation map. There is no groundwater potential in the basin and its main water supply is rainfall. Soil of the basin is commonly limestoneless brown forest soil and it has shallow (20-50 cm) and very shallow (0-20 cm) deep. Basin is exposed to medium and violent erosion. General soil types are loam sandy loam and sandy clayey loam. Slope is so various and its ranges from %0 to %38. Land use of the basin is forestry, pasturage and agricultural.

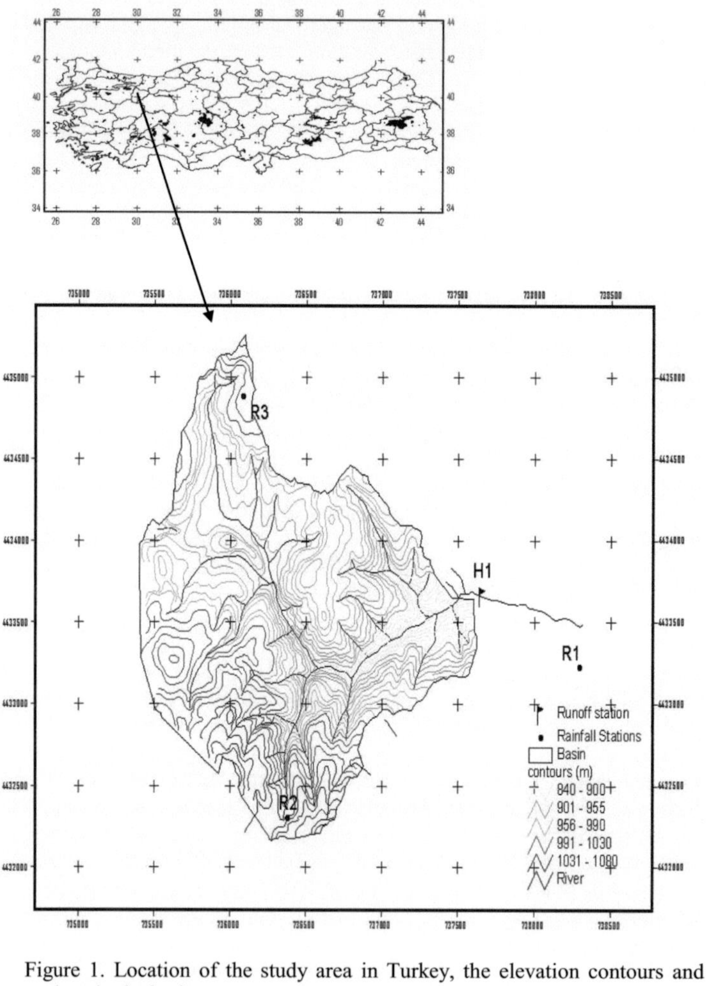

Figure 1. Location of the study area in Turkey, the elevation contours and the recording stations in the basin.

4 Application of the Wepp Model to Sample Area

4.1 Input Data Collection for the Model

The climate data required by WEPP model include daily values for precipitation, temperatures, solar radiation, and wind information. A stand-alone program called CLIGEN (Climate Generator) was used to generate the climate data file (USDA, 1995) for the duration of this study. However, the measured rainfall data for the site were collected from three precipitation station that are located at 840, 1020 and 1050 m height and were used in the simulation (the other climate data required by WEPP were generated by CLIGEN). Taking samples from the area and from the model user manuals collected soil data. The baseline effective hydraulic conductivity, interrill erodibility parameter, rill erodibility parameter, and critical flow hydraulic shear stress values for WEPP were estimated as described in the WEPP user manual (Flanagan et al., 1995).

Data collected during the 1984-2003 period were used to develop a linear relationship between measured sediment yield and measured runoff (discharge flow). The relationship was:

Sediment Yield (t/yr) : $\log Q_s = 0.722029 + 0606778 \log Q$

with a regression coefficient (R^2) of 0.82. The equation was then used to predict annual sediment yields from runoff measured during the 1984-2003 period (Karaş, 2006). Slope data were obtained from topographic survey of the area. The basic characteristics of the soil and vegetation are summarized in Table 1.

Table1. Initial characteristics of the soil and vegetation as entered to the WEPP model.

Parameter	Units	Value
Effective hydraulic conductivity	mm h^{-1}	3.47
Critical Shear	Pa	5.95
Soil depth	m	1.75
Rock fragments	vol. %	7.2
Organic matter content of soil	vol. %	1.25
Clay content of soil	vol. %	20.8
Sand content of soil	vol. %	51.8
Cation exchange capacity	meq. 100 g^{-1}	9.61
Total foliar cover	%	45
Frost-free period	days	300
Minimum temperature to initate growth	^{0}C	6.0

4.2 Application of the Model

The model was used in continuous-simulation mode (rangeland option) to predict soil erosion and sediment yield in the sample area. Watershed version was used for prediction. The WEPP watershed model is recommended to satisfy a maximum size field limitation of 260 ha (Foster and Lane, 1987), or of 40 ha (Baffaut et al., 1997). It is known that Weep can make bigger predictions when the area of a watershed is bigger than 260 ha. Despite this limitation, several test of WEEP capabilities have been conducted on watershed size greater than 100 km^2 (Amore et al., 2004). The basin was divided to seven pieces to identify topography of the whole basin. Then, from these pieces two sub basins, which have area of 230 and 195 ha respectively, were built. Soil input files were generated from values that have been collected before. Climate input file was generated from observed daily measurements of preticipitation, wind and temperature values. Management input file was generated from the usage of the basin.

5 Results and Conclusions

After running the program the observed values and estimated values quietly similar to each other and the results indicate that WEPP model can predict soil loss very good. Average annual runoff was estimated 90.79 and 77.36 mm for two sub basin respectively. Average annual sediment yield was estimated 0.326 and 0.280 t/yr respectively for two subbasins. WEPP prediction for sediment yield was approximately %10 bigger than observed values as seen in Table 2.

In general, WEPP predictions of precipitation, runoff, erosion and sediment yield show reasonable agreement with observed values. Hence the results obtained from model analysis revealed that the model could be used to simulate runoff and sediment in agricultural watersheds.

Table 2. Runoff and sediment yield observed and estimated values

	Observed	Estimated	
		Subbasin 1	Subbasin 2
Area (ha)	425	230	195
Runoff (mm)	122.90	90.79	77.36
Sediment Yield (t/yr)	0.550	0.326	0.280

The present study we conclude that this model could be used for predicting the sediment yield of basins like Kurukavak Creek Basin and could be used for estimating economical life of small dams.

6 References

[1] Amore, E., Modica, C., Nearing, M.A., Santoro, V.C., 2004. Scale effect in USLE and WEPP application for soil erosion computation from three Sicilian basins. Journal of Hydrology 293 (1–4), 100–114.

[2] Baffaut, C., Nearing, M.A., Ascough, J.C., Liu, B., 1997. The WEPP watershed model: II. Sensitivity analysis and discretization on small watersheds. Transactions of the Asae 40 (4), 935–943.

[3] Brazier, R.E., Beven, K.J., Freer, J., Rowan, J.S., 2000. Equifinality and uncertainty in physically based soil erosion models: application of the GLUE methodology to WEPP the

[4] Water Erosion Prediction Project, for sites in the UK and USA. Earth Surf. Process. Landforms 25, 825–845.

[5] Huang et al., 1996; Laflen et al., 2004), in Australia (Rosewell, 2001) and in the United Kingdom (Brazier et al., 2000). Huang, C.H., Bradford, J.M., Laflen, J.M., 1996. Evaluation of the detachment transport coupling concept in the WEPP rill erosion equation. Soil Sci. Soc. Am. J. 60, 734–739.

[6] Flanagan, D.C., Nearing, M.A., 1995. USDA-Water Erosion Prediction Project: Hillslope profile and watershed model documentation. NSERL Report 10, USDA-ARS National Soil Erosion Research Laboratory, West Lafayette, IN, USA.

[7] Foster, G.R., Lane, L.J., 1987. User requirements: USDA-Water Erosion Prediction Project (WEPP). NSERL Report 1, USDA-ARS National Soil Erosion Research Laboratory, West Lafayette, IN, USA.

[8] Karaş,E., 2006. Bilecik-Pazaryeri-Kurukavak Creek Basin characteristics of rainfall-runoff and sediment yield. Part of Project Report Project Number:KHGM-83110E01, 143 (in Turkish)

[9] Laflen, J.M., Flanagan, D.C., Engel, B.A., 2004. Soil erosion and sediment yield prediction accuracy using WEPP. J. Am. Water Resour. Assoc. 40, 289–297.

[10] Nearing, M., (1995), Overview of WEPP. In: Presented at the WEPP/WEPS Symposium. 50th SWCS Annual Meeting. Des Moines, IA.

[11] Nearing, M.A., Foster, G.R., Lane, L.J., Finkner, S.C., (1989), A process-based soil erosion model for USDA–Water Erosion Prediction Project Technology. Transactions of the ASAE 32 (5), 1587–1593.

[12] Nearing, M.A., Deer-Ascough, L., Laflen, J.M., (1990), Sensitivity analysis of the WEPP Hillslope profile erosion model. Transactions of the ASAE 33 (3), 839–849.

[13] Rosewell, C.J., 2001. Evaluation of WEPP for runoff and soil loss prediction in Gunnedah, NSW, Australia. Austr. J. Soil Res. 9,
230–243.

[14] Savabi, M.R., 1993. Modeling subsurface drainage and surface runoff with WEPP. J. Irrig. Drain. Eng. 119, 801–813.

[15] Savabi, M.R., Flanagan, D.C., Hebel, B., Engel, B.A., 1995. Application of WEPP and GIS GRASS to a small watershed in Indiana. J. Soil Water Conserv. 50, 477–483.

[16] US Department of Agriculture — Water Erosion Prediction Project, (1995), In: Flanagan, D.C., Livingston, S.J. (Eds.), NSERL Report No. 11. USDA–ARS National Soil Erosion Research Laboratory, West Lafayette, IN.

[17] Zhang, X.C., Nearing, M.A., Risse, L.M., McGregor, K.V., (1996), Evaluation of WEPP runoff and soil loss predictions using natural runoff plot data. Transactions of the ASAE 39 (3), 855–863.

Land evaluation for irrigation with treated wastewater

Karsten Schacht[1]
Sven Gönster[2], Elisabeth Jüschke[1], Bernd Marschner[1]
[1]Department of Geography, Ruhr-University Bochum, Universitätsstraße 150, 44801 Bochum, Germany, e-Mail: karsten.schacht@rub.de, Telephone: +49(0)234/32-23322, [2]Faculty of Organic Agricultural Sciences, University of Kassel, Germany

Abstract

A risk assessment for wastewater irrigation in the Middle East is carried out regarding several major agricultural soil risks. This is done based on digital soil maps joined with spatial data on soil properties. Particular soil property changes due to irrigation with low water quality are evaluated and discussed. Focussing on soil suitability criteria, risk and suitability maps are presented for supporting sustainable irrigation practices.

Keywords

risk assessment, soil suitability, irrigation, treated wastewater, sewage, Middle East

1 Introduction

In the Middle East, a region with one of the lowest per capita water availabilities worldwide, the steadily growing demand for water leads to political, socio-economic and environmental problems. Hence, there is an increasing need for water reuse and irrigation of agricultural crops with wastewater. This practice carries a variety of risks to sustainable land use and agriculture, such as salinisation of soils, slaking, dispersion and contamination of soils with organic and inorganic pollutants.[1]

As a part of the multilateral project network GLOWA Jordan River the research subgroup "Wastewater Management" focuses on soil conditions and soil properties in the catchment of the Jordan using geographic information systems (GIS) and on site investigations.

For the first time digital soil maps were joined with soil parameters on a supranational scale using GIS. The values of these parameters were set based on a literature analysis and expert interviews. Several risks of primarily agricultural significance regarding treated wastewater irrigation were defined. Based on the local soil parameters, specific soil related risk- and suitability grades regarding the irrigation with treated wastewater were assessed using standard and specially developed methods. As a result of this approach a small scaled digital map is provided together with recommendations for sustainable agricultural production and wastewater management.

2 Data base

The data collection is based on the digital Soil Association Map of Israel.[2] This map covers the territory of Israel and the Golan Heights, the Gaza Strip and the West Bank. The soil associations are defined as "geographic associations of [...] soil units which are distributed in a landscape segment according to a definite pattern related to the physiographic, lithologic and microclimatic conditions".[2] The scales of the underlying soil mapping campaigns mainly conducted in the 1950´s and 1960´s by the Israeli Soil Survey, were 1:20,000 for the northern parts, 1:50.000 for the Golan Heights and 1:100.000 for the southern parts. Further small scaled soil maps are not available for the whole study area. Additional data for Jordan will be joined in the further course of the project.

Through a literature analysis [3] and various expert interviews the averaged values of various soil parameters, divided into two depths ($\leq$ 30 cm and > 30 cm), were allocated to the local soil associations. These parameters include soil texture, pH, organic carbon, bulk density, $CaCO_3$, EC, CEC and soil depth. Further parameters were derived in accordance with German standard methods (field capacity, TW, available field capacity, effective rooting depth, available field capacity in the root zone, hydraulic conductivity).[4]

3 Risk definitions and risk assessment

The environmental risk regarding the irrigation with treated wastewater mainly depends on the local soil properties and the water quality.[5] The quality of the wastewater depends on the water resources, the “pick-up” during usage and the treatment technology. It is therefore regionally variable. For a supranational approach this results in the need for an evaluation independent of water quality. In collaboration with regional partners the currently most important agricultural risks associated with wastewater irrigation were defined (Tab. 1) based on the assumption that treated wastewater has generally a poorer quality than fresh water regarding several properties (e.g. total dissolved solids, dissolved organic matter, salts). Criteria through which the risks can be assessed are set. Hence, methods were determined and developed to evaluate the determinable parameters in the context of the appropriate parameters. Finally, the particular risks are aggregated and displayed as an overall risk. The result of the assessment is the respective risk grade, which is given on a three staged scale: 1 - low risk, 2 - medium risk and 3 - high risk. From this, the respective suitability classes arise conversely; a low risk area has therefore a high suitability for irrigation with treated wastewater.

Table 1: Overview of the defined risks, criteria, methods and the used parameters

Risks		Criteria	Methods	Parameter
A	mobilisation of inorganic adsorbable pollutants	relative bond for inorganic adsorbable pollutants (e.g. heavy metals)	Blume & Brümmer 1991 [6] DVWK 1988 [7] (modified)	texture/percentage of clay, organic matter, pH
B	slaking of the upper soil layers	slaking risk	Ad-Hoc-AG Boden 2005 [4] (modified)	texture, organic matter
C	salinisation of soils	salinisation of soils	own method	texture, bulk density, depths of horizons, depth of root zone, soil depth, field capacity, hydraulic conductivity
D	mobilisation of boron	buffering capacity for boron	own method	texture/percentage of clay, organic matter, pH
E	mobilisation of non-adsorbable substances	buffer capacity for non-adsorbable substances (e.g. nitrate)	DIN 19732 [12] (modified)	texture, bulk density, depth of root zone, field capacity, leaching rate

Risk A: Mobilisation of inorganic adsorbable contaminants

Assuming a worst-case risk assessment, the evaluated criteria is the relative bonding strength of cadmium in the top soil $\leq$(30 cm) [6, 7], as Cd is the first heavy metal mobilised by decreasing pH values. The relative bond is mainly dependent on the amount of clay minerals and organic matter in the soil. The original method has been modified. For the reason of simplification, it includes no translocation processes due to complexation and no previous impacts of the sites.

Risk B: Slaking of the upper layers

The slaking risk of the topsoil is dependent of the silt and fine sand contend. Important for stabilizing the soil structure is also the soil organic matter.[8] The assessment is based on Ad-Hoc AG Boden (2005) on the basis of the texture.[4] The determined risk level is modified in dependency of the organic matter content.

Risk C: Salinisation of soils

Salinity is the process of accumulation of salts in soils or soil horizons and is favoured by low leaching rates. Leaching rates are dependent on the field capacity of the soil and the amount of precipitation. For the same rate of evapotranspiration a sandy soil loses proportionally more water than a clay soil, which leads to an increase in soil solution concentration. Assuming good irrigation practices sandy soils are usually irrigated more frequently, which leads to higher leaching rates.[9] The prevention of salt accumulation in fine-grained and low-permeable soils is most difficult.[10] The lower the field capacity, the lower is the risk of salinisation. Higher drainage abilities are assumed if the saturated hydraulic conductivity is high. Furthermore, it is supposed that higher profile depths (≥ 1 m) ensure better drainage.

The particular risk is evaluated for the effective root zone on the basis of an evaluation matrix with the parameters of saturated hydraulic conductivity and field capacity of the root zone. For profile depths < 1 m the risk level is increased by one unit.

Risk D: Mobilisation of boron

Boron is an important plant micronutrient, but the margin between deficiency and toxic concentration is smaller than for all other nutrients.[11] Borate compounds are conveying into the wastewater as parts of bleaching and disinfection detergents. The binding of boron in soils is largely determined by pH, clay and organic matter content. The risk is evaluated via an assessment matrix.

Risk E: Mobilisation of non-adsorbable substances

The determination of the potential mobilisation is done according to DIN 19732.[12] The assessment is a function of the soil water exchange rate, which is the ratio of the leaching rate and the field capacity in the root zone. The leaching rates have been modelled within other GLOWA Jordan River working groups based on the same joined data base soil associations and soil properties.[13]

Risk aggregation

The aggregation of the particular risks are equally weighted in accordance using the principle of maximum values, as neither a hierarchy nor an averaging of the sub risks can be made (Tab. 2). This results in the presentation of the risks by means of a three-staged scale (Fig. 1).

Table 2: Evaluation matrix for the aggregation of risks A to E

Total assessment risks A - E	**Total risk**	**Total risk grade**	**Suitability**
≥ 3 x "low risk" & no "high risk"	low	1	high
all other cases	moderate	2	moderate
≥ 2 x "high risk"	high	3	low

4 Perspectives

Currently, more particular risks are examined. For example, the effect of the arise of hydrophobic soil surfaces has been associated with wastewater irrigation.[14] This effect is, among others, related to the properties of soil texture, pH and the contents of organic matter and $CaCO_3$. Furthermore the threat of surface water contamination is going to be evaluated with regard to the erosion tendency.

Further detailed substrate mapping campaigns should be made at the field scale to validate and refine the data base. The ultimate goal of the studies is a site specific suitability evaluation for irrigation with wastewater, depending on the specific wastewater quality.

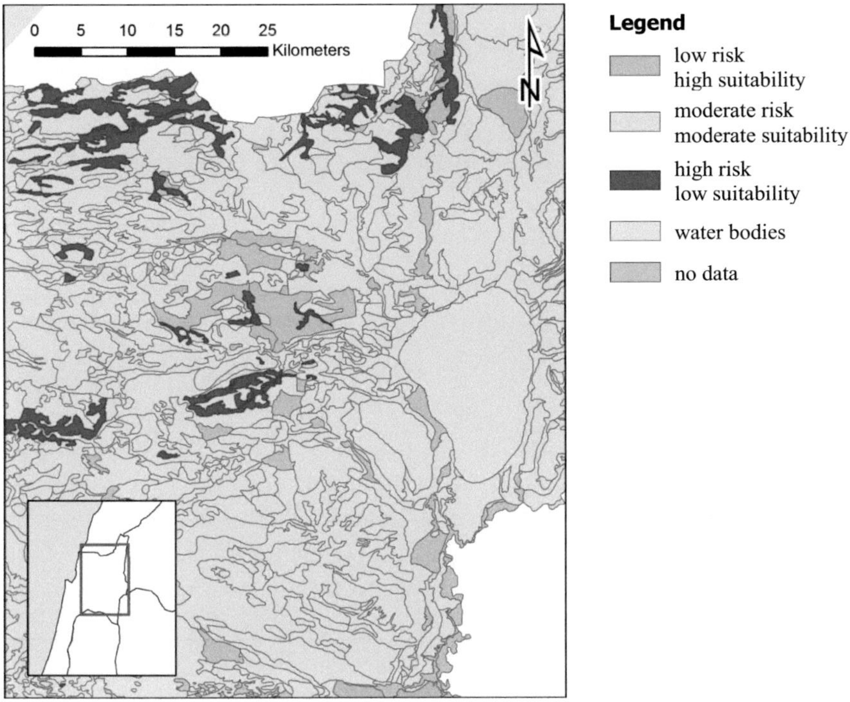

Figure 1: Detail of the aggregated risk and suitability map

5 Acknowledgement

The research project is part of the multilateral GLOWA Jordan River project and is funded by the Federal Ministry of Education and Research (BMBF). We would like to thank all involved project partners.

References

[1] Toze, S. (2006): Reuse of effluent water-benefit and risks. Agricultural Water Management. 80: 147-159.

[2] Dan, J., Yaalon, D. H., Koyumdjisky, H., Raz, Z. (1972): The soil association map of Israel. Israel Journal of Earth Sciences. 21: 29-49.

[3] Singer, A. (2007): The Soils of Israel. Berlin.

[4] Ad-Hoc-AG Boden (2005): Bodenkundliche Kartieranleitung. 5th Edition. Hannover.

[5] Feigin, A., Ravina, I., Shalhevet, J. (1991): Irrigation with treated Sewage effluent. Berlin.

[6] Blume, H.-P., Brümmer, G. (1991): Prediction of heavy metal behaviour in soil by means of simple field tests. Ecotoxicology and Environmental Safety. 22: 164-174.

[7] DVWK - Deutscher Verband für Wasserwirtschaft und Kulturbau e.V. (1988): Filtereigenschaft des Bodens gegenüber Schadstoffen, Teil 1: Beurteilung der Fähigkeit von Böden, zugeführte Schwermetalle zu immobilisieren. DVWK-Merkblatt. 212.

[8] Levy, G.J. (2000): Sodicity. in: Sumner, M. E. [Ed.] (2000): Handbook of Soil Science. Boca Raton.

[9] Shalhevet, J. (1994): Using water of marginal quality for crop production: major issues. Agricultural Water Management. 25: 233-269.

[10] Rhoades, J. D.; Kandiah, A.; Mashali, A. M. (1992): The use of saline waters for crop production. FAO Irrigation and Drainage papers - 48.

[11] Goldberg, S. (1997): Reaction of boron with soils. Plant and Soil. 193 (2): 35-48.

[12] DIN – Deutsches Institut für Normung e. V. (1997): DIN 19732. Bestimmung des standörtlichen Verlagerungspotentials von nichtsorbierbaren Stoffen. Berlin.

[13] L. Menzel, Heidelberg University, personal communication

[14] Tarchitzky, J., Lerner, O., Shani, U., Arye, G., Lowengart-Aycicegi, A., Chen, Y. (2007): Water distribution pattern in treated wastewater irrigated soils: hydrophobicity effect. European Journal of Soil Science. 58: 573-588.

Alternatives for a sustainable management of stormwater in megacities of emerging countries

The case of Santiago de Chile

Gerhard Schleenstein[1]
Dr. Helmut Lehn[2]
*[1]Ingeniería Alemana S.A., Santiago, Chile, E-mail: gs@iasa.cl,
Telefon: +56(0)2/2698158; [2]Karlsruhe Institute of Technology, Institute for Technology Assessment and Systems Analysis, 76021 Karlsruhe, Germany,
E-mail: helmut.lehn@kit.edu, Telefon: +49(0)7247/823977*

Abstract

Santiago de Chile lacks efficient stormwater-management systems, so little precipitation events flood the lower and often poorer areas of the city. This contribution describes the actual and proposed stormwater management systems and gives hints on Low-Impact Development stormwater-management methods which could be implemented in Santiago, considering legal, economical, social and environmental criteria. Furthermore, a first approach on sustainability indicators for water and stormwater management will be described.

Catchwords

Megacity, stormwater management, sustainability, indicators

1 Introduction

The Metropolitan Region of Santiago de Chile has a population of about 6,5 million and, except for the inner city, lacks efficient stormwater-management systems. Precipitation events of low intensity convert streets into little streams that flood the lower areas of the city.

It is known that an unintended consequence of these conventional drainage systems includes increased frequency and magnitude of flooding events in natural waterways downstream. Additionally, in developing and emerging countries increased non-point source pollution must be expected due to waste accumulation in waterways or public spaces, insufficient street cleaning, building materials and wastewater infiltration.

Integrated water management, or Low-Impact Development (LID), emphasizes stormwater-management methods which imitate drainage and flow patterns that existed prior to urban development. LID offers several techniques including stormwater harvest, infiltration and biofiltration or bioretention to store and treat runoff and release it at a controlled rate to reduce impact on streams and treatment systems. While international discussion prompts towards an increased tendency of artificial infiltration [1], [2], local Chilean efforts are less advanced [3].

2 Aims of the investigation

Environmental, social and economic dimensions of water and stormwater management shall be described and quantified within Helmholtz-Association's integrated sustainability concept and the sustainability indicators that are actually developed within the "Risk-Habitat-Megacity" initiative (www.risk-habitat-megacity.ufz.de). A set of about 40 key indicators aims to describe the ecological interaction of the megacity with its catchment area or to assess sustainability of services in the city. Additionally, the investigation aims to identify those LID stormwater-management methods that are recommendable and worthwhile to implement in emerging countries. This will include technical, environmental/ecological, social/institutional and economic cost factors as sustainability criteria in terms of assessing long-term, cost-effective drainage options.

3 Precipitation and water offer

Santiago de Chile has an average annual rainfall ascending from 261.6 mm in the west by 347.2 mm in the city centre and 442.0 mm in the areas of the Andean piedmont, due to orographic precipitation and rain. The semi arid climate is responsible for dry summers and an annual rainfall encompassing about 32 rainy days in the winter season. Assuming complete run-off and standard IDF-curves, a one in five year storm of 60 mm [4] causes a rainfall intensity of 145 L/(s*ha) in five minutes ($r_{5,5}$). Therefore, rainfall intensities are weaker than in Germany, where DIN 1986-100 gives $r_{5,5}$ values of about 350 L/(s*ha).

4 Stormwater management

Aguas Andinas (the principal local water utility) concessionary service area covers 51,870 hectare, serving 4,648,000 habitants. Santiagos sewer network comprises 9,000 km, including about 1,000 km of combined sewers [4]. At present, separate stormwater sewers do not exceed 300 km.

Per capita wastewater contribution is 182 L/(capita*d), so an average wastewater intensity of 0.189 L/(s*ha) can be calculated. Q_{max}/Q_{24} is estimated to be 1.7 at La Farfana WWTP (3.3 million PT). Consequently, an average maximum wastewater intensity of 0.32 L/(s*ha) or a lineal value of 0.00184 L/(s*m) might be good comparative values for further investigations or indicator building. In comparison, rainfall would yield an annual average of about 0.099 L/(s*ha), considering complete run-off.

After years of relative inactiveness, the recent stormwater-masterplan considers conventional end-of-pipe systems and separate sewage systems that channel stormwater quickly and efficiently away from development along the main avenues to receiving bodies of water [4]. These systems would involve an estimated investment cost of around one billion Euro. In contrast, Aguas Andinas, proposes the utilization of idle capacities of the wastewater sewers that cover the entire urban area [5]. Although currently forbidden, the extensive discharge of stormwater into the wastewater sewer network would become legalized. Even though Aguas Andinas' proposition would be two-thirds less expensive and would consider adding supplementary decentralized retention works and storage basins, local authorities reject the initiative, fearing discharge of untreated wastewater into receiving waters.

5 Legal framework

During privatization of the water sector in the 1990s, responsibility for storm water management was not assigned to the water utilities with the clear political intention to improve economical feasibility and assure future gaining. Just in 1997 the law 19.525 instructed the water works authority (DOH) to (1) realize stormwater master plans within a five year timeframe and (2) afterwards invest and maintain primary stormwater sewers. Secondary sewers shall be attended by the public housing authority (SERVIU). Nevertheless, considering the actual DOH budget, the implementation of the projected primary sewers would require a whole century. As there are no fees on stormwater, and dominating neoliberal background in Chilean politics, modifications to the law were discussed by the lawmakers, but are still not realized. On the other hand, a decree from 1926 forces any effluents to be discharged to the public sewerage system, so that any alternative methods legally are no feasible at present. Technical guidelines on alternative stormwater management systems were approved in 1996 but are not obligatory.

6 Environmental impacts

Floods occur regularly as a result of winter precipitation, even of lower intensity as storm water is discharged by the streets. The risk of flooding is increasing because the city is experiencing a dynamic spatial expansion which results in a loss of retention areas and an mayor amount of sealed surfaces that contribute to increased run-off.

Most of the combined sewers probably discharge directly to the Mapocho River, which still receives some 27% of all wastewater generated in Santiago de Chile. Own estimates, based on data reported from La Farfana WWTP, show that stormwater discharge to the sewage network at present is virtually insignificant and a high share of stormwater is conduced in superficial flow through the city and will cause floods in urban lowlands or end up in open channels or in natural receiving waters.

Additionally, increased non-point source pollution must be expected due to waste accumulation in waterways or public spaces, insufficient street cleaning, building materials and wastewater infiltration. Measured contamination is higher than reference values and investigations conducted in developed countries [6]. Local emission standards were exceeded for aluminium, manganese, lead and suspended solids.

Conventional end-of-pipe systems do not contribute to the groundwater recharge, which should be a cause for concern, as the groundwater table in Santiago has descended some 20 to 30 m within the last three decades. Neither the master-plan, nor the proposal of Aguas Andinas has considered decentralized stormwater infiltration in the permeable subsoil which is characterized by non consolidated fluvial and fluvioglacial sediments.

7 Sustainability indicators

National or supra national Sustainable Development Strategies (SDS) generally use a few readily available indicators that describe water resources management but on EU-level. Benchmarking indicator systems (IBNET or DWA) generally focus on financial, technical and process indicators, assuming that water resources management in developed countries is sustainable “per se” [7], [8]. Therefore, the integrative sustainability concept of the Helmholtz Association [10] was applied and contextualized, adopting a combination of sustainability analysis, status analysis and distance to target analysis. Appropriate sustainability criteria and indicators were identified together with regional institutional stakeholders of the water sector (governmental and non governmental). The developed key indicators are grouped in the following seven dimensions and 22 aspects:

Dimension	Aspect
Human life	Minimization of environment-caused adverse impacts on health
	Satisfaction of basic needs
Natural resources	Sparing use of non-renewable resources
	Sparing use of renewable resources
	Equal access to environmental resources
	Maintaining the regeneration capacity of ecological systems
	Internalization of external costs
Social and cultural resources	Social coherence of society
	Preservation of the cultural heritage and cultural diversity
	Preservation of the cultural function of nature
Securing society's productive potential	Development of human and knowledge capital
	Development of manmade capital
	Limitation of public indebtedness
Equal opportunities	Possibility of autonomous subsistence
	Equal access to education, profession, information etc.
	Compensation of extreme differences in income
	Participation
Governance	Society's ability to respond
	Self-Organization
	Balance of Power
Global stewardship	Fair international economic relations
	International co-operation

Tab. 1: Keyindicators.

8 Conclusion and Forecast

Decentralized stormwater management facilities are mostly absent in Santiago. Tab. 2 shows advantages and drawbacks for different stormwater management technologies, judging their applicability in the local context, considering economical, social and environmental issues. Popular reuse strategies for example can not be exported from industrial countries to take-off Chile, due to cost and absent summer precipitation in Santiago. Maintenance intensive infiltration technologies probably will fail due to little developed conscience of house-owners and local authorities.

	Advantages	Drawbacks	Applicability
Sewerage systems			
Combined sewers	Optimization of existing infrastructure	Untreated discharges	++
Separate sewers	Pollution reduction	Maintenance, cost	+
Run off minimization			
Green roofs	Climatisation, Treatment	For new buildings only	-
Infiltration			
Open space infiltration/ Ground unsealing	Little complexity Little cost	Space requirements	++
Trench infiltration			++
Infiltration ditch	Little space requirements	Maintenance, treatment efficiency	++
Basin infiltration	Multi utilization possible	Space requirements	0
Infiltration wells	Very little space requirements	Maintenance, Pre-treatment necessary	0
Reuse			
Domestic use	Substitution of tap water	Cost	-
Drinking water			--
Irrigation	Substitution of raw water	Precipitations and water demand do no coincide	--

Tab. 2: Applicability of stormwater management technologies in Santiago.

Sustainability indicators, target values and evaluation matrices of stormwater management methods for urban Santiago are still being developed, based on the Risk Habitat Megacity's indicator set. The most promising are the following:

- Percentage of the population connected to a save and hygienic sewage system (%).
- Tariff rates for the collection of storm water according to the degree of sealed surface of real properties ($).
- Floods as a cause of death (by 100.000 habitants).
- Losses as results of flooding ($/hab/a).

List of Tables

List of Literature

[1] P. Fox (ed.) Management of Aquifer Recharge for Sustainability. Proceedings of the 6th International Symposium on Managed Artificial Recharge of Groundwater (ISMAR6) Phoenix, Arizona, USA, 28.10-2.11.2007.

[2] The United Nations Educational, Scientific and Cultural Organization - UNESCO. Recharge systems for protecting and enhancing groundwater resources. IHP-VI, Series on Groundwater No. 13, Proceedings of the 5th International Symposium on Management of Aquifer Recharge (ISMAR5), Berlin, Germany. 11-16.06.2005.

[3] E. Brown (ed.). Hacia un Plan Nacional de Gestión Integrada de los Recursos Hídricos (Towards a national plan for integrated water resources management). Taller Nacional-Chile, Cepal, Santiago, Chile. 11-12.12.2003.

[4] Ministerio de Obras Públicas, Transporte y Telecomunicación. Plan Maestro de Aguas Lluvias para el Gran Santiago (Stormwater masterplan for Santiago). Santiago, Chile. 2003.

[5] J. Santamaria, M. Concha, F. Strajilevich. Soluciones de Gestión de Aguas Lluvias Urbanas El Caso de Santiago (Solutions for urban stormwater management - The case of Santiago), XXXI Congreso Interamericano AIDIS, Santiago, Chile, 12–15.10.2008.

[6] J. Montt, P. Rivera, B. Fernández and R.Valenzuela. Caracterización de la Calidad de las Aguas Lluvias Urbanas de Santiago (Quality caracterization of urban stormwater in Santiago). XVI Congreso Chileno de Ingeniería Hidráulica, Santiago, Chile. 13-14.11.2003

[7] W. Schenkel. Nachhaltig, nachhaltiger, noch nachhaltiger – wozu so ein Leitbild in der Wasserwirtschaft. In: Wasserforschung e.V. Nachhaltige Entwicklung in der Wasserwirtschaft. Interdisziplinäre Fachtagung am 27. und 28. Juni in Berlin. Schriftenreihe Wasserforschung Nr. 7. Berlin, Germany. 2001.

[8] B. Weigert, H.-P. Lühr and C. Steinberg (ed.). Nachhaltige Wasserwirtschaft – Entwicklung eines Bewertungs- und Prüfsystems. Schriftenreihe Wasserforschung Nr. 4. Berlin, Germany. 1998.

[9] K. Gantner. Nachhaltigkeit urbaner Regenwasserbewirtschaftungsmethoden (Sustainability of urban stormwater management technologies). PhD thesis, Tecnical University of Berlin, Germany. 2002

[10] J. Kopfmüller, V. Brandl, J. Jörrisen, M. Paetau, G. Banse, R. Coenen and A. Grunwald. Nachhaltige Entwicklung integrativ betrachtet. Konstitutive Elemente, Regeln, Indikatoren. Edition sigma, Berlin, Germany. 2001

Estimating water retention potentials for Saxony

Catchment wide implementation of agricultural measures with an innovative method

Mariusz Merta
Harald Sommer, Heiko Sieker, Walter Schmidt
Ingenieurgesellschaft Prof. Dr. Sieker mbH, 15366 Hoppegarten, Germany,
E-mail: m.merta@sieker.de, Telefon: +49(0)3342/359526

Abstract

Within the framework of the project "Estimating water retention potentials for Saxony" [6] an innovative methodology was developed to determine the water retention potential. The extreme summer flood 2002 in the Elbe River watershed initiated a debate on the role of land-use change as measure for preventive flood protection. To quantify the effect of measures on flood runoff from catchments reliable model calculations are essential. The article presents an example for a model-based assessment of potential water retention caused by land-use changes in Saxony (Germany). The potential of flood control by land-use management measures is highly dependant on the site-specific soil and relief conditions and the rainfall event characteristics. It is shown how the locations more suited for water storage can be identified catchment-wide, how the effective retainable precipitation volume is quantified, and where the potentials for implementation of infiltration-enhancing measures are located. The presented maps are central components for a structured procedure for land use management.

Catchwords

flood protection, water management, infiltration potential, land use planning, expert system

1 Introduction

The number and magnitude of flood events in Europe and other parts of the world in recent years made clear that flood protection is universal societal issue that challenges more than just the areas of water management, disaster preparedness and the insurance industry. Flood Protection Technology (dykes, floodwater retention tanks etc.) and Flood Precaution (construction measures, human preparedness and risk management) are important components of flood protection [5] that have been especially emphasised in Germany since the flooding of the Elbe River in 2002. In comparison, only very limited steps have been taken in the area of land management. Land management is often a synonym for identifying flooding areas. In contrast, the aspect of adjusted land use is practically ignored. This represents a weak point in current flood protection strategies. In the task of precaution through land management, the players and decision makers in Land Use Planning, Agriculture and Forestry must take part in the task of flood prevention.

The goal is quantify the influence of different land use for flood protection with these tools. The target groups are planners and authorities from the areas of land use planning, agriculture, forestry and nature protection who are responsible for the implementation of infiltration-improving measures in catchments with flooding danger.

2 Material & Methods

While selecting the input data, attention was paid to dealing with available data for many catchments. For this reason the need for a development a of transferable method was obvious.

A potential map shall show where flooding reduction areas in Saxony (18.200 km^2) can be found and how large the storage potential in the soil (in m^3) is estimated. Creation of the potential map is based on the connection of two independent expert systems.

The first part of the methodology consists of the interpretation of the digital soil map. Taking the soil saturation, the groundwater depth, and the layering density into consideration, the infiltration potential was determined. For this, the infiltration model according to [3] was used in a modified form for saturated soils according to [2].

Since it can only be assumed in certain cases that the infiltrated precipitation can be stored in the soil for a while – at least for the duration of a flooding event – it was necessary to characterise the surfaces in Saxony in terms of their runoff behaviour. For the analysis of the hydrological situation in the project area the expert system WBS FLAB [7] was applied. This expert system identified areas with dominant runoff components (surface runoff, saturation overland flow, fast interflow, deep percolation) on the basis of landscape characteristics such as soil type, land use and slope.

The combination (GIS-intersection) of both of these GIS-based landscape evaluation systems leads to a catchment-wide approximation of the water retention potential for the soil (Fig.1). An actual water retention potential is assigned only to those surfaces that are dominated by heavily delayed runoff components. The surfaces that are marked by rapid, temporary runoff absolutely have a potential infiltration capability, although underground lateral runoff processes lead to the fact that the water retention potential is not enough to be of use to flood protection. This applies especially for areas with steep slopes and/or thin soil layers.

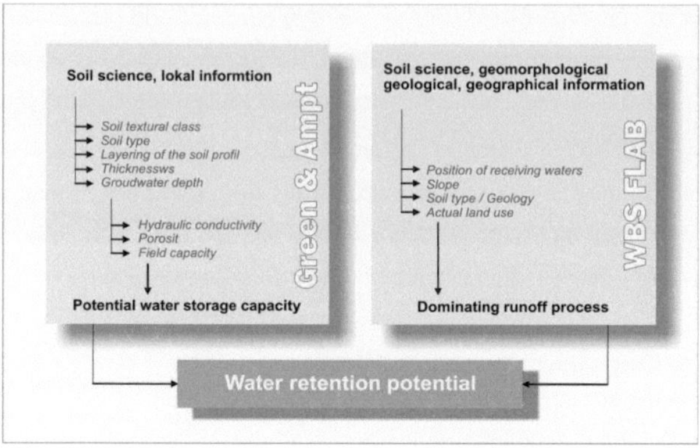

Fig. 1: Methodology for creating the potential map

Based on the analysis of the distributesd storage potential, with the aid of the GIS-based expert system FLEXT [4] a map with measures for the agriculturally used surfaces was created. With the aid of the measures maps, it shall be shown for the Saxony, where and which changes to the land use or land management are especially effective from the point of flood protection. Figure 2 schematically shows how the creation of the measures map was realised and which links exist between the geographical information system (GIS) and FLEXT. The starting point in this methodology is the decision matrix (Fig. 2, IIb). Graphic IIa in Figure 2 symbolises the software programme FLEXT. FLEXT provides a graphical user interface (GUI) that allows the input of rules of the decision matrix. All relevant input parameters of the decision are processed by a decision support system.

In the decision-making process for the agricultural surfaces, measures for modified soil tillage methods, for changing land use and for ground formation were evaluated. The current type of agricultural use, the slope, the water storage potential in the soil and the dominant runoff process in the natural condition (potential map) as well as the distance to the ground surface, the proximity to a water body and the size of the farming units were taken into account as evaluation parameters.

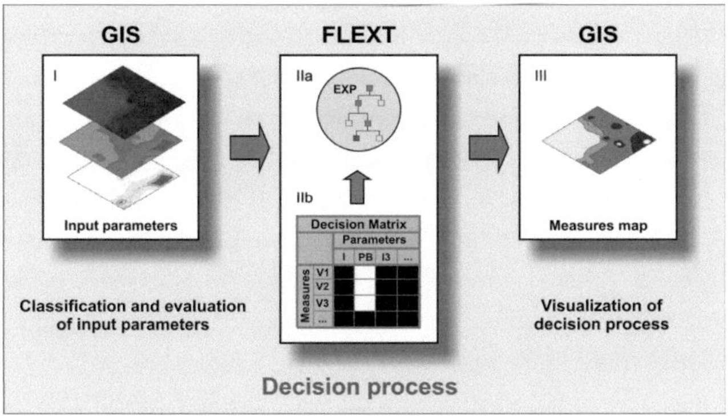

Fig. 2: Structure of Expert System FLEXT

3 Results

Figure 3 shows the results of the calculation of the soil water retention potential. The entire Saxony catchment (area of 18.200 km²) is shown. The potential map displayed is based on a 3-day flood event with a recurrence period of 100 years.

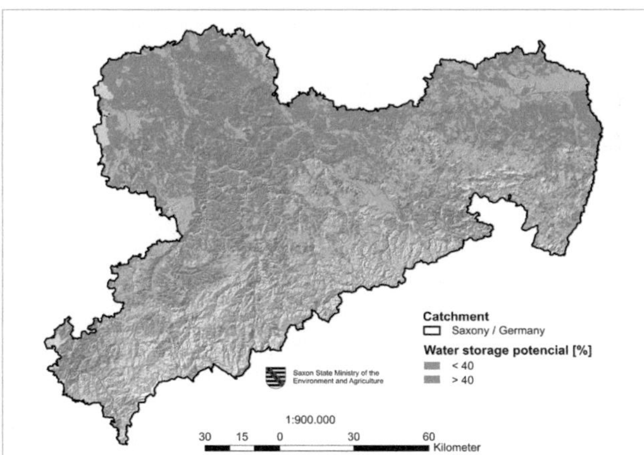

Fig. 3: Water storage potential for the Saxony (100-year rain event [1])

That means that the 72-hour infiltration capacity was contrasted with the statistical regionalised heights of precipitation of the same duration level [1]. It is clear that not all areas in the catchment are suitable for augmented floodwater storage in the soil. This applies especially to the flat and steep areas of the Erz Mountains in the southern part of the catchment as well as to the floodplains that are in direct hydraulic contact with the waterbody. In these areas, decentralised storages and runoff retention measures such as hedges are more effective. In the middle part of the catchment there are nevertheless wide areas on which 40% of the flood-causing precipitation can be retained. In the downstream part of the catchment in flat areas with highly permeable soils, the precipitation could be nearly completely stored in the soil.

Figure 4 is the result of the FLEXT evaluation process for agricultural areas using the example of conservation tillage.

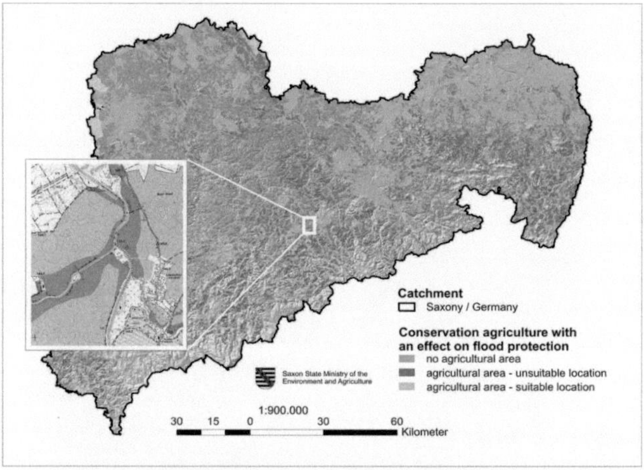

Fig. 4: Measures map for agricultural surfaces

The overview map shows that around a half of all cultivated surfaces (7.370 km²) are suitable for conservation tillage and at the same time, it should be expected that this measure bring positive (side-) effects for flood protection on the designated surfaces since a large portion of the additionally infiltrated precipitation can be temporarily stored in the soil. The detailed excerpt shows the spatial resolution of the measures map.

On this base (Fig. 5) different land use scenarios were developed and evaluated from a flood prevention perspective. One of these scenarios considered conservation tillage increased by 20% within 10 years. Through the implementation of this scenario additionally 8.9 million m^3 of water could be stored in the Saxony catchment.

4 Conclusions & Forecast

With the aid of a GIS-based expert system, maps were created as examples for Saxony. These maps can serve as the foundation for systematic land management.

The potential map gives an overview to spatial planners, administrators and politicians as well as those involved in water management on the areas marked for flood reduction in a catchment. With this information their priority areas, active focuses and plans can be adjusted. At the same time it allows a high spatial resolution for the map in order to identify the areas within partial catchments which contribute to flooding because of their specific local properties. Areas that can store significantly more precipitation than it is the case with conventional farming can be allocated. A comparison with other approaches to landscape retention potential [8] shows that the potential map is different not only in the level of spatial detail but also in chronological resolution. Unlike the annual water balance, an observational time period of 3 days allows for better statements about the runoff behaviour of the landscape in a flood event. Last but not least, provision of equivalent storage volumes and the integration of the dominant runoff process means a clearly better representation and understanding of the processes.

The presented measures maps help the stakeholders in regional and local planning to grasp the measures which fit to a location and thereby to utilise systematic synergy effects (erosion prevention plus flood prevention, drainage plus flood protection). The goal is, even in the area of agriculture, to have the information from the potential and measures maps to be integrated into the planning process.

In the scope of the research project (see acknowledgement) the measures maps are used as the foundation for the setup of realistic future scenarios. The computer-simulated representation of these scenarios, with the aid of precipitation-runoff models, should clearly show the benefit of taking precautions in a specific area and thereby incite that the potentials of land management are better used than ever before.

List of Illustrations, Abbildungsverzeichnis

List of Literature

[1] DWD. Starkniederschlagshöhen für Deutschland KOSTRA. Selbtverlag des Deutschen Wetterdienstes. 2000.

[2] Flerchinger, G.N., F.J. Watts and G.L. Bloomsburg. Explicit Solution to Green-Ampt Equation for Nonuniform Soils, Journal of Irrigation and Drainage Engineering (ASCE) JIDEDH Vol. 114, No. 3, p 561-565. 1988.

[3] Green, W.A. and G.A. Ampt. Studies on soil physics I. The flow of air and waterthrough soils. Journal of Agricultural Science 4: 1-24. 1911.

[4] Jin Z. Development of a Transparent Knowledge-Based Spatial Decision Support System for Decentralised Stormwater Management Planning“ Dissertation, Universität Hannover. 2005.

[5] LAWA. Länderarbeitsgemeinschaft Wasser, Instrumente und Handlungsempfehlungen zur Umsetzung der Leitlinien für einen zukunftsweisenden Hochwasserschutz, Düsseldorf. 2004.

[6] Merta, M. & H. Sieker (2009): Wasserrückhaltepotenziale in Sachsen – Teilprojekt I und II. Abschlussbericht zum Projekt des Sächsischen Landesamts für Umwelt, Landwirtschaft und Geologie. 2009.

[7] Peschke, G., C. Etzenberg, G. Müller and S. Zimmermann. Das wissensbasierte System FLAB. Ein Instrument zur rechnergestützten Bestimmung von Landschaftseinheiten gleicher Abflussbildung. IHI-Schriften, Heft 10 . Internationales Hochschulinstitut Zittau. 1999.

[8] Röder, M. and C. Beyer. Abflussbildung und vorbeugender Hochwasserschutz in der Landes- und Regionalplanung. Dargestellt am Beispiel Sachsens. Naturschutz und Landschaftsplanung 34 (7): S. 197-202. 2002.

[9] Sieker, F., M. Reich, W. Schmidt, J. Jasper, D. Wilcke, S. Rüter, S. Zacharias and M. Salzmann. Vorbeugender Hochwasserschutz durch Wasserrückhalt in der Fläche unter besonderer Berücksichtigung naturschutzfachlicher Aspekte - am Beispiel des Flusseinzugsgebietes der Mulde. Abschlussbericht zum DBU-Projekt AZ 21467. 2007.

Requirements for rainfall retention and storage in cold climate

Laura Inha
Päivi Paavilainen, Pekka Pietilä (Tampere University of Technology), Tapio Katko (Tampere University of Technology)
Ramboll Finland, PL 718, Pakkahuoneenaukio 2, FIN-33101 Tampere, Finland,
E-mail: laura.inha@ramboll.fi, Telefon: +3580207556916

Abstract

Climate change is expected to make winters in Finland milder and cause a larger proportion of precipitation to fall as rain than as snow during winter. Since winter time temperatures will still remain around zero degrees, the soil will still be frozen and there will be at least an occasional accumulation of snow. That will make many of the commonly proposed rainfall retention and storage systems, such as green roofs, low swales, infiltration trenches and permeable pavings, useless, or at least very difficult to use and maintain.

The objective of this study was to find out methods for managing rainfall retention and storage in relation to soil properties and vegetation in cold climates with special attention to urban areas. The idea was not to concentrate on technical installation manuals and sizing criteria but to introduce methods which have been proven to work in real life cases.

In cold regions, the most reliable and functional storage systems are built underground, preferably below the frost line. When heavy rain falls on ice covered urban areas, the runoff can be substantial. Therefore, traditional flood route and flooding area planning will continue to play a vital role in cold season stormwater management. In spite of alternative methods, such as green roofs, stormwater networks still need to be dimensioned for large amounts of water, which means that alternative methods will not necessarily be economically more advantageous in cold regions.

All in all, maintenance is very important with regards to all retention and storage systems both prior to and during winter. Since it is often very difficult to find the space to locate these systems in built-up city environments (without major financial and structural investments), it is vital to consider rainfall retention and storage needs during the planning phase.

Catchwords

stormwater, retention, urban flood, cold climate, climate change

1 Introduction

Global warming has remarkable effects on nature and human environment. One of the effects is the change in rainfall. It has been estimated that due to climate change the amounts of rainfall will increase in the Polar Regions whereas in the arid areas the rainfall will decrease [1].

1.1 Climate change in Finland

During the years 2000–2005, in the Finnish research "Heavy rains and floods in urban areas", weather radar and rain gauge observations were studied to find out how climate change will change the short duration, intensive heavy rains and how those changes will affect the urban areas. The research showed that by the year 2100 in a period from May to September the overall amount of rainfall will increase approximately by 10–15 % and the intensity in single heavy rain events will increase up to 10–40 %. The increase is estimated to be fairly linear, which means that an event now occurring once in three years will in the future occur every second year. Even though both the intensity and rainfall amounts of the heavy rain events in the summer time increase, the overall amount of rainfall will mainly increase in winter time. [2].

1.2 Current situation of rainfall retention and storage in Finland

Research of infiltration and retention techniques started in the late 1970's in Finland when Technical Research Centre of Finland started various test sites [3]. The experiences with stormwater retention and infiltration have been positive and the structures still function today [4]. However, after the test projects were completed in late 1980's, there has been a downshift within similar research through the 1990's.

In the early 2000's during the research "Heavy rains and urban floods" it was noticed that flood risks, planning and construction need to be considered more carefully. More innovative solutions were wanted especially in planning and construction of infrastructure. The quality of stormwater also started to raise concern. [2]

Nowadays an increasing number of cities and municipalities have their own stormwater management strategies and plans. A general, national management plan for stormwater is also being prepared.

2 Requirements for rainfall retention and storage in cold climate

Stormwater management during the winter time is challenging in cold climates, when the soil is frozen and often covered with snow and ice, but occasional rain events and snow melt can still take place. [2] Frozen soil lowers the capacity of the stormwater structures. When the precipitation falls as rain instead of snow, there is also a flood risk.

When the winter precipitation falls as snow, the challenge is storing the snow in a way that it will not block the stormwater surface systems. If the surface systems have been blocked, the spring time melting waters will cause a risk of flooding.

2.1 The functioning of the retention and storage systems in cold climate

Winter conditions lower temporarily especially the reduction, storage and transportation capacity of surface structures for stormwater management. These properties are, however, restored as soon as snow and ice melt away.

1.2.1 Transportation

Transportation methods can be divided into surface and pipe systems. Surface systems include for example open ditches, swales and canals. Besides transporting, some of these methods can also reduce and detain stormwaters.

One of such methods is a green swale, which combines transport and infiltration during the summer months and functions as a transport channel during the frozen period. It is possible to use the green swales for storing snow, for which swales should be designed so that the other slope is less steep. This way the bottom of the swale remains clear of snow and enables drainage when melting occurs. [5]

1.2.2 Reduction

Stormwater reduction methods decrease the amount of stormwater by infiltrating it into groundwater. Surface infiltration is not possible in winter time because of ground frost, frozen surface and ice cover [5]. When the frost melts in the spring time, the soil can get very moist which also reduces the infiltration capacity. [6] If the water does not infiltrate, it should be transferred away by drainage [5].

Besides frost, the winter time gritting and de-icing can also cause problems: the gritting sand fills the pores of permeable surfaces whereas de-icing salts have adverse effects on ground water. Even mechanical clearing of the snow has its disadvantages, as it easily damages the surface of the permeable paving. [6, 8]

Frozen green roofs lack storage and infiltration capacity during winter. However, in spring both features return.

Underground structures survive the winter better [7]. Subsurface infiltration fields and underground basins were built in the corners of street and park areas in the town of Kouvola, Finland. After more than 20 years in use these infiltration fields still function well. The ones that are located in places where the snow cover is left untouched for the winter work best and have not experienced problems with ground frost. [4]

1.2.3 Retention

Stormwater retention means storing the water for a while to slow down the flow rate and decrease the load in receiving systems. Retention methods include for example different kinds of wetlands, ponds and both underground and surface pools and basins.

Generally the retention structures function well at winter time even though winter conditions lower their capacity temporarily. The quality of effluent also decreases during winter due to the diminishing microbiological activity and decreasing particle settling speed in cold water. [6] The use of retention structures as snow storage in built-up areas should be avoided, since they should be available to handle rainfall and snow melting in spring. [5]

Frozen pond surface does not only decrease storage capacity, but also increases water flow rate when the ice is melting. The increased flow rate can detach already sedimented particles from the bottom. To solve this problem an idea of a combination of a pond and an infiltration swale has been introduced. The combination will be dry during the winter season, which enables the infiltration of the first flush to the subsoil of the pond. For greater flow rates a dam and outlet pipe system can be installed. [9]

Another opportunity to get more storage capacity for ponds during winter is an outlet pipe system, which keeps the water level lower during the winter than during the summer. In spring the outlet pipe must be closed before melting to enable the water to rise again. Normal water level makes the sedimentation of suspended solids and other particles possible. Same solution also applies to wetlands, where more regulation capacity can also be achieved by directing the flow by building low steps to the bottom of the wetland. [9]

The inlet and outlet pipes of ponds and basins are advised to be constructed below the frost line to prevent pipe freezing [6]. However, the frost often reaches deeper than the receiving ditches, so the installation would require pumping of stormwaters, which is questionable both in an environmental and economical sense.

Hence to avoid frozen pipes and to ensure the functioning of the whole retention structure, maintenance is very important before winter. If leaves and other trash are not removed before freezing, the stormwater will flow past the structure and cause a risk of flooding. [5] Winter is a good time to build and maintain the structures since the frozen ground enables the access to ponds and wetlands also with heavy vehicles.

3 Conclusions and forecast

3.1 Storm and melting water management in new areas

Planning of the storm and melting water management for new land use development sites is relatively straightforward when the management structures are integrated in the new land use plans. In ideal situation the storm and melting water management structures are designed simultaneously with the early land use planning phase, and suitable sites can be allotted for retention and treatment. Often the examples shown for managing stormwaters of urban areas are from small, newly built housing estates, where there are relatively good opportunities to build rain gardens etc. to receive the stormwaters. However, the real challenge is managing sites such as large retail parks and industrial areas.

Cold climate makes such places especially difficult to handle, as the large storage and parking areas require continuous gritting and anti-icing with salt and large amounts of smudgy snow piles up during the winter months. The dirty melting waters formed in biologically inactive period, grit and anti-freeze salts have the dual aspect of clogging management systems and rendering the stormwaters more difficult to treat.

Because of the commercial functions, the land use on the lots is also restricted by the need to make the most of the money invested in the property. Thus the on-site solutions usually are underground or rooftop ones, which often are the most costly to build. E.g. green roof prices can be up to five times more expensive than standard roof solutions, and the price of underground structures can easily reach 35 000 € and over for a site the size of just one hectare. In some cases the municipalities can actually be reluctant to set very stringent stormwater control demands to the land owners, as they run the risk of losing investments to neighbouring municipalities. Hence the stormwater management on commercial properties is often removed at least to some extent to the adjacent common areas, where more economical superficial structures can be located – often at the city's expense.

3.2 Storm and melting water management in built up areas

Stormwater management plans of built up old areas need to consider the whole watershed just like the plans for completely new areas. The special problem of these built up sites is that there normally are very few places suitable for stormwater management structures, and the management methods must be creatively fitted to existing infrastructure. Theoretically speaking it would be best to withhold and manage the storm and melting waters right on site where they are formed, but it is very difficult to obligate existing premises to carry out stormwater management measures, as the neighbouring premises should carry out similar measures as well, and as there was no such obligation when the premises were originally built. Thus the measures are most often paid by the city and located in recreational areas, but even when trying to shift the existing or threatening flooding problems from more valuable properties to relatively lightly built green areas, the attitude towards leading even only occasional additional stormwaters into a park is often at least initially negative.

The subsurface methods, which are the most reliable and functional in cold climate, are often the most suitable solutions in existing areas, although some special restrictions apply. Infiltration systems are restricted primarily by the soil type, which in Finnish conditions is often rather challenging for infiltration – the impervious bedrock is often found right below the surface and the most commonly occurring soil types conduct water very poorly. The infiltration systems can affect the soil freezing in adjacent structures (e.g. roads or house foundations), and this has to be taken into account in all cold climates. The lack of space in built-up areas also applies below surface: the subsurface systems need to be fitted to other subsurface infrastructure like water, wastewater, district heating, gas, electricity and IT networks. As the joining points to existing stormwater networks are pre-determined, sometimes the installation of subsurface stormwater structures would require pumping, which renders the solution questionable in the environmental and economical sense.

The central routes of stormwaters have luckily often remained as open channels in the housing estate areas outside the urban centres. The easiest management solution for these sites lies with the clever landscaping of the central stream, where retention volume can be achieved by terracing the stream cross section or creating controlled flooding fields on low lying areas. Often in these cases it is possible to greatly increase the recreational and even ecological value of the urban streams, as the circumstances to water flora and fauna can be improved and lost species reintroduced to the streams by improving the essential habitats. The typical challenge is the management of dry periods in the summer time. Maintaining even a small dry period flow requires a steady input from groundwater, which only takes place when the watershed includes enough green areas. Additional input of water from drinking water network has even been introduced in some cases in high quality areas.

List of Literature

[1] Ilmastonmuutos. Suomen ympäristökeskus. (The website of Finland's environmental administration)

[2] Aaltonen, J., Hohti, H., Jylhä, K., Karvonen, T., Kilpeläinen, T., Koistinen, J., Kotro, J., Kuitunen, T., Ollila, M., Parvio, A., Pulkkinen, S., Silander, J., Tiihonen, T., Tuomenvirta, H., Vajda, A. Heavy rains and urban floods (Rankkasateet ja taajamatulvat, RATU). 2008.

[3] Huttunen, K. et al. Kevennetty kunnallistekniikka. Putkijohtojen routasuojaus ja lämmöneristäminen Elimäen Napan koerakennuskohteissa. VTT, Tiedotteita 519. 1985.

[4] Inha, L. Kokemuksia 1980-luvun kevennetyn kunnallistekniikan koerakentamishankkeista. 2010.

[5] Hyöty, P. Hulevesien luonnonmukaisen hallinnan menetelmät. Suunnitteluohje. 2007.

[6] Ahponen, H. Luonnonmukaisten hulevedenkäsittelymenetelmien ja aluesuunnittelun keinoin kohti parempaa taajamahydrologiaa. In: Vakkilainen, P., Kotola, J., Nurminen, J. Rakennetun ympäristön valumavedet ja niiden hallinta. 2005.

[7] Larm, T. Dagvattnets sammansättning, recipient påverkan och behandling. 1994.

[8] Caroco, D. & Claytor, R. Stormwater BMP Design Supplement for Cold Climates. 1997.

[9] Jormola, J. Hulevesien käsittelyperiaatteita ja -menetelmiä. In: Niemelä, J., Helle, I., Jormola J. Purovesistöjen merkitys kaupunkiluonnon monimuotoisuudelle. Loppuraportti. 2004.

Infiltration and Surface Runoff in Urban Areas

Hydrological Characteristics and Parameter Recommendations

Marc Illgen

DAHLEM Beratende Ingenieure, Poststrasse 9, D-64293 Darmstadt, Germany

Email: m.illgen@dahlem-ingenieure.de, Fon: +49(0)6151/8595-23

Abstract

The stormwater runoff and infiltration performance of pavement structures has been systematically evaluated within an extensive research project. Over 230 field and lab scale experiments have been completed and analyzed for surface runoff and infiltration characteristics as well as for water content changes inside the construction. The test series comprises several types of conventional and explicitly permeable structures under various boundary conditions. In addition, the field data have been superimposed on the results of former infiltration tests accomplished by several researchers to a unique database containing over 350 single infiltration tests and enabling substantive statistical analyses. Based on these widespread monitoring results, the primary characteristics of the specific infiltration phenomenon on pavements have been determined. The analyses yield to an identification of the governing constraints, to a quantification of their impacts on the infiltration capacity of the entire pavement construction and finally to the development of an advanced conceptual approach for modelling runoff and infiltration processes on pavement structures. Moreover, profound recommendations for type specific infiltration rates, runoff coefficients and parameter values for conventional applications have been worked out, which facilitate planners a better estimation of the stormwater runoff contributed by the manifold types of more or less permeable pavements.

Catchwords

infiltration rate, infiltrometer, pavement, runoff coefficient, surface runoff

1 Background and Objectives

The stormwater runoff is the central and dimension determining parameter in almost all planning tasks in the fields of urban drainage and flood mitigation. Hence, the common design methods are obliged to reproduce the specific runoff and infiltration processes of the entire spread of urban surfaces adequately – whether simplified computations in combination with runoff coefficients or more complex numerical models are applied. Thereby, the application of surface specific runoff and infiltration parameters is vital to both computational methods. However, up to now sound and reliable parameter recommendations do not exist in regard to common permeable surfaces [1, 2].

Moreover, the conventional urban drainage models are predominantly using methods that have been adopted from the simulation of natural watersheds. Even though permeable surfaces are receiving more and more attention and the expectations regarding the accuracy of model applications are continuously rising, these makeshift methods have not been reconsidered or significantly improved for years [3, 4]. The lack of both, validated parameter recommendations as well as adequate methods for modelling stormwater runoff and infiltration processes of urban areas are a result of evident knowledge deficits regarding the particular infiltration performance of pavements and can lead to considerable uncertainties of hydrological computations.

Driven by these deficits, the infiltration performance of various pavings has been systematically evaluated within a comprehensive research project over several years [5]. The primary objective of the investigation was to generate a broad and reliable database that enables a phenomenological analysis of the infiltration process in order to elaborate reliable parameter values for conventional computational methods as well as to develop an advanced simulation module for urban drainage modelling.

2 Material and Methods

Valuable investigations have been realized in recent years, such as [6-10]. However, these investigations are predominantly focusing on particular aspects, are restricted to one or few single sites only and do not enable entire water balances, mostly. Overall, more detailed investigations that include the evaluation of the entire infiltration and percolation processes through top and base layer into the soil layer as well as clogging effects and other leading constraints are rare. Moreover, while great effort has been devoted on the investigation of exceptionally permeable eco-pavements in recent years, the infiltration performance of conventional flagstone, block and interlocking pavements has been neglected entirely, even though these pavements are the most prevalent ones. Consequently, further investigations comprising all types of pavement and offering a detailed view inside the construction and on the major process chain are necessary to obtain a reliable database for the development of an enhanced computational method.

In order to achieve these purposes, a holistic methodical approach has been chosen that combines infiltration tests on existing pavements after several years of use with methodical and more conclusive lab scale experiments under determined boundary conditions together with the application of a finite element soil hydraulic model. Fig. 1 gives an overview of the basic approach including scope and methodology of the particular tasks and the types of pavement represented in the study. A detailed description of the methodical approach and the particular analytical steps of the overall investigation can be found in [11, 12].

	METHODICAL APPROACH					
method of investigation	field measurements		lab test series		application HYDRUS-2D	
scope of investigation	statistical analysis of infiltration rates		phenomenological analysis of infiltration process			data extension & verification
data origin	external data	own measurements & applications				
included types of pavement	flagstone pavement	conventional block and interlocking pavement		open-jointed eco-pavement	porous eco-pavement	concrete grid eco-pavement
+ width of joints	3-5 mm	3-5 mm	6-10 mm	> 4 mm	3-5 mm	-|-
+ open surface	0,5-2%	3-6%	6-12%	10-30%	3-6%	> 35%
+ main aggregates	sand	sand	sand / gravel	crushed stone	sand	soil / gravel

Fig 1: Methodical approach and included types of pavement.

3 Runoff and Infiltration Characteristics of paved Surfaces

The infiltration performance of a pavement structure is strongly affected by a multitude of site-specific constraints. Accordingly, infiltration rates can exceedingly vary from one site to another – irrespective of the particular type of pavement. But even on a site-scale the infiltration capacity of a particular pavement reveals a high spatial variability. On the one hand, remarkably high infiltration rates have been found at many locations, even on pavements which are generally considered to be hardly permeable and despite long-lasting periods of operation. On the other hand, point-wise or localized extremely reduced infiltration capacities have been recorded as well.

A decline of the infiltration capacity is basically caused by a reduction of the water conductivity of the entire pavement construction. This reduction can be either a result of fine material accumulating into the slots or voids (clogging or colmation) or of a compaction of the aggregates in the different layers of the construction (e.g. due to mechanical impacts by cars), where clogging effects seem to dominate. Even though reduced infiltration capacities can be expected for sites of higher age or higher traffic loading, considerable mechanical impacts or an extensive utilization of a pavement do not necessarily lead to a distinct reduction of the infiltration capacity over time.

Beyond that, the overall infiltration performance of a paved surface is characterized by high infiltration rates within the first few minutes of a storm event while the wide pores in the mineral aggregates in the joints are filled with water. Hence, the high values do not represent the hydraulic conductivity of the pavement construction, which may be much lower. The further percolation of the water to the deeper horizons of the base layer starts as soon as particular water contents in the joints have been reached. Already after 5-10 minutes, the infiltration rate has settled down on an almost constant value. Consequently, the particular duration of a rain event has only for short-term events a significant impact on the infiltration rate.

The infiltration rate of a particular pavement (with a certain stage of colmation) depends on the actual rain intensity. Thereby, increasing rain intensities cause considerably rising infiltration rates. This phenomenon is presumably induced by the inhomogeneity of the infiltration capacity along the joints. Whereas for minor rain intensities the infiltration capacities are not exhausted at every section of the pavement, some minor permeable sections already generate considerable surface runoff. For higher rain intensities the higher capacities become effective and lead to an increasing infiltration capacity of the entire structure. Hence, a starting surface runoff does not indicate the maximum infiltration rate of paved surface. It merely represents that phase of a rain event when the rain intensity exceeds the infiltration capacity of the least permeable section.

Beside clogging and mechanical impacts, structural conditions such as opening ratio, surface slope and grain size distribution of the joint aggregates are predominantly affecting the particular infiltration capacity of a pavement. Altogether, the infiltration performance of a particular pavement on a certain site is controlled by many site-specific boundary conditions, but offers notedly stochastic attributes. The infiltration capacity may vary in a huge range of value and is not predictable for a particular site.

Due to the comprehensive set of field data compiled within the presented research project, eminently valuable data on the statistical distribution of the infiltration capacity of several pavement types can be presented for the very first time. The cumulative frequency curves of the five major types of pavement are shown in Fig. 2. The curves are suggested to be representative for the particular pavement types as they contain field data obtained at multitudinous sites.

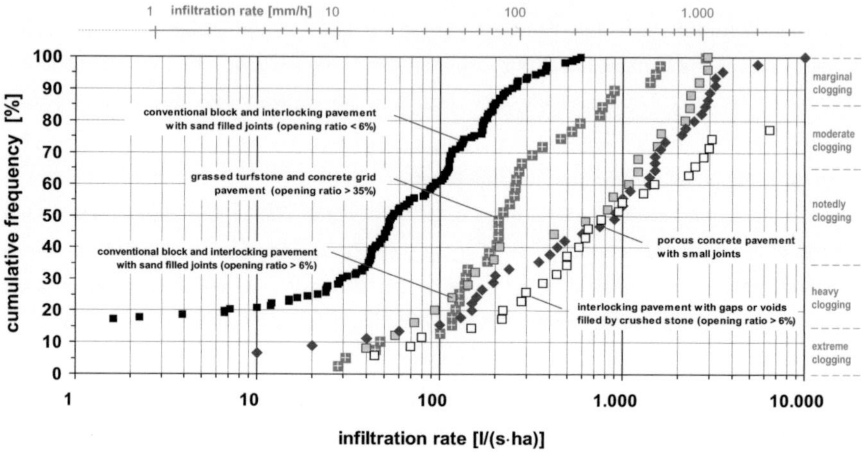

Fig 2: Cumulative frequency of final infiltration rates of common pavement types.

The figure illustrates the enormous variability of the infiltration capacity and its distinctive stochastic character. In a logarithmic representation, the ranges of value between 10th and 90th percentiles of the infiltration rate cover more than two decades for almost all types of pavement. The median infiltration rates amount between 55 l/(s·ha) or 22 mm/hr for conventional block and interlocking pavement with small sand-filled joints and 750 l/(s·ha) or 270 mm/hr for the more pervious types of pavement. For grassed grid pavers a median of 200 l/(s·ha) or 72 mm/hr has been found. Fig. 2 also contains a rough categorization of several stages of clogging correlated with the cumulative frequency of the infiltration rate.

A comprehensive discussion of the monitoring results has been published in [11, 12].

4 Recommendations for common Planning Tasks

Based on the results of the numerous field and lab experiments, far reaching recommendations concerning appropriate infiltrations rates, runoff coefficients and model parameters have been worked out. For instance, practical nomograms representing runoff coefficients relating to particular states of clogging, rain intensities and surface slopes have been developed for all common types of pavement. Some few examples are given in Fig. 3. They emphasize the strong correlation between runoff coefficient and rain intensity, which is neglected all too often.

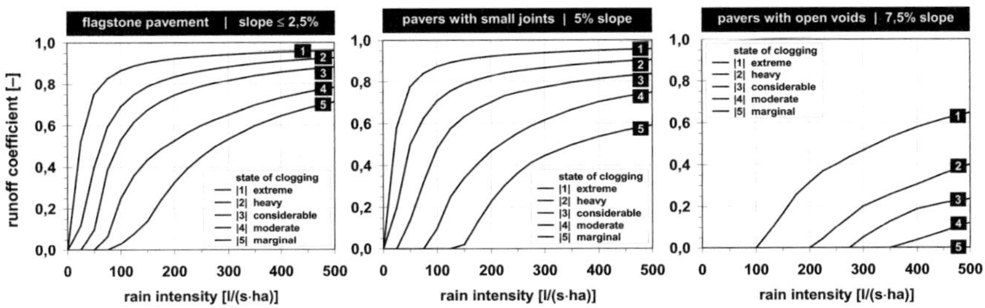

Fig. 3: Nomograms of recommended runoff coefficients for several pavement types.

The entire set of parameter recommendations (nomograms, tables) can be found in [5].

5 Development of an advanced Modelling Approach

In addition to the parameter recommendations, an advanced conceptual approach for modelling runoff and infiltration processes of permeable pavements has been developed. The approach is based on a simplified bi-directional layer model and may substitute the outdated approaches presently implemented in the common urban drainage models. The composition of the model is strictly following the physical construction scheme of a pavement structure, generally consisting of a pervious top layer, an extremely permeable base layer and a more or less permeable subgrade. A basic scheme of the approach is shown in Fig. 4.

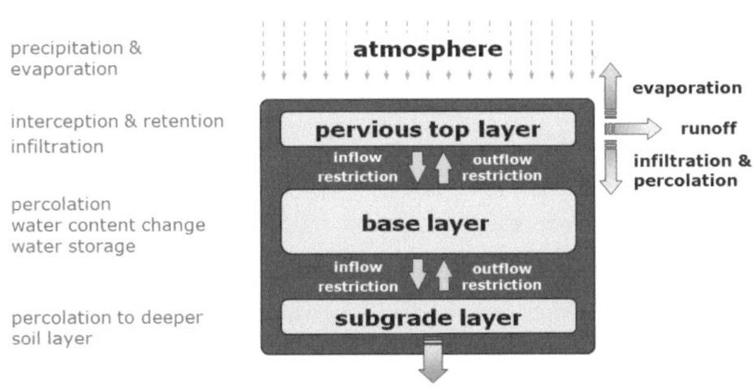

Fig. 4: Basic scheme of the bi-directional layer model and basic infiltration algorithm.

Whereas wetting and depression storage are represented by common initial losses, the infiltration process is described by a rather simple but eminently effective infiltration algorithm. Within this approach any paved area segment is sub-divided into a number of virtual coextensive sub-segments. Instead of applying a uniform infiltration rate for the entire area element, graded infiltration rates are allocated to the particular sub-segments (Fig. 5). The infiltration rates of the virtual sub-segments correlate with the frequency curves shown in Fig. 2 and represent the stochastic character of the infiltration capacity.

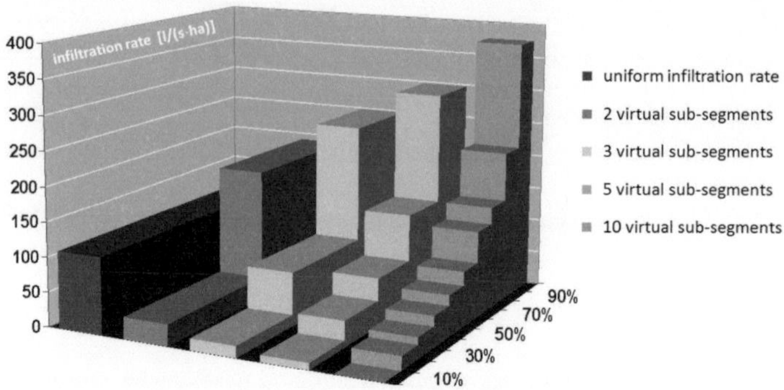

Fig. 5: Sub-division into sub-segments with varying infiltration capacity (e.g. 3 to 10 segments)

In contrast to conventional approaches, this infiltration algorithm accounts for the enormous variability of the infiltration capacity and reflects the increasing infiltration rate with increasing rain intensity. Overall, the approach enables a much more realistic description of the infiltration and runoff performance of paved surfaces in urban environments.

6 Summary and Conclusion

Based on the results of the numerous field and lab experiments, the primary characteristics of the specific infiltration phenomenon on permeable pavements have been investigated within an extensive research project. The comprehensive database has lead to an identification of governing boundary conditions and major constraints as well as to a quantification of their impacts on the infiltration capacity of the entire pavement construction. In addition, an advanced conceptual approach for modelling the infiltration performance of pavements structures has been developed and far-reaching recommendations for reasonable infiltration rates, runoff coefficients and model parameters have been worked out which facilitate planners a better estimation of the stormwater runoff on permeable pavements.

List of Literature

[1] IAHR/IWA (2008). Joint Specialist Group on Urban Drainage Newsletter, No. 21, February 2008, http://www.iwahq.org., visited 29 February 2008.

[2] DWA (2007): Oberflächenabfluss von durchlässig befestigten Flächen (Stormwater runoff of permeable paved sufaces), Arbeitsbericht der DWA-AG ES-2.6, KA Abwasser Abfall, 54. Jhrg., Nr. 8, 784-788.

[3] Ferguson, B. K. (2005). Porous Pavements. Taylor and Francis Group. New York.

[4] James, W. (2000). Applied Modeling of Urban Water Systems. Monograph 8. CHI, Guelph.

[5] Illgen, M. (2009). Infiltration processes on paved urban areas and their hydrological quantification (in German: Das Versickerungsverhalten durchlässig befestigter Siedlungsflächen und seine urbanhydrologische Quantifizierung). [Online] available at: <http://kluedo.ub.uni-kl.de/volltexte/2009/2326/pdf/Dissertation_Illgen_TUKL_2009.pdf> [accessed 23.07.2010].

[6] Bean E.Z., Hunt W. and Bidelspach D.A. (2007). Field survey of permeable pavement surface infiltration rates. J. Irrig. and Drain. Engrg., 133(3) 249-255.

[7] Borgwardt S. (2006). Long-term in-situ infiltration performance of permeable concrete block pavement. Proc. 8th Int. Conf. on Concrete Block Paving, San Francisco, USA.

[8] Brattebo, B.O. and Booth, D.B. (2003). Long-term stormwater quantity and quality performance of permeable pavement systems. Wat. Res., 37(18), 4369-4376.

[9] Davies J.W., Pratt C.J. and Scott M.A. (2002). Laboratory study of permeable pavement systems to support hydraulic modeling. Proc. 9th Int. Conf. on Urban Drainage, Portland, USA.

[10] Pratt, C. J., Mantle, J. D. G., and Schofield, P. A. (1995). UK Research into the Performance of Permeable Pavement, Reservoir Structures in Controlling Stormwater Discharge Quanitity and Quality. Wat. Sci. Techn., 32(1), 63-69.

[11] Illgen, M. (2008). Infiltration and surface runoff processes on pavements: physical phenomena and modelling. In: Proc. 11th Int. Conf. on Urban Drainage, Edinburgh, Scotland, UK, 31 Aug – 5 Sept 2008, [Online] available at: <http://www.11icud.org> [accessed 23.07.2010].

[12] Illgen, M., Harting, K., Schmitt, T. G., Welker, A. (2007). Runoff and infiltration characteristics of pavement structures – review of an extensive monitoring program. Wat. Sci. Techn., 56(10), 130-140.

Multi-criteria Framework for Performance Benchmarking of the Water Sectors in the Nile Basin

Dr. Ahmed S. ELZayat & Dr. Hatem M. Ali
Environmental Engineering Group & Water Quality Management Unit
Cairo, Egypt

Abstract

In various national and regional environments, the performance of water sectors depend on the appropriateness and adequacy of not only water institutions but also the indirect water-related institutions governing the economic, political, and social spheres. In view of the interests of multiple water entities in the Nile basin, the mixed impacts of water, channeled through different pathways, make it difficult to measure the potential of sustainable development for the water sector. As a result, the objective of this paper is to develop a multi-criteria framework to benchmark the performance of the water sectors in the Nile basin with different areas in various national perspectives. Based on a multi-criteria decision method, the study incorporated an evaluation of the national water sectors, started by analyzing the life decade challenge of sanitation coverage as it is insufficient to apply one single performance criterion.

In view of the central role of water in providing food, income, livelihood, and amenity benefits, the proposed method is based on the direct or indirect impacts of water generated through socio-economical, educational, health and environmental categories. The Preference Ranking Organization Method for Enrichment Evaluation (PROMETHEE) is used in the study to outrank the performance of Nile basin countries in achieving their water strategies because of its clear involvement of criteria importance and efficient computational procedure. The multi-criteria framework, interrelated pathways of water impacts and performance outranking of the riparian countries are intended as a 'reality assessment' for Nile communities, in order to fulfill the commitment which can be effectively used for sector capacity-building efforts.

Keywords: Nile basin, Water sector performance, multi-criteria framework, the PROMETHEE outranking procedures.

Making global water resources sustainable: Virtual water trade

Nico Grove
Institute for Information, Organization and Management,
Munich School of Management, Ludwig-Maximilians-Universität Munich,
grove@lmu.de

Abstract

Looking at the world's many challenges in the twenty-first century food supply is one of paramount importance. One issue that is hardly looked at is the political economy of *virtual water* in the region. *Virtual water* is the hidden amount of water that is embedded in all commodities. The paper focuses on the identification and elimination of welfare losses from inefficient use of the resource water, represented in actual worldwide *virtual water streams*. The main area of research is now to identify the causes for the inefficient *virtual water* flows and develop a concept combining reallocation incentives and a resource efficient use of fresh water resources.

Keywords

Virtual water, regulation, resource efficiency, tragedy of commons, trade, emission trade

1 Introduction

In terms of the planet's water availability, certain areas have more resources than others. Times are changing and scenarios predicting continuous conflicts over oil could easily change to water – the water wars. Especially overpopulation is putting severe pressure on certain world regions such as Africa, Latin America and Asia[1].

The increased and rapid urbanization that will result will be one of the greatest challenges for decision-makers. While the drinking water supply infrastructures alone will cost trillions of dollars, the real challenge will be ensuring an adequate food supply through sufficient agricultural water or *virtual water*. While every human being requires approximately 3 liter of drinking water per day, approximately 100 liter are used for domestic

[1] See WHO and UNICEF (2005).

purposes in showers, washing machines or cookers[2]. However, the bulk of individual water consumption is through food consumption. More than 1,2 million liter of water are hidden in the products each of us consumes every day – virtual water[3].

This study focuses on the identification of welfare losses from inefficient water use, represented in worldwide virtual water streams. Existing contributions to virtual water are predominantly led by Allen, Hoekstra and Chapagain, who identified the problem of virtual water resource inefficiency. These and other contributions focus on identifying virtual water inefficiencies. But in contrast to existing regulatory regimes like for carbon emissions, fresh water is a locally concentrated resource, plentiful in one area and untraceable in another. Therefore, a framework based on reduction[4] (e.g., carbon trade) cannot work for virtual water. Moreover, this local component – including the option to use much more water – must form part of a regulatory approach to virtual water. A viable framework would need to be global in scope and make provision for the improvement of local fresh water use efficiency – *virtual water trading*. This study will address this matter and will seek to move towards a regulatory regime for virtual water flows.

2 The notion of virtual water

The notion that water is a prerequisite for life on earth is not new. While approximately 70% of the planet is covered with water, human beings as well as other land animals and insects can consume only 1% of our fresh water resources, which is accessible without industrialized processing such as desalination or filtration. Another 2% is contained within the ice caps, while the remaining 97% is all sea water, which cannot be used for farming or personal purposes without prior, expensive treatment[5].

What makes the situation even more complex is that most of our water – virtual water – is not visible to us. This *hidden water* in everything we consume was labeled *virtual water* by Tony Allan in 1993[6]. Despite being invisible in the product, virtual water is the key to the holistic understanding of water consumption in the 21st century. Virtual water requires a paradigm shift in the political and economic thinking of those in power. The challenge is hidden in the details. Certain crops consume more water than others. While

[2] See Howard and Bartram (2003).

[3] See Allan (2001) and Hoekstra and Chapagain (2008)

[4] See Gleick (1993), p. 79.

[5] Data from World Bank (2004) and Watergap (2003).

[6] See Allan (2001).

15 000 liter of virtual water go into the production of 1 kg of wheat, 1 kg of potatoes requires only 900 liter (17 times less). Barley also requires 1 300 liter per kg, whereas rice – Asia's favorite carbohydrate source – requires 3 400 liter of water per kg[7].

As a matter of fact, developed countries have done much to conserve fresh water resources. However, the figures prove that water saving is not the issue in water-abundant nations. The greatest effect can be achieved by reducing or reallocating virtual water consumption, as in the end, today more than 18% of the world's population (1,1 billion people) does not have access to safe drinking water[8].

3 The tragedy of virtual water

Citizens of developed countries tend to take access to drinking water for granted. We may place access to water in the same class of public goods, characterized by non-excludability and non-rivalry[9]. However, fresh water is perfectly excludable and rivalry. Some people could die of thirst next to an irrigated field fenced with barbed wire and protected by armed guards. The same is true of virtual water. A country's virtual water trade balance could be negative while the majority of its population lacks access to clean drinking water. These issues cannot be solved by regional fresh water savings. They need to be anchored internationally by improving the efficiency of local fresh water use and, thus, virtual water trade.

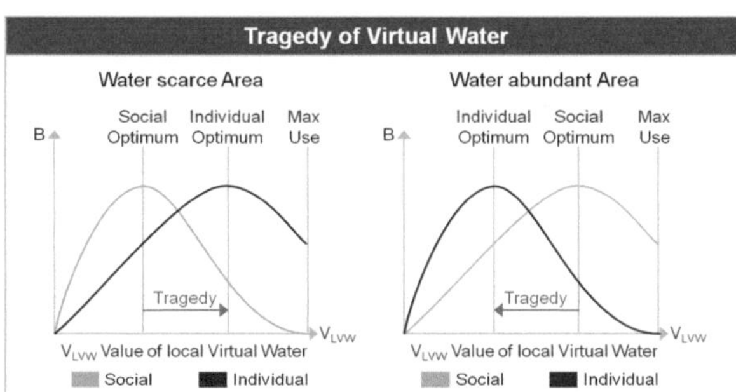

Fig.1: The tragedy of the commons in water-abundant and water-scarce areas

---

[7] See, for example, Allan (1997) and Zuygmunt (2007).

[8] See WHO and UNICEF (2005), p. 4.

[9] See Samuelson (1954), p. 387f.

Negative externalities of location-dependent inefficient fresh water use arise when more fresh water is used than what is available in that area in terms of acceptable, sustainable usage levels. In what has become known as *the tragedy of the commons*, Hardin described a public pasture, where every farmer will try to keep as much cattle as possible on this commons. According to Hardin, every farmer's adding of another animal to the pasture has a positive (+1 for the individual) and a negative component (-1 for all). In the end, all farmers will continue to send cattle to the common to graze until it is depleted due to overgrazing[10].

Hardin's description fits the virtual water phenomenon. However, another perspective is added. Besides overexploitation of fresh water in a water-scarce area, fresh water is used less where it is abundantly available, as depicted in Figure 1. The horizontal axis indicates the value of local virtual water (V_{LVW}) contained in products produced, in relation to the fresh water consumption. The vertical axis indicates the individuals' and/or overall social benefit (B). Local water resources are first exploited, then exported from the area. However, the social optimum lies below that level, as the water should be used as drinking water and for the production of goods that require less blue water. For a water-abundant area, the opposite is true. Herein lies the tragedy[11] of virtual water.

4 Redistribution goals

While at present the agricultural and industrial production process does not internalize virtual water consumption in the production process, the production function does include the cost of local fresh water sourcing. Therefore, an economic frontier depends on the availability of fresh water in a specific region. Keeping cattle in a desert would not be economically viable. However, production does not take worldwide resource efficiency into account. As shown, overall welfare would be much higher if goods that require large quantities of fresh water are grown in water-abundant areas. These goods can be exported to water-scarce areas in exchange for products requiring low quantities of fresh water[12].

Hence, I developed an idealistic virtual water flow regulatory model, depicted in Figure 2, based on Ricardo's notion of comparative advantages[13]. The export of goods containing large quantities of virtual water has to be supported in the case of water-abundant

[10] See Hardin (1968), p. 1244.

[11] See Akerlof (1997).

[12] See Chapagain and Hoekstra (2008a), p. 43.

[13] See Ricardo (1821), p. 41f.

areas and restricted in the case of water-scarce areas. Similarly, the export of goods containing low quantities of virtual water must be supported in the case of water-scarce areas. In order to keep the international supply of goods containing low quantities of virtual water at a constant level, their export from water-abundant areas must be restricted. The model's ultimate target is to reverse the actual virtual water trade balance between water-scarce and water-abundant areas, leading to a negative virtual water trade balance for water-abundant areas and a positive virtual water trade balance for water-scarce areas.

The theoretical model developed contains options for the integration of location-dependent resource-efficient virtual water consumption into the production function in order to increase overall welfare.

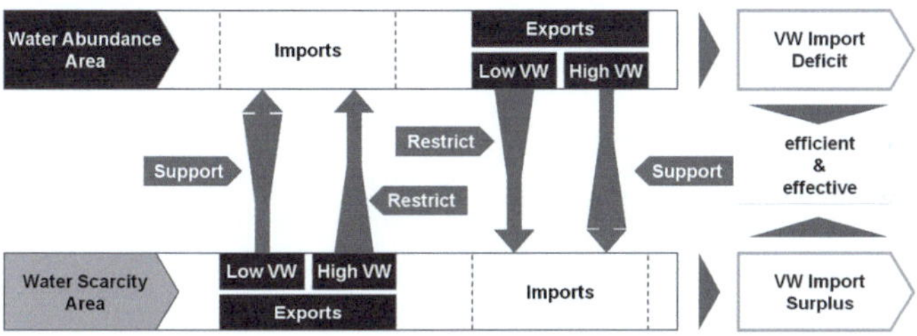

Fig.2: Idealistic model of virtual water flow regulation

5 Perspectives of regulatory regimes

A major goal of a new regulatory regime is to increase local resource-efficient virtual water consumption. Some of considered regulatory mechanisms fulfill these goals better than others. The acceptance and applicability of a virtual water regulatory framework and its instruments depend on sociopolitical and economic dimensions and considerations[14]. The following criteria must be fulfilled: reallocation potential, supporting developed countries, supporting developing countries, transparency, controllability[15], and implementation.

Applying these evaluation criteria to the aforementioned regulatory mechanisms allows for feasibility assessment. The results of the previous chapter's applicability analysis have been summarized in Figure 3.

[14] See Laffont and Triole (1993), p. 592.

[15] See Newbery (2005), p. 20.

	Rules and Laws	Consumer Tax	Producer Tax	Trade
Reallocation potential	○	◔	◑	◕
Supporting industrialized nations	◔	◔	◕	◔
Supporting dev. countries	○	◔	◑	◕
Intend to defraud	◕	◕	◑	◔
Control	○	◕	◔	◕
Implementation	●	◕	◔	◑
TOTAL	◔	◑	◑	◕

Fig.3: The valuation of regulation measures for virtual water

6 Conclusion and Recommendations

The actual use of fresh water without taking into account the local relative value leads to massive negative impacts on overall welfare. In March 2008, 32 states in the UN Human Rights Council agreed to establish a right to water and sanitation, which sought to make drinking water a universal human right[16]. Such agreements prove the (stated) willingness by governments to cooperate over water. The natural next step would be to include virtual water in this international agreement. Virtual water is also a unique opportunity for developing countries with booming populations to combine economic prosperity and environmental sustainability.

In the end, the idea of regulating virtual water flows might seem a bit unrealistic. But the case of carbon trade has developed swiftly from inception to internationally accepted system. If this concept were to spark a debate on and work towards a regulated political economy of virtual water, it would fulfill its intention.

List of illustrations

[16] See United Nations (2008).

List of literature

[1] WHO/UNICEF. WHO/UNICEF Joint Monitoring Programme for Water Supply and Sanitation - Water for life : making it happen. WHO Library, Geneva, 2005.

[2] Howard, G./Bartram, J. Domestic Water Quantity, Service, Level and Health. Geneva: World Health Organization, 2003.

[3] Allan, A.J. The Middle East water question: hydropolitics and the global economy. Tauris Publishers, London, 2001.

Hoekstra, A. Y./Chapagain, A. K. Globalization of Water – Sharing the Planet's Freshwater Resources, Blackwell, 2008.

[4] Gleick, P. H. Water and Conflict: Fresh Water Resources and International Security, in: International Security Vol. 18 No. 1, 1993.

[5] World Bank. WDI, GDF & ADI Online Databases. available online at: http://web.worldbank.org/, 2010.

Watergap. Global modelling of water resources, water use and water quality.Tool by the Institute of Physical Geography, University of Frankfurt, available online at: http://www.geo.uni-frankfurt.de/ipg/ag/dl/forschung/WaterGAP/index.html, 2010.

[6] See above.

[7] Allan, A. J. Virtual Water`: a long term solution for water short Middle Eastern economies? paper presented at 1997 British Association Festival of Science, Roger Stevens Lecture Theatre, University of Leeds, Water and Development Session - TUE.51, 14.45, 9 September 1997.

Zygmunt, J. Hidden Waters. Report on behalf of waterwise, London, 2007.

[8] See above.

[9] Samuelson, P.A. The Pure Theory of Public Expenditure. Review of Economics and Statistics 36 (4): pp. 387–389, 1954.

[10] Hardin, G. The Tragedy of the commons. Jounal of Science Vol. 162. No. 3859, pp. 1243 – 1248, 1968.

[11] Akerlof, G. A. Social Distance and Social Decisions. Econometrica, Vol. 65, No. 5 (Sep., 1997), pp. 1005-1027.

[12] See above.

[13] Ricardo, D. On the Principles of Political Economy and Taxation. John Murray, London, 3rd edition, 1821.

[14] Laffont, J.-J./Tirole J. A Theory of Incentives in Procurement and Regulation. Massachusetts: MIT Press, 1993.

[15] Newbery, D. M. Why Tax Energy? Towards a More Rational Policy. The Energy Journal, Vol. 26, No. 3, p1-39, 2005.

[16] United Nations Promotion and Protection of all Human Rights, civil, political, economic, social and cultural Rights, including the Right to Development. General Assembly, GE.08-11903 (E) 200308, 2008.

Energy Efficiency as the Limiting Constraint for Water Supply

Case Study Bribin Water Distribution System, Java, Indonesia

Philipp Klingel
Nico Hassel, Peter Oberle, Franz Nestmann
Karlsruhe Institute for Technology, Institute for Water and River Basin Management, 76131 Karlsruhe, Germany, E-mail: klingel@kit.edu, Phone: +49(0)721/6084561

Abstract

The availability and the inefficient use of energy resources can be limiting constraints for the implementation of a functional water distribution - especially in developing countries. This is the case for the Bribin water distribution system in Java, Indonesia. This article discusses the improvement of energy efficiency in water distribution systems in order to establish a reliable water supply using the Bribin system as a case study. Therefore two system and operation scenarios are analysed and compared: the existing system and a restructured and optimised system. Thereby, the energy demand is determined for both scenarios and the main potentials to increase the energy efficiency are pointed out.

Keywords

Water supply, water distribution, energy efficiency, hydro power, Bribin, Gunung Kidul, Indonesia

1 Introduction

The transport of water from the point of availability through water distribution networks to the consumers requires energy. In particular, energy is needed to exploit water bodies, to treat the water, to pump water up to higher elevations and to pressurise pipe networks. Therefore, the availability and the efficient use of energy resources play a key role in realising a secure water supply. Furthermore, considering the large amounts of energy consumed to distribute water, the maximisation of energy efficiency always should be an objective in order to reduce carbon dioxide emission. Major potentials to increase energy efficiency are found in (1) the design of the system components, (2) the system concept,

(3) the potential for energy recovery, (4) the real water losses and (5) the use of local regenerative energy resources.

The head loss in pipes rises exponentially with increasing discharge [2]. If the actual discharge through a pipe is larger than the design discharge, the energy losses and therefore the needed pressure head of pumps are unnecessarily high. In addition, the incorrect design or operation of pumps leads to avoidable low degrees of efficiency [2]. Furthermore, given potential energy within the distribution system is often not utilised extensively or even wasted, e.g. pumping against a partially closed valve. In addition, existing potential might be recovered with small turbine units or reverse running pumps [5]. Given the water lost through leakages is equal to a certain percentage of the input volume, the same percentage of the energy needed to produce and transport the input volume is wasted. Therefore, reducing water losses is always connected with an increase of energy efficiency [6]. If power cuts, expanses for energy or shortcomings of fuel are a constraint, the use of local regenerative energy resources, e.g. hydro power, may be an alternative.

The Gunung Sewu region of Java in Indonesia is a karst area characterised by a moved topography with hundreds of narrowly located and steep hills (see fig. 1). Water is scarce due to a long running dry season and high infiltration rates. Apart from collected rain water during the rainy season water is only available from the underground karst cave system. Therefore, the approx. 80.000 inhabitants have to be supplied by a regional water distribution system which is fed by the underground river Bribin.

The given topography results in a generally high energy demand which is aggravated by the inefficient use of energy. Diesel shortage and power cuts lead to frequent pump failures resulting in supply interruptions. Therefore, the limited availability of sufficient energy is likely to be seen as the limitation of the water supply. However, increasing the energy efficiency of the water distribution might close the gap between energy demand and energy availability. Within the frame of an Integrated Water Resources Management project funded by the German Ministry for Education and Research (BMBF) an optimisation of the system concept and components was planned based on the analysis of the existing system [4].

This article discusses the improvement of the energy efficiency in water distribution systems in order to establish a reliable water supply using the Bribin water distribution system as a case study. In the following the methodology of investigation is described before two possible system scenarios are analysed. Finally, the scenarios will be compared and conclusions will be drawn.

Fig.1: Topography of the Gunung Sewu

2 Methodology

The Bribin water distribution system consists of a transmission main system and distribution networks. Water is pumped through the approx. 30 km long transmission main system to several elevated tanks. The distribution networks transport water by gravity from the elevated tanks to the consumers of the supply area. If the elevated tanks are taken as given parameters, the distribution networks do not affect the energy balance. Thus, this study exclusively discusses the transmission main system. Two scenarios are analysed:

Scenario A: the existing system

Scenario B: the planned restructured system

Scenario A represents the existing system (see section 3). Scenario B represents the improved system which is at the time of writing at a planning stage (see section 4).

For the scenarios numerical models were set up and calibrated based on measurements of flows and pressures [3]. The numerical models were used for the system analysis and for determining the improvement measures and the operation strategies of scenario B.

For each scenario the transmission main system can be divided into network sections connecting two tanks. Water is transported either by pumps or by gravity. For each pumping section the energy demand per day P_d [kWh/d] and year P_a [kWh/a] is calculated. Summarising the yearly energy demand of the sections gives the total energy demand per year for the system operation P_{tot} [kWh/a]. To determine the energy demand of a section the discharge, the water losses, the system and pump characteristics, the elevation difference and the operation strategy are considered. The degree of efficiency of generators and engines are neglected ($\eta_e = 1.0$) in both scenarios due to missing data.

Since it is difficult to compare scenarios without taking the different volumes of distributed water into account, the indicator I_{del} [kWh/m³] is introduced, which relates the energy demand to the amount of water delivered to the tanks. Furthermore, the supplied quantity per capita is given for each scenario in order to point up the quality of service.

3 Scenario A – Existing system

The only connected source of the existing system is the underground river Bribin. An open cave entrance is used to reach the water of the river (see fig. 2, *Bribin*). The main distribution system consists of six pump stations (*PS*) feeding six elevated tanks (*ET*) (see fig. 2). To run the pumps, electricity from diesel generators or the public grid is used.

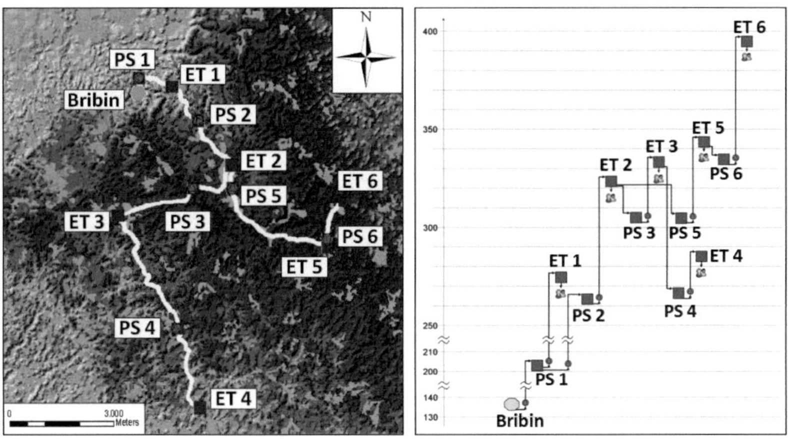

Fig. 2: Existing transmission main system (scenario A)

The system is operated rather intuitively. Thus, an equal distribution of the available water is not maintained. The tanks at the far end of the system (e.g. *ET4* or *ET6*) receive less water than the ones closer to the water extraction (e.g. *ET1* or *ET2*) [7]. Exact records of the current system operation are not available. Therefore, in scenario A it is assumed that the system is operated in a way that the input volume of approx. 2,178 m³/d is distributed evenly to the consumers. The pump pressure heads and discharges were determined through measurements. The degrees of efficiency were found in data sheets of the manufacturers.

Due to the intuitive operation (tank overflows, emptying of pipes) and the generally bad condition of the infrastructure real water losses are high. Own measurements have shown for the real losses an average of approx. 58 % of the input volume. Thus, just approx. 911 m³/d of water reach the elevated tanks.

Table 1 summarises the energy demand of scenario A. Approx. 813 MWh/a are needed for the operation of the existing transmission main system. Related to the total volume transported to the tanks, for the distribution of 1.0 m³ approx. 2.5 kWh are needed (I_{del} = 2.45 kWh/m³). The distributed volume corresponds to a per capita consumption of approx 12 l/cap/d if water losses within the distribution networks connected to the elevated tanks are neglected (which in reality is not the case).

Section	Pressure head	Discharge	Efficiency degree	Daily operation	Energy demand P_d	Energy demand P_a
	[m]	[l/s]	[-]	[h/d]	[kWh/d]	[kWh/a]
Bribin – PS1	93.8	40.0	0.75	11.00	539	196,735
	73.0	15.0	0.80	11.00	147	53,801
PS1 – ET1	79.4	16.3	0.80	0.32	5	1,829
PS1 – PS2	79.1	76.1	0.76	7.88	611	223,266
PS2 – ET2	63.8	38.1	0.50	9.25	440	160,750
	76.0	15.6	0.79	9.25	136	49,589
PS3 – ET3	38.3	38.0	0.50	8.10	231	84,297
PS4 – ET4	37.5	24.0	0.74	3.56	42	27,904
PS5 – ET5	39.8	15.9	0.52	3.32	40	15,480
PS6 – ET6	61.3	10.9	0.51	2.83	36	13,250
Sum (P_{tot})						813,443

Tab. 1: Energy demand of scenario A - the existing transmission main system

4 Scenario B – Planned system

Within a joint research project lead by the Karlsruhe Institute of Technology funded by the BMBF the world's first underground hydro power plant was developed and implemented in the Bribin cave to pump water to the surface for consumption [5]. Reverse running pumps serve as turbine units which are directly connected to centrifugal pumps. Depending on the discharge of the Bribin river up to 63 l/s can be pumped over

220 m elevation to a new storage tank (see fig 3, *ET7*). A planned restructuring of the existing system integrates this new water extraction *Sindon* and optimises the system layout and components (see fig. 3). Additionally, three decentral energy recovery units are planned within the gravitational sections *ET7-ET2*, *ET7-ET3* and *ET7-ET4*.

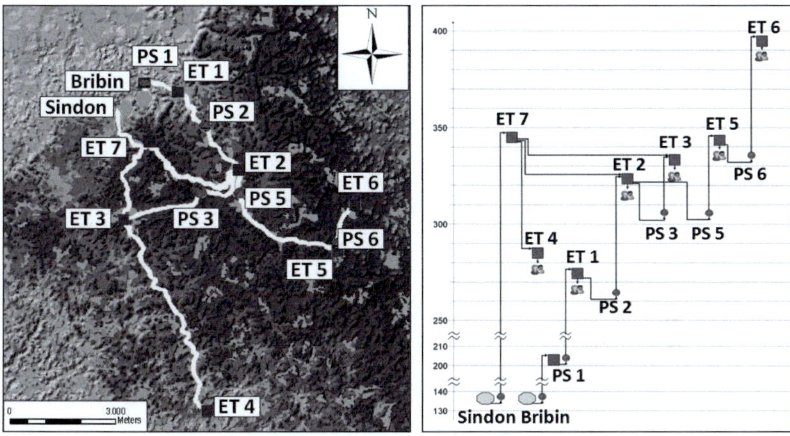

Fig. 3: Planned transmission main system (scenario B)

According to the planned operation strategy the total water production (*Bribin* and *Sindon*) of 5,637 m³/d will be distributed evenly to the consumers. Furthermore, the input volume will be maintained constantly. In times of lower discharge in the Bribin river and therefore a production of *Sindon* less than the maximum, additional water will be extracted in the old production site *Bribin*. In this study, the river discharges are assumed to be constant over a month. For each month an operation scenario is considered.

The restructuring includes rehabilitation of the system components and the installation of a Supervisory Control and Data Acquisition system which will be implemented by the industry partner IDS GmbH. All pipes are over ground, which eases the leak detection and repair processes. Thus, real water losses within the transmission main system are expected to be minimised to at least 15 % (Q_{del} = 4,791 m³/d).

Table 2 summarises the average energy demand of scenario B. The energy demand of the new water extraction *Sindon* is covered completely by its integrated hydro power plant and therefore can be excluded from the calculation. Thus, approx. 659 MWh/a are needed for the operation of the existing transmission main system. Approx. 6.0 % of the total energy demand can be provided for further use through decentral energy recovery. Related to the total volume transported to the tanks, for 1.0 m³ approx. 0.4 kWh are needed (I_{del} = 0.38 kWh/m³). The distributed volume corresponds to a per capita consumption of approx. 62 l/cap/d if water losses within the distribution networks are neglected.

Section	Pressure head	Discharge	Efficiency degree	Daily operation	Energy demand P_d	Energy demand P_a
Unit	[m]	[l/s]	[-]	[h/d]	[kWh/d]	[kWh/a]
Bribin – PS1	75.5	65.0	0.77	8.64	539	196,675
PS1 – ET1	80.4	38.0	0.78	13.27	510	186,184
ET1 – ET2	70.2	52.0	0.80	8.60	386	140,928
ET2 – ET3	57.3	8.9	0.75	17.24	114	41,712
ET2 – ET5	51.7	38.0	0.78	7.00	173	63,265
ET5 – ET6	25.5	14.3	0.76	17.70	83	30,290
Sum (P_{tot})						659,055

Tab. 2: Energy demand of scenario B - the planned system

5 Comparison

Table 3 compares the results of scenario A and B which are described in section 3 and 4. The energy demand of scenario B is significantly less than the energy demand of scenario A even so in scenario B more than five times (1 / 5.26) the delivered water of scenario A is distributed. Comparing the indicators I_{del} leads to the corresponding result. In scenario B, for the distribution of one cubic meter less than a sixth (1 / 6.45) of the energy of scenario A is needed.

Scenario	Q_{inp}	Q_{del}	Q_{cap}	P_{tot}	I_{del}
	[m³/a]	[m³/a]	[l/cap/d]	[kWh/a]	[kWh/m³]
A	794,970	332,350	11.8	813,443	2.45
B	2,057,674	1,749,023	61.6	659,055	0.38
A / B	1 / 2.56	1 / 5.26	1 / 5.26	1 / 0.81	1 / 0.16
B / A	1 / 0.39	1 / 0.19	1 / 0.19	1 / 1.23	1 / 6.45

Tab. 3: Comparison of scenario A and B

Amongst all measures the major factors for the better energy efficiency of scenario B are the hydro power plant *Sindon* and the optimised system concept.

6 Conclusion

Increasing the energy efficiency is a necessary boundary condition to realise a reliable water supply especially in developing countries. This paper reflects the specific situation of the Bribin water distribution system situated in a karst area on Java. The existing system was compared with a planned restructured and optimised system. The analysis showed that the planned improvement measures (scenario B) lead to an immense saving of energy. The major factors influencing the energy demand of the case study are the minimisation of real water losses, the optimisation of the system concept and operation and the use of regenerative hydro power to extract and pump water. The analysis is limited to the comparison of energy demands. In order to investigate the feasibility and sustainability of the scenarios, cost for improvement measures, system operation, maintenance etc. have to be considered. Therefore further studies will be carried out.

List of Literature

[1] S. Baumgarten and W. Guder. Pumpen als Turbine. Technik kompakt, KSB, Nr. 11. 2005.

[2] R. Karger, K. Cord-Landwehr and F. Hoffmann. Wasserversorgung, Teubner Verlag, 12. Auflage. 2005.

[3] P. Klingel and A. Knobloch. Anwendung hydraulischer Simulationsmodelle zur Analyse und Planung von Wasserverteilungssystemen, WasserWirtschaft 99, Heft 7-8. 2009.

[4] F. Nestmann, P. Oberle, M. Ikhwan, T. Lux and U. Scholz. Bewirtschaftung unterirdischer Fließgewässer in Karstgebieten, WasserWirtschaft 99, Heft 7-8. 2009.

[5] P. Oberle, M. Ikhwan, F. Nestmann, J. Fritz and T. Ravisundar. Angepasste Wasserfördertechnologie unter Einsatz von Pumpen im Turbinenbetrieb, WasserWirtschaft 99, Heft 7-8. 2009.

[6] J. Parker, P. Fanner and M. Feldman. Increasing Energy Efficiency of Water Supply - How to introduce carbon credits to the water sector?, International Water Association, Task Force on Water Utility Efficiency. 2008.

[7] U. Scholz, B. Unger and T. Lux. Sozioökonomische Analyse der potenziellen Wassernutzer, Bericht – unveröffentlicht, Institut für Geographie, Universität Gießen. 2004.

Mitigating the water footprint of export cut flowers from the Lake Naivasha Basin, Kenya

M.M. Mekonnen
A.Y. Hoekstra

Twente Water Centre, University of Twente, Enschede, The Netherlands
m.m.mekonnen@utwente.nl, Tel: +31(0)53/4892080

Abstract

Kenya's cut-flower industry has been praised as an economic success as it contributed an annual average of US$ 141 million foreign exchange (7% of Kenyan export value) over the period 1996-2005. On the other hand, the commercial farms have been blamed for causing a drop in the lake level and for putting the lake's biodiversity at risk. The water footprint of one rose flower is estimated to be 7-13 litres. The total virtual water export related to export of cut flowers from the Lake Naivasha Basin was 16 Mm^3/yr during the period 1996-2005 (45% blue water; 22% green water; 33% grey water). Although the commercial farms around the lake have contributed to the decline in the lake level through water abstractions, smallholder farms in the upper catchment are mainly responsible for the lake pollution due to nutrient load. In this study we propose "a sustainable-flower agreement"" which involves a water-sustainability agreement between major agents along the cut-flower supply chain and includes a premium to the final product at the retailer end of the supply chain.

1. Introduction

Lake Naivasha is Kenya's second largest freshwater lake without surface outlet and the natural fluctuation in water levels over the last 100 years has been in excess of 12 meters (Mavuti and Harper, 2005). The lake remains fresh due to a significant exchange of groundwater. The lake has international value as a Ramsar wetland. Since the late 1990s, flower farms started to expand at a faster rate (Becht et al., 2005). The total irrigated commercial farm area around Lake Naivasha is about 4450 ha. Cut flowers account for about 43% of the irrigated area, followed by vegetables with 41% and fodder with 15% (Musota, 2008).

The objective of this study is to quantify the water footprint within the Lake Naivasha Basin related to horticulture, in particular the flower farms, and assess the potential for mitigating this footprint by involving cut-flower traders, retailers and consumers overseas. In addition, we will explore the idea of a voluntary sustainable-flower agreement between major agents along the flower supply-chain that involves a water-sustainability premium to be paid by the consumers in the countries importing flowers from Kenya.

2. Method and data

The green, blue and grey components of the water footprint of the products were calculated following the method of Hoekstra and Chapagain (2008) and Hoekstra et al. (2009). The water footprint of a crop (m^3/ton) is calculated as the ratio of the volume of water (m^3/ha) consumed or polluted during the entire period of crop growth to the corresponding crop yield (ton/ha). The crop water requirements (*CWR*) and irrigation requirement for the different vegetables and cut flower grown around Lake Naivasha were estimated using FAO's CROPWAT model. For the other 22 crops grown in upper catchment of the Lake Naivasha Basin, a crop water use model (Mekonnen and Hoekstra, 2010) was used to carry out daily soil water balance and calculate the green, blue and grey water footprint at grid level. Virtual water export (m^3/yr) related to trade in cut flowers and vegetable products were calculated by multiplying the trade volumes (tons/yr) by their respective water footprint (m^3/ton).

The Lake Naivasha Basin can be schematised into two parts: the upper catchment part with smallholder farms and the part around Lake Naivasha with big farms producing for export. Grid data on type and size of farms around Lake Naivasha was obtained from the ITC Naivasha database (Becht, 2007). For crops grown in the upper catchment, the crop growing areas were obtained from Monfreda et al. (2008). Data on irrigated area per crop was obtained mainly from Portmann et al. (2008). Grid-based soil moisture data was taken from ISRIC-WISE (Batjes, 2006). The crop parameters (crop coefficients and start and length of cropping seasons) for cut flowers and other crops were taken from Orr and Chapagain (2006) and Chapagain and Hoekstra (2004). Average monthly reference evapotranspiration were obtained from FAO (2008). Monthly values for precipitation, wet days, minimum and maximum temperature were obtained from CRU (Mitchell and Jones, 2005). Daily precipitation values were generated from these monthly average values using the CRU-dGen daily weather generator model (Schuol and Abbaspour, 2007). The commodity trade flows have been taken from the International Trade Centre (ITC, 2007)

3. Water use within the Lake Naivasha Basin related to cut-flower production

Table 1 show the water footprint related to crop production for two groups of crops: fully irrigated crops grown by commercial farms mainly for export and concentrated around Lake Naivasha, and other crops which are cultivated by small farmers in the upper catchment. The total water footprint related to crop production sums up to 102 Mm^3/yr. About 69% of the water footprint is related to green water, 13% grey water and 19% blue water. The commercial crops contribute 41% to the total water footprint related to crop production. About 98% of the blue water footprint can be attributed to the commercial farms around the lake, while about 61% of the grey water footprint can be attributed to agriculture activities carried out in the upper catchment.

Land use	Area cultivated		Water footprint (1000 m^3/yr)			
	Area (ha)	Irrigated (%)	Green	Blue	Grey	Total
Commercial farms around the lake						
Cut flower	1911	100	3640	7576	5627	16842
Vegetables	1824	100	7887	7375	1834	17097
Fodder	665	100	3716	3194	452	7362
Macadamia	50	100	278	303	34	615
Total of commercial farms	4450	100	15521	18448	7947	41916
Farms in the upper catchment of the basin						
Cereals	12125	1	34776	82	1655	36513
Pulses	2199	0	3958	0	2673	6631
Others	3813	7	15876	382	809	17067
Total of upper catchment farms	18137	2	54609	465	5137	60211
Grand total	22587	21%	70130	18913	13084	102127

Table 1. Water footprint of crops grown in the Lake Naivasha Basin. Year 2006.

Cut flowers take the major share of the water footprint related to crop production around Lake Naivasha accounting over 50% of both the blue and total water footprint. The production water footprint related to cut flowers is about 16.8 Mm^3/yr (Table 1). Flowers grown in greenhouse are assumed to be fully supplied with irrigation water while flowers on open field get both rain water and irrigation water. The average water footprint of cut flowers grown in Naivasha is 367 m^3/ton. About 45% (165 m^3/ton) of this water footprint refers to blue water, 22% (79 m^3/ton) to green water and 33% (123 m^3/ton) refers to the polluted water due to leaching of nitrogen fertilizer into the water systems.

Depending on the yield and weight of a rose flower stem, the water footprint per stem varies from 7 to 13 litre/stem. If we assume an average rose flower stem weights about 25 gram, its green water footprint would be 2 litre/stem, its blue water footprint 4 litre/stem and its grey water footprint 3 litre/stem or a total water footprint of 9 litre per stem.

The average virtual water export from the Lake Naivasha Basin related to export of cut flowers was 16 Mm^3/yr in the period 1996-2005 (45% blue water, 22% green water and 33% grey water). The European Union remains Kenya's principal market for cut flowers. The Netherlands is the principal market accounting for 69% of the total export, followed by the UK with 18% and Germany with 7%. The virtual water export related to vegetable products was 8.5 Mm^3/yr. Therefore, for the period 1996-2005 the total virtual water exported related to export in cut flowers and vegetable products was 24.5 Mm^3/yr.

The cut flower industry is an important export sector, contributing about US$ 352 in 2005 alone and an annual average of US$ 141 million foreign exchange [7% of Kenyan export value] over the period 1996-2005.

Most of the year except the high flow month of May the blue water footprint is above the usable runoff as shown in Figure 1. Such high variability in the water regime is very relevant for the water allocation decisions.

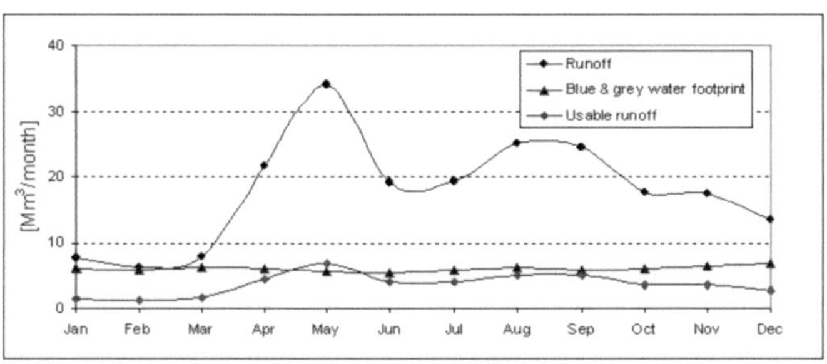

Figure 1. Long term average monthly runoff, blue water footprint and environmental flow of the Lake Naivasha Basin. Source: runoff from the ITC Naivasha database – Becht, 2007; blue water footprint from own calculation.

4. Reducing the water footprint in the Lake Naivasha Basin: involving consumers, retailers and traders along the supply chain

Implementation of the 2007 Water Resource Management Rules is actually hampered by reluctance of many water users to follow the regulation and difficulties government encounters in enforcing the regulation. The current water pricing policy has several weaknesses including: illegal water abstractions and in practice it is difficult for the government to actually check whether farmers have actually installed water meters as legally required. The flower farms feel they are already overtaxed and burdened with a number of remittances and some even have threatened to relocate to Ethiopia if local authorities force them to pay more tax. Attracted by a number of incentives five major flower companies have already made the switch to Ethiopia with more to follow (ARB, 2007). Despite the fact that farmers have indicated that the newly introduced water tariff is too high, the tariff actually does by far not cover full economic cost of the water. As a result, the funds generated by the current water pricing scheme are very small. The level of water price increase that would be required to have a significant impact on demand would be politically very difficult to enforce.

Under such conditions, the implementation of full-cost pricing at the source is not feasible. A unilateral implementation of stringent pricing strategy by a country would affect the competitiveness of its local companies in the global market (Hoekstra, 2006). To address this problem, Hoekstra (2006) have proposed national governments negotiate on an international protocol on water pricing.

As an alternative to the international protocol involving national governments we propose an alternative that can be implemented with a focus on sustainable water use in flower farming around Lake Naivasha alone. The proposal involves a water-sustainability agreement between major agents along the cut-flower supply chain and includes a premium to the final product at the retailer end of the supply chain.

The agreement should include two key ingredients: a fund-raising mechanism at the consumer-end of the supply chain, which will raise the funds for making water use by flowers sustainable, and a labelling or certification scheme, which will provide the guarantee that the funds are properly spent and that the flower production actually moves in the direction of sustainable water use.

The premium collected when selling cut flowers from the Lake Naivasha Basin to consumers should be used to invest in better watershed management and, most in particular, in reducing the water footprint of the flower farmers. Adding a water sustainability premium to the final product at the consumer end of the supply-chain, will yield more than 200 times the amount of money raised than through local water pricing at the current water abstraction fee in Kenya.

There is a need to provide institutional infrastructure through which the funds could flow back to the basin and be used in environmental protection, watershed management, support of small farmers to improve their water management and community development.

Collecting a water-sustainability premium at the lower end of the supply chain needs to go hand in hand with a mechanism for certification of the farmers that deliver the premium-flowers and for labelling the premium-flowers. Labelling can be interpreted here in physical sense – where indeed a consumer-oriented label is attached to a flower – but it can also get the shape of 'attached information' to whole batches of flowers.

There are different options for which parties could engage in making a water-sustainability agreement. In its most modest form, an agreement could involve one retailer, one trader and one of the major farmers. In a more ambitious setting, several retailers, traders and farmers could be involved. Retailers, traders or farmers could also be represented by their respective branch organisations. In the case of the flower farmers this could be the Lake Naivasha Growers Group or the Kenyan Flower Council. In the Netherlands, the flower market is organised by FloraHolland, which may take a central role in facilitating an agreement.

Success of the water-sustainability premium depends on all stakeholders' commitment to reach agreement and implement it. Further, a clearly defined certification procedure and institutional arrangements for the flow of fund back to the basin is required.

5. Discussion

Horticulture export is the second export earner after tea by contributing 27% to the total exports value in the period 1996-2005. Cut flowers export account over 50% of Kenyan horticultural export value and its overall contribution to the country's export earnings is growing rapidly. Kenya's cut flower industry is too big to fail. If it dies out the consequences would be disastrous. Beside the contribution to the countries export earning and GDP, the commercial farms provide employment, housing, schools and hospitals provided free to employees and their families. Losing the cut flower business means over 25,000 workers and their dependence will be facing the prospect of losing everything. On the other hand the treatment of Lake Naivasha as a 'common pool' resource will have dire consequences both on the Lake's sustainability and business image of the commercial farms. Therefore, sustainable management of Lake Naivasha requires:

- Decide on the allowable drop in the lake water level and the amount of water that can sustainably be abstracted from the lake,
- Price water at its full-cost to create incentive for the sustainable use of water. Related with this is the creation of tradable water permit,

- Encourage the use of greenhouse flower production coupled with rainwater harvesting,
- Discourage production of water-intensive products such as beans and low value products such as fodder and grass around the lake and encourage the use of rain water for the production of fodder and grass in the upper catchment,
- Reduce the flow of sediments and agricultural nutrients to the lake both from commercial farms around the lake and farms in the upper catchment.

The water sustainability premium to final product in retail shop would generate more funds for watershed management and create awareness among consumers on the value of water. In addition, the water sustainability premium approach will reduce the risk of Kenya's losing its business and the effect on the farms competitiveness of unilateral water pricing by Kenya. In addition, an added value of the water-sustainability premium includes an aspect of fairness through cost sharing among producers and consumers, it can enhance the green image of the commercial farms and increase chances in the market for sustainable products. Successful implementation of the water-sustainability premium to final products at the retailer depends on the commitment of all stakeholders: governments, civil society organizations, private companies and consumers.

References

[1] ARB (2007) Flowers: Kenya, Africa Research Bulletin 43(11): 17197A-17198A.

[2] Batjes, N. H. (2006) ISRIC-WISE derived soil properties on a 5 by 5 arc-minutes global grid. Report 2006/02, ISRIC – World Soil Information, Wageningen, The Netherlands.

[3] Becht, R. (2007) Environmental effects of the floricultural industry on the Lake Naivasha basin, unpublished paper, ITC Naivasha Database, Enschede, The Netherlands.

[4] Becht, R., Odada, O., and Higgins, S. (2005) Lake Naivasha: Experience and lessons learned brief. In: ILEC, 2005, Managing lakes and their basins for sustainable use: A report for lake basin managers and stakeholders, International Lake Environment Committee Foundation. Kusatsu, Japan.

[5] Chapagain, A.K. and Hoekstra, A.Y. (2004) Water footprints of nations, Value of Water Research Report Series No. 16, UNESCO-IHE, Delft, The Netherlands.

[6] FAO (2008) Global map of monthly reference evapotranspiration - 10 arc minutes. http://www.fao.org/geonetwork/srv/en/resources.get?id=7416&fname=ref_evap_fao_10min.zip&access=private (15 Oct, 2008)

[7] Hockstra, A.Y., Chapagain, A.K., Aldaya, M.M., Mekonnen, M.M. (2009) Water footprint manual: state of the art, Water Footprint Network, Enschede, The Netherlands.

[8] Hoekstra, A.Y. (2006) The global dimension of water governance: nine reasons for global arrangements in order to cope with local water problems, Value of Water Research Report Series No. 20, UNESCO-IHE, Delft.

[9] Hoekstra, A.Y. and Chapagain, A.K. (2008) Globalization of water: Sharing the planet's freshwater resources, Blackwell Publishing, Oxford, UK.

[10] ITC (2007) SITA version 1996–2005 in SITC, [DVD-ROM]. International Trade Centre, Geneva.

[11] Mavuti, K.M., Harper, D.M. (2005) The ecological state of Lake Naivasha, Kenya, 2005: Turning 25 years research into an effective Ramsar monitoring programme (Accessed 25 Sept., 2008) http://iodeweb1.vliz.be/odin/bitstream/1834/2127/1/WLCK-30-34.pdf

[12] Mekonnen, M.M. and Hoekstra, A.Y. (2010) A global and high-resolution assessment of the green, blue and grey water footprint of wheat, Value of Water Research Report Series No.42, UNESCO-IHE, Delft, The Netherlands.

[13] Mitchell, T. D. and Jones, P. D. (2005) An improved method of constructing a database of monthly climate observations and associated high-resolution grids. International Journal of Climatology, 25, 693-712.

[14] Monfreda, C., Ramankutty, N., Foley, J.A. (2008) Farming the planet: 2. Geographic distribution of crop areas, yields, physiological types, and net primary production in the year 2000, Global Biogeochemical Cycles, Vol.22, GB1022, doi:10.1029/2007GB002947.

[15] Musota, R. (2008) Using WEAP and scenarios to assess sustainability of water resources in a basin. Case study for Lake Naivasha catchment-Kenya, M.Sc. Thesis, ITC, Enschede, The Netherlands.

[16] Orr, S. and Chapagain, A. (2006) Virtual water: a case study of green beans and flowers exported to the UK from Africa, Fresh Insights No. 3.

[17] Portmann, F., Siebert, S., Bauer, C. & Döll, P. (2008) Global data set of monthly growing areas of 26 irrigated crops. Frankfurt Hydrology Paper 06, Institute of Physical Geography, University of Frankfurt, Frankfurt am Main, Germany,

[18] Schuol, J. and Abbaspour, K.C. (2007) Using monthly weather statistics to generate daily data in a SWAT model application to West Africa. Ecological Modelling 201: 301–311.

Capacity Development for Integrated Water Resources Management (IWRM) in a transition country

-Improving River Basin Management in the Western Bug Basin, Ukraine-

M. Leidel[1]
S. Niemann[2], N. Hagemann[3]
[1]*Technische Universität Dresden, Department of Hydrosciences, 01062 Dresden, Germany, Email: marco.leidel@tu-dresden.de, Phone: +49 (0)351 46331282*

[2]*Helmholtz Centre for Environmental Research – UFZ, Department Aquatic Ecosystems Analysis and Management (ASAM), Brückstraße 3a, 39114 Magdeburg, Germany, Email: steffen.niemann@ufz.de, Phone: +49 (0)391 8109653*

[3]*Helmholtz Centre for Environmental Research – UFZ, Department of Economics, Permoserstr. 15, 04318 Leipzig, Germany, Email: nina.hagemann@ufz.de, Phone: +49 (0)341 2351734*

Abstract

Even if the paradigm of Integrated Water Resources Management (IWRM) is representing nowadays the basis for the water sector, the implementation is still in its infancy.
IWRM deals with dynamic adaptive systems that comprise human, environmental and technical factors and their interactions, so that implementing IWRM means more than integrating all facets of water management. Good (water) governance and thus a strong civil society as well as stakeholder participation are important additional factors for making IWRM operational. Yet, many countries struggling for the implementation of IWRM, and especially transition countries like Ukraine, do to a certain extent lack such factors. Therefore, the respective society needs capacities, in order to start such processes and keep them running. Hence, Capacity Development (CD) is required that strengthens the capabilities of the water management itself, but also the society as a whole.
As one of the first steps towards improving River Basin Management in the Western Bug River Basin, besides an analysis of the water system, an analysis of the political and social situation was conducted. According to those analyses, a capacity assessment was accomplished and tailor-made measures to strengthen the available competencies are developed.
Based on the outcomes and continuous work of the International Water Research Alliance Saxony (IWAS), a project funded by the German Federal Ministry of Education and Research (BMBF), this paper shows one possible approach to support the implementation process of IWRM.

Keywords

Integrated Water Resources Management (IWRM), Implementation, River Basin Management, Capacity Development, Good Governance, Ukraine

1 Introduction

The pressure on the world's water resources is steadily increasing and therefore, the competition for the limited water resources is rising as well. To meet these challenges, a water management approach is required that is, putting it simply, balancing the human needs and the ecological requirements. That means that the prevailing sectoral approach to water management has to be shifted towards a more interdisciplinary and therefore holistic one. The concept of Integrated Water Resources Management (IWRM) acknowledges this, however, it has to be recognized that IWRM is a process of change and not a perfect system that can be implemented at once. The prevalent IWRM-guidelines have to be adapted to the specific country or region in which IWRM is applied, i.e. it has to be customized for the specific and unique requirements of the concerned region. Therefore, IWRM has to put special attention to a well elaborated and adapted Capacity Development concept. However, there is no doubt that the planning and implementation is a difficult and longsome process. In 2007, UN-Water and the Global Water Partnership (GWP) [1] claimed that a roadmap accompanying established national water plans will be needed, so that the focus would shift from the planning to the concrete implementation process.

2 Capacity and Capacity Development for Integrated Water Resources Management

Yet, the worldwide implementation of IWRM is still dissatisfying according to UN-Water [2]. Making IWRM operational and effective needs more than integrating all aspects of water management. According to [3], it is assumed that measures for solving existing water problems can be only sustainable and effective in the long term, if the generated knowledge about possible solutions will be deeply rooted within the particular region/ country. Thus, the respective society needs capacities, in order to start the IWRM process and to keep it running. According to [4], capacity can be defined as the capability of a society to identify, understand and address problems, to learn from experience and to accumulate knowledge for future issues. Thus, CD is „*a much broader concept that refers not merely to the acquisition of skills (individual), but also to the capability to use them (institutional), which requires access to the necessary resources, the right framework conditions, etc.*" [5]. Accordingly, the international CD-research often distinguishes between different levels of CD measures, namely between the individual and organisational level and the *enabling environment* (following the multi-level approach of [4], [6] and 5th World Water Forum (2009). *Figure 1* shows that such three levels are in many ways intertwined. Thus, CD- measures addressing one particular level do often include and integrate consecutive influences on other ones.

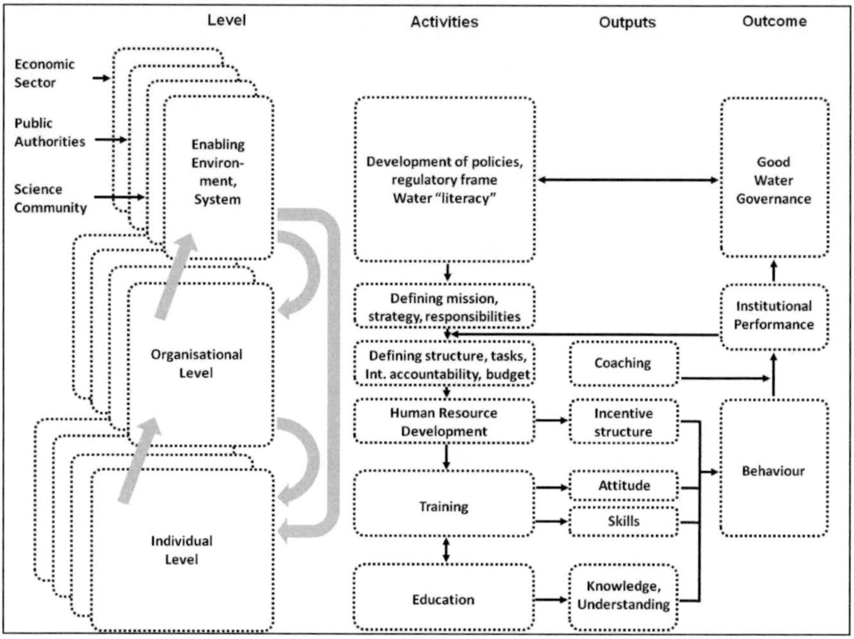

Modified after VAN HOFWEGEN (2004), 5th World Water Forum (2009) and ALAERTS (2009)

Fig.1: Multi-level approach

A reasonable procedure how to elaborate sustainable CD was developed by UNDP [7]. Regarding this process, a general target definition together with a stakeholder and institutional analysis should be conducted as a first step. Those steps are prerequisites for the capacity assessment. According to such an assessment, the identification of already existing and missing competencies towards an improved water management is possible.

Eventually, the capacity assessment leads to the development of a CD-strategy including the planning of CD-measures (CD responses). The next step is the implementation of the proposed CD-measures followed by monitoring and evaluation activities. The evaluation should then, if necessary, lead to the adaptation of the CD-strategy and eventually another CD-cycle has to proceed.

3 Capacity Development in the International Water Alliance Saxony (IWAS)

Reflecting the situation elaborated above, the concept of IWRM with a strong CD constitutes also the framework for the International Water Research Alliance Saxony – IWAS[1], which was initiated by the Helmholtz Centre for Environmental Research – UFZ, the Technische Universität Dresden and the Stadtentwässerung Dresden and is funded by the German Federal Ministry of Education and Research (BMBF). In order to face the challenge set by the UN in the MDGs[2], more than 50 scientists develop, together with partners from industry and with political decision-makers, specific solutions to particular water related problems in five regions worldwide. A variety of water-related problems is addressed, for instance the improvement of the surface water quality with regard to the EU- Water Framework Directive in East European countries.

Additionally, CD is of major importance within IWAS, since the significance of CD for a sustainable implementation of the developed system solutions, adaptive strategies, measures and methods in the respective regions was recognized. Further, IWAS acknowledges that CD is much more than education and training. Therefore, general guidelines for the Capacity Development and the knowledge transfer within IWAS have been elaborated, which constitute the basis for individual CD strategies in the regional projects. The concepts for the individual model regions are adapted to the respective goals and necessities according to the above mentioned CD-process from the UNDP [7].

Moreover, measures are elaborated, which are applicable worldwide, respectively in all model regions. One of those measures is the development of an IWRM E-learning module together with the German National Committee for the International Hydrological Programme of UNESCO and the Hydrology and Water Resources Programme of World Meteorological Programme (WMO). The target groups of this module will be the water related administration, academic instructors and post-graduate students. The core of the

[1] http://www.iwas-sachsen.ufz.de
[2] Millennium Development Goals

module includes the most important concepts and facts about IWRM. Yet, since IWRM is always regional specific, there will be an additional part of the module, which will reflect this fact and thus will show regional specific aspects of IWRM.

3.1 CD for improving River Basin Management in the Western Bug Basin, Ukraine

Among others, Eastern Europe and especially the Western Bug River Basin, Ukraine, is one out of five model regions within IWAS. In order to adapt the general CD guidelines to the Ukrainian circumstances, an evaluation of the social and political situation, including an analysis of relevant stakeholders, was conducted as one of the first steps towards improving River Basin Management (RBM). Such an evaluation is especially important for a transition country like Ukraine that still faces institutional shortcomings such as partial absence of the rule of law and overlapping institutional responsibilities. It has to be taken into account that the post communist transformation path differs significantly from other countries in several aspects [8]. Therefore there are no experiences for CD strategies to be built upon so far. Several international projects have already been initiated to analyse environmental problems in Ukraine but they missed to take the local situation such as the social, political and cultural environment into account and ended only in technical assistance [9]. One major challenge for RBM in Ukraine is the unstable political situation, which results in short term oriented political actions and also involves frequent changes in staff. As a consequence, reliable partners have to be found to ensure sustainable long term cooperation. Civil society actors as multipliers within the IWRM-process are not sufficiently developed and especially environmental NGOs lack substantial support. In general, people are not much engaged in civil society activities [10]. The lack of awareness and interest in environmental problems at societal but also political level shows the urgent need of an all-encompassing CD process for improving RBM.

Based on the outcomes of this societal analysis, a capacity assessment workshop (needs assessment) with key stakeholders was conducted, in order to identify expected difficulties as well as already existing competencies towards an operational IWRM. According to that situation analysis, tailor made measures for an improved river basin management are currently developed at the individual level integrating all three aspects of learning (knowledge, skills and attitude). The target groups for the proposed measures are the scientific community, as well as the public authorities (Environmental administration, water authorities, River Basin Organisation) and the economic sector (municipal water and wastewater companies). The currently developed and already partially executed CD measures are of very different kinds, such as the vocational training of people working in the water sector, knowledge exchange, and the exchange of scientific staff. Another measure within this model region is the support of scientific education and curricula development, e.g. the development of series of lectures for IWRM in transition countries. Those lectures will be designed according to the above mentioned IWRM E-learning module that is jointly developed with UNESCO/WMO.

Crucial for the success of such measures is the elaboration of functioning networks in both the academic and the practical arena in order to ensure the mutual exchange of experiences. Thus, for instance, a network of Ukrainian wastewater companies ("vodakanal") is currently established comparable to German water associations.

4 Summary and perspectives

The concept of IWRM is currently the frame for water management. However, the implementation process is, except for some success stories like the European Water Framework Directive (WFD), not accomplished yet.

Therefore, the essential point should be the development of measures for implementing operational IWRM. One important fact for implementing IWRM is CD since knowledge for starting and maintaining the IWRM-process is often not adequate. For this reason, a well devised CD process is needed that is adjusted with the IWRM process. Capacity development responses that accord with such strategies are therefore by all means rational no matter how big or small they are.

Especially in Ukraine, knowledge on environmental problems and problem solving approaches is missing. Therefore, the local stakeholders have to be trained and educated for improving RBM. However, the institutional structures and the enabling environment have to be improved as well, since otherwise, the gained knowledge of the stakeholders may not be applied and implemented. Thus, the current political, economic and social situation needs to be taken into account, as well as the specific characteristics of a transition country.

Hence, it can be stated that IWRM and capacity development cannot stand alone if we pursue sustainable and long-lasting development.

List of Literature

[1] **UN-Water and GWP (2007):** UN-Water and Global Water Partnership (GWP) Roadmapping for Advancing Integrated Water Resources Management (IWRM) Processes.

[2] **UN-Water (2008):** Status Report on IWRM and Water Efficiency Plans for CSD16.

[3] **Leidel, M., Niemann, S. (2009):** Internationale WasserforschungsAllianz Sachsen (IWAS)-Capacity Development Leitlinien. http://www.iwas-sachsen.ufz.de/data/IWAS-CD-Leitlinien11658.pdf.

[4] **Alaerts, G.J. (2009):** Knowledge and capacity development (KCD) as tool for institutional strengthening and change. In: Alaerts, G.J., Dickinson, N (Eds.) Water for a changing world - Developing local knowledge and capacity. Delft.

[5] **Vollmer, R., Ardakanian, R., Hare, M., Leentvaar, J., van der Schaaf, Ch., Wirkus, L., UN-Water Decade Programme on Capacity Development (2009):** Institutional Capacity Development in Transboundary Water Management. Paris.

[6] **Van Hofwegen, P. (2004):** Capacity-building for water and irrigation sector management with application in Indonesia. FAO Water Reports 26. Rome: FAO.

[7] **UN Development Programme (2008):** Capacity Development Practice Note.

[8] **Maćków, J. (1999):** Der Wandel des kommunistischen Totalitarismus und die postkommunistische Systemtransformation: Periodisierung, Problematik und Begriffe. In: Zeitschrift für Politikwissenschaft 9, 4/1999: 1347-1380.

[9] **Nazarov, N., Cook, H., Woodgate G. (2004):** Water pollution in Ukraine: the search for possible solutions. In: International Journal of water resources development 20 (2). 205-218).

[10] **Hutcheson, D. S., Korosteleva, E. A. (2006):** Patterns of Participation in Post-Soviet Politics 4: 23-46.

Capacity building activities of SMART Research Program as an important tool for an integrated water resource management strategy for the Jordan River Valley (JRV)

Wasim ALI (wasim.ali@agk.uka.de), Heinz. HOETZL[1] (Heinz.hoetzl@agk.uka.de), Leif WOLF [1](leif.wolf@agk.uka.de), Manfred VANAFFERDEN [2] (manfred.afferden@ufz.de), Jaime CARDONA [2] (jaime.cardona@ufz.de, David RIEPL [1] (David.Riepl@agk.uni-karlsruhe.de),

1) Karlsruhe Institute of technology (former Karlsruhe University) Karlsruhe, Germany,
2) UFZ, Leipzig, Germany

Corresponding author: **Associate Prof. Dr. Wasim Ali**, SMART Research Project, Institute of Applied Geosciences, Hydrogeology Department, KIT Karlsruhe Institute of Technology, former Karlsruhe University, Germany, Tel: 0049 721 608 2758, Fax: 0049 721 606 279, E-mail: (wasim.ali@kit.edu)

Key words: Capacity building and Training activities of SMART project, PhD program, Water fun and other activities, Water Resources Management Concept, Lower Jordan Rift Valley (LJV)

Abstract

One of the main objectives of SMART research program, funded by the German Federal Ministry of Science and Research (BMBF), is to achieve an integrated water resource management strategy concept for the Jordan River Valley (JRV).

The wastewater training program in Jordan and Palestine was based on demonstration, technology transfer in the field of Decentralized Wastewater Systems Solutions (DWWSS) and was designed, developed and tested as a pilot teaching unit for primary schools in Ramallah, Palestine and Al-Salt, in Jordan. The training enhances the reflection process about wastewater components and gives arguments to the participants to understand the possibilities of wastewater treatment and its reuse purposes in decentralized systems.

Another training program dealing with protection zones for springs and wells was carried out in Ramallah, Palestine. This workshop for water experts stressed the vital need of protection zones regulations to protect extracting ground water wells and springs from pollution.
To enhance intensive scientific and cultural exchange of the students from the participating countries, within the SMART project a PhD program was launched in SMART 1 (2006 – 2009) and in SMART 2 (2010 – 2013). The young scientists profited from this vital and important program. This program was in SMART 1 a very successful tool for technology transfer.

For promoting and sharing the results with the associated project partners and informing not only internal users, but also the public (e.g. stakeholders in the region), a project website was established. The website provides current information about research and implementation activities and results of the work packages, information concerning the project meetings and Project Status Reports.

INTRODUCTION

One of the main activities of SMART research project (s. Fig. 1) is to achieve an integrated water resource management strategy concept for the Jordan River Valley (JRV). Both Work package 11 in phase I and WP 9 in phase II deals with training and capacity building activities, which play a major role in the sustainable management of available water resources concepts. The objectives of the two WPs are in addition to the organizing of project workshops open to interested people, technology and know-how transfer and training activities in the LJV region.

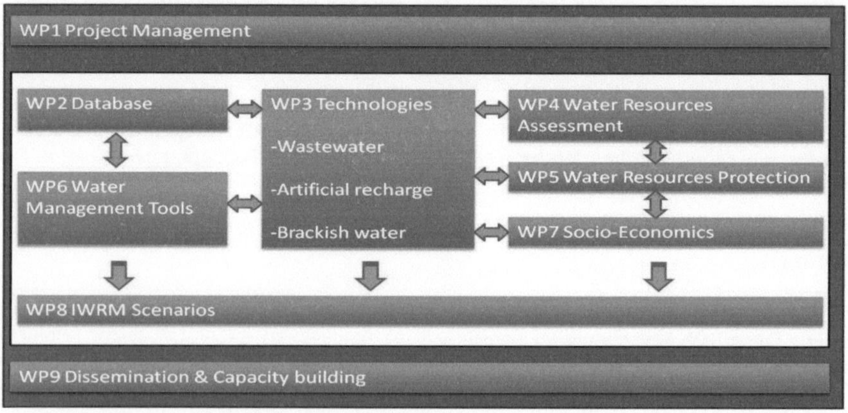

Figure 1: Work packages in SMART Research Project (Phase II), *from project proposal SMART II, 2009*

The training and capacity building program in Jordan and Palestine was based on demonstration, technology transfer in the field of Decentralized Wastewater Systems Solutions (DWWSS) and was designed, developed and tested as a pilot teaching unit for primary schools in Ramallah, Palestine and Al- Salt, Jordan. The training enhances the reflection process about wastewater components and gives arguments to the participants to understand the possibilities of wastewater treatment and its reuse purposes in decentralized systems. Another training program about protection zones for springs and wells was carried out in Ramallah, Palestine. This workshop for water experts stressed the vital need of

protection zones regulations to protect extracting ground water wells and springs from pollution. A PhD program in was launched To enhance intensive scientific and cultural exchange of the students from the participating countries, within the SMART project. The young scientists profited from this vital and important program. This program was in addition a very successful tool for technology transfer.
For promoting and sharing the results with the associated project partners and informing not only internal users, but also the public (e.g. stakeholders), a project website was established. The website provides current information about research and implementation activities and results of the work packages, information concerning the project meetings and Project Status Reports.

Workshop Stakeholder meeting, Ramallah, Palestine: Training und Capacity building need (November 19. 2008)

The focus of that meetings was to define the needs of a training and capacity-building program and to introduce a school training course on protection zones for springs and wells and "Water fun" course for school about the reuse of waste water. This workshop took place on November 19. 2008. Local stakeholders from the Ministry of Education, the Palestinian Water Authority PWA and the Palestinian Hydrologic Group PHG attended the workshop (s. Photos, fig. 2). The meeting was an opportunity to introduce SMART project to important authorities in Palestine as well as to initiate an approach to the teachers designated to attend the training. During the meeting the training program "Water Fun" within the framework of the SMART project was introduced. The meeting provided also the opportunity to contextualize the capacity-building program into the local water resources problematic. During this meeting the PHG organized a presentation concerning the wastewater problematic in Palestine and the role of alternative technologies to treat grey-water.

Figure 2: Photo left : Palestinian Water Authority meeting, Ramallah 19th November 2008 with participants from Sakeholder and SMART project From right to left side, Mr. Azed Asstor delegate from the Ministry of Education of Palestine, Dr. Marwan Ghanen , Palestinian Hydrologic Group PHG, Dr. Wasim Ali , SMART project, University of Karlsruhe, Ing. Aymen Jarrar, Palestinian Water Authority PWA, Photo right: Dr. Manfred Van Afferden, UFZ, Leipzig, Germany explain the activities in grey water field (*photos © Wasim Ali and Manfred Van Afferden)*

PhD students program

A PhD program was launched between the years 2006 – 2009 allowing students from the participating countries within the SMART project, phase I to participate. The young scientists profited from this vital and important program. This program was in addition a very successful tool for technology transfer.

A. PhD Thems
B. PhD student meeting

PhD Thems

The following table 1. include the names of the PhD students and the themes of their PhD-work.

Table 1: names of PhD students participating in SMART PhD Program and and the these of their PhD-work

Family Names	First name	Themes
ALFARA	AMANI	USING IWRM MODELS AND CONCEPTS TO EXPLORE THE FUTURE OF TREATED WASTEWATER IN JORDAN VALLEY AGRICULTURE
AL-MADBOUH	SUHA	SOCIAL ACCEPTABILITY OF WASTEWATER REUSE IN AGRICULTURE: THE CASE OF THE LOWER JORDAN VALLEY (LJV), PALESTINE
AL-KHOURY	WILIAM	RAINFALL-RUNOFF RELATIONSHIP IN MICROSCALE WADIES IN A SEMI ARID ENVIRONMENT / A CASE STUDY FROM WADI KAFREIN IN JORDAN
AL-SAQARAT	BETTY	MICROSTRUCTURES IN THE UPPER CRETACEOUS ROCKS OF JORDAN AND THEIR DEFORMATION MECHANISMS USING GEOTECHNICAL, MINERALOGICAL AND GIS TECHNIQUES
AWAD	IBRAHIM	TOWARD EFFICIENT, EQUITABLE AND SUSTAINABLE MUNICIPAL WATER SUPPLIES FOR DOMESTIC PURPOSES IN THE WEST BANK: A CONTINGENT VALUATION ANALYSIS
INBAR	NIMROD	THE EVAPORITIC SUBSURFACE BODY IN KINNAROT BASIN: STRUCTURE AND SUB-SURFACE HYDROGEOLOGICAL RELATIONSHIP
JASEM	HIND	GROUNDWATER VULNERABILITY IN WADI KAFRAIN CATCHMENT AREA AND SURROUNDINGS
LASTER	YUVAL	WATER-ENERGY NEXUS AND THE NEW WATER & ENERGY GEOGRAPHY IN ARID AND SEMI ARID ZONES
NIVALA	JAIME	SUBSURFACE-FLOW TREATMENT WETLANDS: THE EFFECT OF DESIGN CONFIGURATION ON TREATMENT PERFORMANCE, HYDRAULIC EFFICIENCY, AND WATER LOSS
RIEPL	DAVID	KNOWLEDGE MANAGEMENT FOR IWRM DECISION SUPPORT
SAHAWNEH	JULIA	STRUCTURAL CONTROL OF HYDROLOGY, HYDROGEOLOGY AND HYDRODROCHEMISTRY ALONG THE EASTERN ESCARPMENT OF THE JORDAN RIFT VALLEY
SAMHAN	SUBHI	OCCURRENCE AND TRANSPORT OF TRACE METAL AS ANTHROPOGENIC POLLUTANTS IN SEDIMENTS AT AL-QILT CATCHMENTS / JORDAN VALLEY, WEST BANK

SCHMIDT	NATALIE	ELIMINATION OF ANTHROPOGENIC TRACE ORGANICS AND PATHOGENIC MICROORGANISMS DURING SOIL PASSAGE
SORGE	SABINE	OPERATIONS AND FINANCING MODELS FOR DECENTRALISED WASTEWATER TREATMENT TECHNOLOGIES INCLUDING WASTEWATER REUSE IN ARID/SEMI-ARID COUNTRIES WITH MARKET DISTORTION
ZEMANN	MORITZ	TRANSPORT AND DISPERSION OF MOBILE TRACE ORGANICS AND PHARMACEUTICALS IN AQUIFERS AFTER INTENDED AND UNINTENDED INFILTRATION

The majority of the PhD student started with their PhD work with the beginning of SMART project. Phase I in September 2006. Some started later in year 2007 and only 1 PhD student started late in year 2008.

B PhD student meeting, 30. March – 01. April 2009 in Karlsruhe

A PhD students meeting of SMART project took place in Karlsruhe between the 30. March to 04. April 2009 including:

- Workshop (2 days): 31. March - 1. April 2008
- Field trip (2 days): 2.-3. April 2009

The field trip for 2 days took place in South Germany (Black forest, Kaiserstuhl, Landau tunnel test site, Danube springs, Rheinwaterfall) and the northern part of Switzerland. The workshop started with presentations in the morning session from invited guests dealing with research activities similar to SMART project activities. The field trip for 2 days included visits to South Germany (Black forest, Kaiserstuhl, Landau tunnel test site, Danube springs, Rhein waterfall and the northern part of Switzerland. Prof. Heinz Hoetzl guided with the assistance of Dr. Wasim Ali (both KIT) the geological, hydrogeological, structural and engineering- geological field trip.

Figure 3: Photo left : different PhD students discussing their results during the PhD meeting Photo right: Prof. Heinz Hötzl main scientific coordinator of SMART project and Dr. Wasim Ali (PhD program coordinator) both (KIT) explaining the Karst system of the springs of the Danube River in South Germany (*photos © Wasim Ali)*

Water Fun training for schools

In the framework of SMART project (phase I, 2006 - 2009) a training program, based on demonstration, technology transfer and capacity building in the field of Decentralized Wastewater Systems Solutions (DWWSS) was designed, developed and tested as a pilot teaching unit for primary schools in Ramallah, Palestine and Salt Jordan.

As a dissemination strategy, the training program gives the possibility to pupils to conduct experiments in water/wastewater analysis and to build wastewater filters and constructed wetlands models. The training enhances the reflection process about wastewater components and gives arguments to the participants to understand the possibilities of wastewater treatment and its reuse purposes in decentralized systems. Additionally by the experiments, pupils understood wastewater as a resource; especially in regions with water big shortages. The central methodological objective of the teaching Unit "Water Fun" is to achieve Human Resource Development (HRD), which represents one of the cross-section elements to achieve economic and social development. The teaching unit is oriented to introduce concepts of water value and wastewater reuse by the hands-on contact with wastewater treatment eco-technologies and waste/water analysis experiments.

The pilot training activity denominated Teaching Unit *"Water Fun"* was conducted in November 2008 in Ramallah, Palestine in a primary school for a group of 21 students between 10 and 11 years old. Teachers from different Palestinian primary schools and Palestinian water administration authorities took part in the activity.

The teaching material based on the primary school teaching unit "Constructed Wetland" (Van Afferden et al. 2008), that was developed for European primary schools within the European Community funded project "Play with Water". This teaching unit was translated and modified according to suggestions from our local Palestinian partners. One of the important modifications was related to the teaching Unit's name. Originally the teaching unit was presented in Ramallah under the name "Play with Water". After the pilot activities it was realized that modifications in the name were necessary. Considering the water constrains in the Lower Jordan Rift Valley, children should avoid to "playing" with water. For this reason the name "Water Fun" will be adopted for future activities.
The development and implementation of the teaching unit "*Water Fun"* enhances the capacity building dimension in IWRM. As a key element the package explores human resources development as a strategy to achieve water development management in the Lower Jordan River Basin.

The main importance of Capacity building is directly associated with the increasing need and value of water. Particularly in Palestine and Jordan. In Palestine only less than 6% of the population is connected to treatment, plants (Mahmoud et al. 2003) the implementation of such CB programs in schools have a direct impact on the water value attitudes and by this CB will facilitate the future implementation of a sustainable water recourses management. The teaching unit sensitizes pupils but teachers as well for the basic and essential concepts related to water value and scarcity. Most of the water analysis experiments promote reflections and behaviour changes concerning water consumption and enhance a vision of considering wastewater as a resource and not as a waste.

The teaching unit "Water Fun" as a capacity building strategy represents an adequate tool to empower local communities and organizations in the implementation of water policy and management initiatives trough education. The direct impact of the CB strategy on the IWRM in the research area is associated with a common lack of basic knowledge of water resources management in the population. Thus, the development of water resources knowledge at the primary school level is a sustainable strategy to establish effective water management programs on the long-run.

Figure 4: Teaching unit *Water Fun* applied in Ramallah, Palestine at 19.11.2008. Left, children from Al-Shurook school build a constructed wetland model. Right , teachers from different Ramallah schools attend the Water Fun training in the Palestinian Water Authority PWA (photos © Manfred Van Afferden)

REFERENCES

ALI, W. & H. Hötzl (2009): SMART Project, Work-package 11, Report 2009, 24 pages, University of Karlsruhe, available at: http://www.iwrm-smart.org

Cardona, J., Ali, W., Mueller, R. , Samhan, S. & Van Afferden, M.(2009): Capacity building in wastewater treatment and reuse as a strategic component for sustainable water management in the Lower Jordan Rift Valley, 7 th International Conference of the European Water Resources Association, Limassol, Cyprus

Mahmoud, N., Amarneh, M. N., Al-Sa'ed, R., Zeeman, G., Gijzen, H., and Lettinga, G. (2003). "Sewage characterisation as a tool for the application of anaerobic treatment in Palestine." Environmental Pollution, 126(1), 115.

Van Afferden, M., Reimann, S., Müller-Bogotá, A., Peretzki, K., and Müller, R. (2008). "Primary school teaching unit "Constructed Wetland"." University of Waedenswil, Switzerland. pp 51, available at: http://www.play-with-water.ch

ACKNOWLEDGEMENT

The authors are thankful for financial support by the BMBF in Germany. They are also thankful for the assistance and help of all engaged universities, authorities, schools in the project region.

Using the SAS (Story and Simulation-) approach to support the development of sustainable water management strategies under climate change in the Jordan River region

Janina Onigkeit, Joseph Alcamo and others
Center for Environmental Systems Resarch, Kassel University, Germany
onigkeit@usf.uni-kassel.de

Abstract

Developing management strategies to cope with increasing scarcity of water resources in the future requires an adequate handling of uncertainty. Within the GLOWA Jordan River project we used an innovative procedure called the "SAS approach" (**S**tory **a**nd **S**imulation) to carry out a scenario exercise dealing with the future situation of water and the environment under climate change conditions in the Jordan River region. SAS is a stakeholder-driven and iterative procedure that requires an engagement of both, stakeholders and project scientists in the scenario building process. The aim of this approach is to integrate regional knowledge in a narrative form provided by stakeholders from different backgrounds and quantitative information provided by scientific models in a balanced, systematic and transparent way. As members of a scenario panel, stakeholders and other experts from Israeli, Jordanian and Palestinian water-related agencies and ministries as well as representatives of NGOs played an active role in the scenario development.

In a first step, the "GLOWA Jordan River scenarios of regional development under climate change" have been developed: As most important factors regarding the future water situation in the region, the members of the scenario panel identified the economic development and the extent to which regional water resources will be shared in the future. Combinations of extreme developments of these two factors result in four scenario storylines, which provide a range of uncertainty that helps to work out water management options under a variety of future conditions.

The four scenarios cover the future development up to 2050 in Israel, Jordan and the Palestinian Authority and can be characterized as follows:
(1) "Willingness & ability" – A scenario under which economic growth and multi-lateral water sharing leads to a flourishing region due to lasting peace and world-wide economic growth.
(2) "Modest hopes" – A scenario under which despite of unilateral dividing of water economic growth prevails due to heavy regional investments of outside donors to prevent deterioration of the political situation.
(3) "Suffering of the weak & environment" – A scenario under which economic recession in combination with unilateral dividing of water resources leads to a stagnation or worsening of the regional development.
(4) "Poverty & peace" – A scenario under which economic recession prevails, but multi-lateral water sharing reflects an improvement of the political situation.

These four scenarios serve as background for the development of strategies for a sustainable management of water resources which contain qualitative as well as quantitative elements. Covering a wide range of plausible and consistent futures the scenarios allow for an elaboration of a diversity of water-related management options. Furthermore, they help to develop strategies which might be useful to prepare for the challenges the Jordan River region might face in the coming decades.

The aim of this paper is to describe the SAS methodology including examples of the quantitative results and storylines and discuss the strengths and weaknesses of the SAS approach with respect to the support of developing water management strategies in the Jordan River region.

WEAP for IWRM in the Jordan River Region

Bridging between scientific complexity and application

Christopher Bonzi[1], Holger Hoff[2], Johannes Stork[3], Ali Subah[3], Leif Wolf[4], Katja Tielbörger[1]

1) University of Tübingen, Germany 2) Stockholm Environmental Institute, Sweden 3) Ministry of Water and Irrigation, Jordan 4) Commonwealth Scientific and Industrial Research Organisation (CSIRO), Australia

E-mail: christopher.bonzi@uni-tuebingen.de

Abstract

The need for integrated modelling to find sustainable water management solutions is not new. However, with integration comes the challenge of bridging between scientific uncertainty and complexity and practical application, which is a particular challenge in situations of limited data availability, institutional capacity and political barriers. We address this challenge by using the node-link type model WEAP as integration tool within a highly complex setting, the Jordan River Basin. WEAP provides both a consistent analytical and data management framework for addressing supply and demand management, as well as a user-friendly interactive graphical display for exploring simulation results. Therefore, WEAP is well suited for transdisciplinary applications in integrated water and land management. By integrating data and information from various sources, including process-based simulation models, it supports decision making on a sound scientific basis. Furthermore, WEAP can be seen as tool for data management and preservation which is, by itself, a contribution toward integrated water resources management.

Key words

Integrated Water Resources Management, WEAP, Jordan River, GLOWA Jordan River, SMART

1 Introduction

Integrated Water Resources Management (IWRM) acknowledges the interdependency and complexity of the water cycle within the river basin and landscape context, going beyond traditional sectoral approaches (Odendaal, 2002). Today, IWRM is a key concept, promoted by international agencies, governmental and academic institution all over the globe.

This paper addresses one of the major challenges of IWRM: how to bridge between scientific uncertainty and complexity and the need for practical application and tools. Here, we present an example by using WEAP (Water Evaluation and Planning tool) in the Jordan River Basin. By focusing on one of the most complex and politicised settings worldwide, we will illustrate both the use of WEAP as integration and application tool.

2 Complexity and uncertainty versus applicability – The IWRM-dilemma

IWRM is "a process which promotes the coordinated development and management of water, land and related resources, in order to maximize the resultant economic and social welfare in an equitable manner without compromising the sustainability of vital ecosystems" (GWP-TAC, 2000). This holistic description implies that sufficient knowledge from various disciplines is integrated as well as having a range of stakeholders identified and included in order to implement sustainable management solutions. Consequently, a major inherent characteristic of the IWRM concept is its complexity, spanning a wide range of disciplines, from climate sciences, land-use, hydrological, socio-economic and policy analyses. Communicating such complex information and the inherent uncertainty to decision makers is almost impossible. In this context, there are two challenges that need to be addressed:

1) Bridging between science and application: Water managers need practical tools which are flexible enough to address various kinds of questions. Ideally these tools can handle and reduce the complexity of a scientific IWRM approach and help facilitate active discussions between scientists, water managers and stakeholders in general.

2) Data availability: Sound and comprehensive data of sufficient quality, resolution and coverage is the prerequisite for calibration and validation of models and for analysing different management options under local and global changes. In the Jordan River Basin, trans-boundary and inter-sectoral competition aggravates this problem.

3 Which tool for IWRM?

WEAP, the Water Evaluation and Planning tool (www.weap21.org), has been chosen by several institutions at local, national and regional scale in the JRB, supported by the GLOWA and SMART science projects. Key advantages of WEAP are its low license fees, a growing number of documented case studies and its flexibility regarding the level of complexity used (for a detailed description see Yates et al. 2005).

WEAP is based on a water balancing approach, designed to be used by authorities and institutions for integrated water resources planning. If there is a need for more advanced modelling features, WEAP offers a range of modules and interfaces (e.g. for coupling with the MODFLOW groundwater model, the Qual2K water quality model, the MABIA crop water model and the MYWAS economic optimization model). The user can evaluate water demand and supply scenarios over extended planning horizons and test different adaptive management strategies. WEAP can explore physical changes to the system, such as new reservoirs, transfers or desalination as well as socio-economic changes e.g. water pricing and demand management. It provides a consistent analytical and data management framework and an interactive graphical display to engage stakeholders, decision-makers and scientists jointly in the planning process.

4 The Jordan River Basin (JRB)

The Jordan River basin is subject to extreme water scarcity, exceeding all typical scarcity parameters by far (Philips et al., 2009). Its natural water resources are overexploited, most evident from the lack of freshwater flowing in the lower Jordan and the shrinking Dead Sea. Water scarcity is expected to worsen in future, as precipitation is projected to decrease (Kunstmann et al., 2007) while rapid population growth drives demand. While the riparian of the Jordan River basin have initiated some demand management measures, key responses are still focused on the supply side.

Distribution and access to water are persistently disputed, due to the complex political situation. Therefore, co-operative water resource management has very high priority and should build on principles of equitable efficient and sustainable resource use. A recent survey in Israel, Jordan and the West Bank (Twite and Kölsch, unpubl.) suggests that water governance may be a key problem, e.g. management is currently implemented unilaterally, with no or very little cooperation between sectors and political entities. In addition, there is no or little data on water resources and demands in the catchment openly accessible.

5 WEAP in the Jordan River Basin

The German Federal Ministry for Education and Research (BMBF) is supporting a program for IWRM in regions with water problems, such as the GLOWA Jordan River (Global Change in the Hydrological Cycle) and SMART (Sustainable Management of Available Water Resources with Innovative Technologies) projects.

GLOWA Jordan River (GLOWA JR)

The GLOWA JR project addresses the question "How can the benefits from the region's water be maximized for humans and ecosystems, under global change?" (www.glowa-jordan-river.de). The project conducts research across many different disciplines and supports an active transboundary dialogue between science and stakeholders in the Jordan River region. German, Israeli, Jordanian and Palestinian institutions have consolidated information from data bases and through a modelling framework. Conceptual WEAP models of the regional and sub-regional water systems have made use of these. An Upper Jordan River Catchment (UJRC) WEAP model on resources, consumptions and hydrology was developed to evaluate different water allocations under climate change (Trondalen, 2009). Jordan Valley WEAP models address future water-balances in the main Jordanian water system and the West Bank. As an overarching transboundary planning tool, a regional WEAP is being developed, integrating aggregated data and model outputs from the different subprojects.

All WEAP models integrate data from different GLOWA JR sub-projects and address the question, how green and blue water can be allocated and managed sustainably to meet the growing demands in the Jordan River region under different scenarios (Fig. 1a). Water managers and GLOWA scientists are jointly developing and evaluating scientifically sound, yet realistic regional scenarios in WEAP, based on the so-called story-and-simulation (SAS) approach (Alcamo et al., 2008). Preliminary results from integrating the socio-economic SAS scenarios with hydro-climatological scenarios (Kunstmann et al., 2007, Menzel et al., 2007) are illustrated in Fig. 1b. This shows how the WEAP tool integrates and presents the output from scientific models for practical water resources planning. The management scenarios tested include, for example, new water options such as desalination or land use options with lower water demand (e.g. Tielbörger et al., in press).

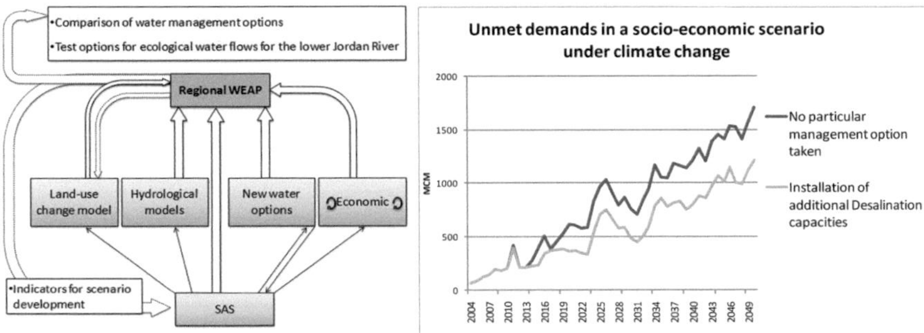

Figure 1: Left: Flow chart illustrating data exchange between GLOWA JR sub-projects and the regional WEAP (e.g. land-use change inputs to WEAP from the LandShift model); Right: Display of calculated unmet demand (preliminary results) in the project region based on socio-economic assumptions and a climate change scenario.

SMART

Within the SMART project, a WEAP simulation was used to analyze the agricultural water situation in the Jordan Valley (Alfarra et al., 2009). Operational rules of the reservoirs were determined using inverse parameter estimation techniques since the documentation of these rules was limited. The model was used to test the impact of various water allocation scenarios on agricultural production in the Jordan Valley.

The SMART project follows a slightly contrasting approach to the IWRM problem. While allocation models and water balance model like WEAP are also supported for upscaling, a focus is on the demonstration of practical solutions like decentralised wastewater treatment, groundwater protection zones or artificial recharge measures. A significant amount of the research effort is invested in local and regional groundwater modelling, with the example of a transregional numerical groundwater flow model incorporating all three riparian of the Lower Jordan Valley.

Using WEAP as central tool in the National Water Master Plan (NWMP)

The Ministry of Water & Irrigation (MWI) in Jordan is developing a Digital National Water Master Plan, to examine conventional and non-conventional water resources and discuss quantitative and qualitative management issues, as well as institutional and regulatory issues. WEAP was selected for strategic water resource planning in the NWMP-directorate. The GLOWA and SMART projects support the MWI scientifically and jointly with GTZ and BGR (German Development Cooperation) also through training workshops.

6 Discussion

WEAP has been shown to support integrated and effective water resources management, by consistently addressing different supply and demand side interventions and scenarios. Scientific model outputs have been integrated with other local data and information, thereby helping water managers to take decisions based on a sound understanding of the basin's water resources, in the context of internal and external environmental and socio-economic drivers and feedbacks – thus bridging between science an application. For example, results from detailed eco-hydrological process models (Menzel et al., 2007) have been integrated with demand side projections into a WEAP's system representation at reduced complexity. The usefulness of the WEAP tool for application becomes apparent in the fact that WEAP has already been adopted by key decision makers such as the Jordanian Ministry of Water and Irrigation.

Sharing of data and information is an important precondition to enable integrated water resources management. With the implementation of a regional WEAP model, containing simplified and aggregated water demand and supply data, we also reached a new consensus database. This is a major step forward compared to the only other multi-lateral water database for the Jordan River region (www.exact-me.org), which is static and non-spatially explicit. An advantage of using a water balance approach like WEAP in data management is that inconsistencies in assumptions and input data are more readily identified. The next step is a database which stores all relevant meta-information and main conclusions from the domain models and monitoring data such as envisaged in Riepl et al. (2009).

This article lists the applications of WEAP in the Jordan Valley and reports on its current applications by the local stakeholders. Results on the reliability and predictive capabilities of the WEAP models are given in the respective technical documentations (e.g. Alfarra 2010; Hoff et al., in preparation). WEAP successfully bridges between the complex IWRM concept and underlying scientific information and the needs for application in the JRB, with impact beyond the duration of the respective research projects.

List of Literature

[1] J. Alcamo, 2008. The SAS Approach: Combining Qualitative and Quantitative Knowledge in Environmental Scenarios. In: Environmental Futures: The Practice of Environmental Scenario Analysis. Developments in Integrated Environmental Assessment - Volume 2, Elsevier, Amsterdam, pp. 123-150. 2008

[2] Alfarra, A., Kemp-Benedict, E., Hötzl, H., Seder, N., Sonneveld, B., Wolf, L., 2009. Simulation of Water Supply and Demand in the Jordan Valley Region.: 7th Intern. Conf. of the European Water Resource Association, 2009, Limassol, Cyprus.

[3] GWP-TAC, 2000. IWRM. *TAC Background Paper No. 4*, Stockholm: Global Water Partnership.

[4] Kunstmann H., Suppan P., Heckl A., Rimmer A., 2007. Regional climate change in the ME and impact on hydrology in the Upper Jordan catchment, IAHS Publ. 313, 141-149

[5] Menzel, L., Teichert, E., Weiß, M., 2007. Climate change impact on the water resources of the semi-arid Jordan basin 3rd Int. Conf. on Climate and Water, Helsinki, 3.-6. September 2007

[6] Odendaal, P.E., 2002. IWRM, with special reference to sustainable Urban Water Management. In: *CEMSA 2002 Conference,* Johannesburg, South Africa.

[7] Philips, D., Jägerskog, A., Turton, A., 2009. The Jordan River Basin: 3. Options for satisfying the current and future water demand of the five riparians. Water International. Vol 34, No2, pp. 170-188

[8] Riepl, D., Wolf, L., Hötzl, H., 2009. Potential of Semantic Wiki tools to organize interdisciplinary IWRM approaches. Proceedings of the 7th International Conference of the European Water Resource Association, 2009, Limassol, Cyprus.

[9] Tielbörger, K., Fleischer, A., Menzel, L., Metz, J. & Sternberg, M., in press: The aesthetics of water and land - a promising concept for managing scarce water resources under climate change. /Philosophical Transactions of the Royal Society/ A (doi:10.1098/rsta.2010.0143).

[10] Trondalen, M., 2009. Climate Changes, Water Security and Possible Remedies for the Middle East. UN World Water Assessment Programme; Side publication series. UNESCO-PCCP.

[11] Yates D., Sieber J., Purkey D., Huber-Lee A., 2005. WEAP21 - A demand-, priority-, and preference-driven water planning model: Part 1, Model characteristics, Water International, 30, 487-500

Water Resources Management in the Arid and Semiarid Zones of India: Issues under a Changing Climate and Environment

K. Shadananan Nair
Nansen Environmental Research Centre India, Gopal Residency, Thottekkat Road, Kochi – 682011, Kerala, India, E-mail: nair59@yahoo.com, Tel: +91 484 2383351

Abstract

Changing climate and increasing anthropogenic impacts are major challenges in water resources management in the drier zones of India. Maintaining food and water securities is becoming very complicated issue in the arid and semi-arid parts of India that cover almost half of its central and north-western parts. Monsoons undergo wide interannual variability, with its impacts on large sectors of the rural society, as the life and economy are closely liked to agriculture, the largest consumer of water. Water resources now face a new challenge associated with rapid industrialization and urbanization. Increasing shortage of reliable water leads to several socio-economic issues. Policies and regulations related to water and climate are inappropriate and their implementation often fails due to social and political reasons.

Catchwords

Arid, semi-arid, India, water resources, climate change

1 Introduction

Globally, water resources are under stress from pollution, overuse, misuse and improper management and conservation. Global climate anomalies and resulting extremes in local weather exacerbate the situation. The impact of weather extremes or human interference is felt more in dry regions already facing multiple issues associated with water shortage. Demand in water has been escalating in India with fast rising population. Population has already crossed one Billion and it is expected to stabilize only by the year 2050 at 1.5 Billion. Increasing domestic, agricultural, and industrial water requirements are serious challenges as the economy is still not strong enough to implement large development projects and to implement measures to adapt with changing climate. Indian economy and life of millions have been traditionally linked to agriculture. Monsoons undergo wide inetrannual variation associated with global anomalies. Wide disparity in rainfall is reflected in local climate and water resources. As a result, vast areas of interior India are arid and semi-arid. Most of these regions are rich in agriculture where another agricultural revolution is possible, if some more water is available for irrigation. Extremes in climate in such regions disrupt all measures to attain food security. Proper assessment and careful management of water resources in such regions are utmost important. India will soon face a very serious water scarcity, unless new resources are identified and effective conservation and management measures are not resorted. Two-third of India's available freshwater is lost due to evaporation and runoff into the Sea, while there exists a freshwater crisis in many parts (UN Newsletter, 1999). Study of the World Watch Institute predicts that the Ganges and other rivers vital for farming will run dry for a part or all of the dry season, 30 years from now (Science update, 97). India has a wide range of geography and climate. The Thar Desert covers the western part, whereas in the eastern part, in the same latitudes lies the world's heaviest rainfall region. The climatic condition and geography influences to a great extent the water resources and its utilization.

Though India may be broadly regarded as belonging to the monsoon type of climate, it possesses almost all the major types of climate (Subrahmanyam, 1982) ranging from Perhumid to Arid. More than one-third of the country is having dry climate. Rain shadow effect of the Western Ghats Mountains makes a considerable part of interior South India semi–arid. This climate extends to the central west coast and western half of north interior India. Semi-arid parts of South India are rich agricultural lands where water shortage is a constraint in food production.

2 Rainfall and water availability

Arid and semi - arid regions of India have been delineated (Fig 1) using the method introduced by Thornthwaite (1948) and later modified by Thornthwaite & Mather (1955). Annual water surpluses, deficiencies and the per capita water availability in different states in these zones have been assessed. Possible changes in these parameters in two decades from now have been estimated, taking into consideration the population increase and change in temperature and rainfall pattern, as predicted by climate models. Necessary data for the period 1901- 2000 have been procured from the India Meteorological Department and the CDIAC, USA. Other relevant informations have been collected from various published records of the Government of India.

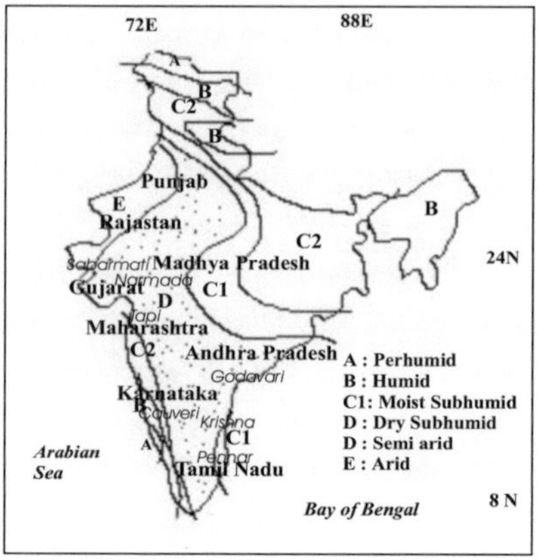

Fig 1 Climate types of India

Though the rainfall of India (nearly 105cm) is slightly above global average, there are wide inter-annual, seasonal and spatial variations. There are regions receiving less than 10cm (arid northwest) and more than 1000cm (perhumid northeast) in a year, both lying in the same latitude. Though the southwest monsoon (June – September) is the principal rainy season in the semi-arid region, the southeast of Peninsular India receives good rainfall from winter monsoon (October - December). In the semi - arid region, rainfall varies between 50 and 100cm. Seasonality of rainfall increases from south to north, indicating more seasonal shortage of water in the north. Disparity in rainfall is so high that in certain years floods and droughts of various intensities hit different parts of the country. Rainfall seasonality is changing in many parts of interior India.

States	Surplus from rainfall (cm)		Per capita availability (M^3)	
	Present level	By 2030	Present level	By 2030
1. Andhra Pradesh	33	26	135	50
2. Gujarat	48	40	250	100
3. Interior Karnataka	50	41	255	101
4. West Madhya Pradesh	42	33	235	93
5. Maharashtra	51	35	231	83
6. East Punjab	10	7	28	8
7. Rajasthan	0	0	0	0
8. Interior Tamil Nadu	49	40	213	102

Tab. 1: Changes in annual water surplus and per capita water availability in dry zones

Seven major rivers in the selected region drain an area ranging from 20,000 to 3,00,000 Km^2 with basinwise percapita 360 to 3109 m^3. These figures represent optimum values. Actual availability of safe water is considerably less than this. Many of these rivers are highly polluted from industries and agriculture. Considering the percapita availability from rainfall alone (after meeting evapotranspiration and soil moisture recharge), even today it is far below sufficiency (Table 1). State of Rajasthan has no surplus water during any season. After two decades from now the percapita will be drastically reduced to almost half. Water surplus occurs only during active rainy season. As per the water balance model, any slight increase in temperature with no corresponding increase in rainfall will make the climate to shift towards a drier category and will have serious impact on the water resources. There is a possibility that the regions now experiencing semi-arid climate may become arid in future. Adding to the land and water degradation, availability of reliable water will be even less than the present estimates. This can initiate new water conflicts. Water related conflicts have always delayed development activities, affecting all facets of life. Increasing tendency in urbanization in the selected zone may soon disrupt all water distribution systems. Groundwater use in many states is far above replenishable limits and has pulled down the water table considerably. In the states of Tamil Nadu and Gujarat, overextraction of groundwater has already affected the water quality. Human activities in the catchments and on the banks of rivers have reduced water flow in many rivers of South India.

3 Issues related to water resources management

Only a portion of the seasonal water surplus is reliably available for human use due to degradation of water resources, especially during dry months when the demand is more. Increased demands in agricultural, industrial and domestic sectors lead to considerable imbalances in the quantity and quality of river water. Climate change and variability are likely to worsen the existing situation by further limiting the water availability. Potential changes in temperature and precipitation pattern may adversely affect soil moisture condition, annual runoff and ground water recharge. Despite extensive efforts to improve the use and management of water resources, water demands continue to rise, contamination degrades water quality, and natural hazards, such as floods and droughts, affect human activities and cause extensive human suffering and economic losses. Population growth and urbanization are among the key factors underlying the enormous growth in the demand of water and the increase of environmental degradation.

Changing climate pattern is of serious concern for the dry regions of India. Rivers in north India are fed by the snowmelt runoff from the Himalayan ranges. Various reports predict an increased recession of Himalayan glaciers and increasing danger from glacial lake outburst floods in an increased temperature and increased seasonal variability in precipitation. Reduction in average flow of snow-fed rivers and an increase in peak flows and sediment yield would have major impacts on hydropower generation, urban water supply, and agriculture. The drier regions are threatened by desertification under a changing climate. Any fall in amount or increase in seasonality of monsoons would lead to soil moisture deficiency or floods in the riverine environments and affect water availability and runoff rates. There can be tremendous increase in the transport of sediment load of the Ganges, which already carries an extraordinarily heavy sediment load. The Ganges delta is very densely populated, and the effects of climate change could cause serious drainage and sedimentation problems, in addition to bank erosion and land loss. These impacts clearly would have immense socio-economic costs.

Extreme water conditions and quality issues add to several socio-economic issues in the heterogeneous society of India. Increasing demand is characterized by increasing competition. The social, environmental and political issues related to the rivers are worsening in the dry areas. There exists dispute over river water sharing, either national or international, around almost all major rivers. In India, domestic water disputes are more severe and difficult to resolve compared to international, because of various political and social reasons. Dispute over the Cauvery waters among the southern states is more than a century old and is becoming worse. Similar dispute exist further north on the Godavari waters also. In Narmada, the issue is about the huge project that needs evacuation of thousands of people and has severe environmental consequences. Water-related environmental problems are becoming transboundary in scope as local pollution problems spread across borders due to the pressure of population growth, increased fertilizer and pesticide use, more industries and inadequate pollution controls. When the availability decreases, existing disputes worsen and new disputes arise. Even the updated national water policy of India doesn't provide any clear solution for conflict resolution.

Shortage of water affects agriculture and rural job opportunities, inviting societal unrest. Rural unemployment promotes urban migration, exerting much pressure on the cities that are not well-designed to accommodate such a large population. Rural unemployment promotes terrorism. Almost one-third of the country is under the spell of extremists.

There are several challenges ahead that may upset current water management practices. A second green revolution has become necessary to maintain food security. But, change in Government policy favouring industrialisation and reducing investment in agricultural sector is a threat to millions living below poverty line. Another major environmental and social issue is gradually spreading with the implementation of the INR560 billion project to interlink rivers. Though the objective is control of floods and droughts by water transfer, the environmental consequences have not been considered seriously. Water rich States strongly oppose the implementation. In addition to these, limited water management capacity, fragmented organizational structures, and inadequate planning, management, and conservation are among the contributing factors to water crisis. Corruption, misappropriation of money, non-cooperation among government departments, slow government machinery, and vested regional and political interests always retard major projects for river basin development, water conservation, irrigation and even public water supply.

4 Conclusions and recommendations

Western parts of Peninsular India have rich water potential. Several rivers originating in the Western Ghats wastefully flow west to the Arabian Sea before they could be effectively harnessed. Some rivers that flow east are the source of water for all purposes in the semi - arid zone. Tapping waters of the west flowing rivers and diverting them to the east can solve almost all water related issues. Eastern part of Madhya Pradesh also has better water resources. If there is an amicable solution to the issues related to water resources project, it could lead to overall development of central and western India. Soil in the many parts of the dry zones is rich and water resources management here is important in view of increasing needs of food for the exploding population. Though water is becoming a serious and challenging issue, measures adopted to face it are inadequate and slow. India needs an appropriate and frequently updated water policy and an efficient mechanism to implement it. Policy development should include climate change impact and socio-economic conditions in different states. Water management should be brought under a central authority. Legal and institutional mechanisms are to be improved to cope with the need of the time. Constitutional amendment becomes necessary to solve water problems in large and transboundary basins. At present, water is a state matter and the central government has only limited role in its control. This is why the water disputes remain unsolved. There should be a consensus among major political parties to overcome this. Autonomous River Basin Organizations generally better coordinate basin welfare programmes. But, in the typical socio-political set up in India, impartiality and demographic and socio-economic justice should be ensured.

Lack of government funds has halter progress of many projects. An option is private sector participation. But, private sector financing should be seen to be complementary and in no way a substitute for the state's responsibility in providing basic necessities.

Development schemes should give due care to the millions below poverty line. Water should continue to be provided free to the extreme poor and to the marginal farmers. Also, social situation in India may not permit pricing easily. The fundamental right of access to safe water and sanitation with environmental protection is to be ensured. Strict control is to be made to avoid overuse and wasteful use. Proper awareness in conservation and management could help this. Satisfactory water allocation with consensus among different users is a key factor in the development of society and in maintaining harmony. Involvement of NGOs is better performed in India than the Government in providing basic necessities like water. People's participation should be an integral part of all aspects of water resources management. Water is to be used in an environmentally sustainable manner in order to maximize its economic and social benefits. Addressing water problems requires an inter-sectoral approach that recognizes their links with land use, agriculture, technology and health. Reformation in agricultural sector is urgently needed in maximum and efficient use of water and in minimising input of pollutants. Policy packages using a mutually reinforcing mix of institutional and policy reform, and legal, economic and management instruments will be needed. Wise and non-vested political decision to implement policy guidelines and the rules and regulations could help overcoming the water crisis in the dry zones of India.

List of Illustrations

List of Tables

List of Literature

[1] Science Update,. 1997. Water Scarcity seen in future. Publication of the US Consulat*e*, 3-4.

[2] Subrahmanyam, V.P. (1982) Some Climatic Problems of Water Balance of the Indian Region. Climatological Notes. No. **32**, 66-80.

[3] Thornthwaite, C.J. (1948) An Approach Towards the Rational Classification of Climate. Geogr.Rev. Vol. **38** No.1.

[4] Thornthwaite, C.J. & Mather, J.R. (1955) The Water Balance. Publ. in Climate. Lab. of Climate. Vol. **8** No.1

[5] UN Newsletter, 1999. UN and Water, **54** (11), 3-8.

Earth crises, counter measures, and their effects on Integrated Water Resource Management

Earth crises and counter measures - overview, geoengineering, and approaches for sustainable solutions within an interdisciplinary dialogue

MDgt Dr.-Ing. Martin Grambow

Bavarian Ministry for environment and health, Department of water management,

Rosenkavalierplatz 2, 81925 Munich, Germany; E-mail: martin.grambow@stmug.bayern.de,

Phone: +49(0)89 9214 4300

Abstract:

The climate conference in Copenhagen did not give reasons for hope of a prompt solution to either the climate crisis or any other closely related problem. In the search for solutions, one method has gained more and more attention: geoengineering. However, this method is quite questionable due to the potential damages and risks it might cause for future generations – risks that we deem acceptable so that we can continue to enjoy our current lifestyle. To develop a deeper understanding of the problems and tasks of Integrated Water Resource Management (IWRM), a group of scientists developed a model of the four major earth crises at a conference on „Earth System Engineering" at Kreuth in 2008. Water management is one of the four large challenges we need to deal with if we want to solve the earth crises. But, similarly to the other crises, there is no single sustainable solution; instead, a sustainable solution needs to take all the other earth crises into account as well.

Keywords:

adaptation, carbon dioxide, climate, earth crises, environment, geoengineering, IESP, intergenerational justice, reduction, sustainability.

1 Introduction

The problems associated with the climate crisis seem to be far from being solved, in spite of intense international debates that have been spurred by scientific studies of the International Panel on Climate Change (IPCC), in spite of *Al Gore* and *Angela Merkel*, in spite of international efforts such as the Kyoto protocol, the UN meeting at Copenhagen in 2009, or efforts on behalf of the climate that have been taken worldwide, but especially in the EU. And in spite of the fact that „the large catastrophe is continuing to send its apostels“, as *Peter Sloterdijk* describes poignantly. He lists the events in the first two-thirds of 2010 as evidence for the apostels: the winter storm Xynthia in Europe at the end of February, the intense flooding in central Europe from May to early June that caused the second largest flooding in Germany (the largest was in 2002), heavy storms with strong winds and hail in the US in mid-May that altogether caused damages of about 7,1 million euro, intense monsoon rains in Pakistan in July and August that rendered more than 15 million people homeless, and at the same time a large number of wildfires that burnt about 200,000 ha in Russia after a heat-wave that lasted for several weeks. Finally, the northeast of China was troubled by heavy flooding in the end of August. According to estimates by *Claudia Kemfert*, professor for energy economy, the economic costs for unabated climate change will be about 800 billion euro by 2050 in Germany alone.

2 Earth Crises and Counter Measures

In September 2008, Prof. Peter Wilderer initiated a dialogue[1] in Wildbad Kreuth, entitled „Earth System Engineering“ (ESE). This dialogue gave the basis for the development of a model on the most important threats to the earth system functions (Figure 1). The participants of this meeting founded the organization „International Experts for Earth System Preservation“ (IESP), to be able to follow up on the basic premises they had developed.

This meeting indicated yet again that the climate crisis is so complex that the resulting problems can only be solved if solutions take all the various global challenges and threats into account. Population growth, excessive use of resources, emissions, and asymmetries in the economic system has resulted in four earth crises. These crises can be understood as dynamically developing critical processes which seriously endanger our current way of life. They entail:

(1) The consequences of **climate change** and the closely related question on **energy** supply. Climate change has a vast amount of serious effects. Long-term disturbances in life-supporting functions of our earth system include not only easily noticeable droughts and floodings, but also substantial changes in the biosphere, especially in oceans and forests.

(2) A reduction in the **supply of water and food** due to the overuse and pollution of water and soil. Climate change worsens this crisis, especially through increased intensity and frequency of droughts and floodings.

[1] Participants included: IASS (Institute for Advanced Studies of Sustainability), IAS (Institute for Advanced Study of the TU Munich), the EU-Chapter of the Club of Rome in Brussels, and UNESCO.

Causes		Consequences / Crises
Population size	Complex interactions	Climate change and energy use
Resource use (e.g. energy, phosphate, land)		Reduced supply of water and food
Pollution, emission		Inequality, social and financial instability
Unsustainable economy (turbo-capitalism)		Loss of biodiversity

Figure 1: The four earth crisis according to IESP

(3) Social tension due to increased **social injustice**. The resulting instability and threats are magnified by the financial crisis or rather by the fractures in the global economic system. At the same time the consequences of climate change, especially the reduced water and food supply, is affecting the poor much more than the wealthy, further increasing the tension between social groups.

(4) The **loss of biodiversity.** It is well known that our earth is a „life supporting and stabilizing system“ which – next to many other functions – is able to keep global temperature between 0°C and 20°C. This temperature is regulated in parts by the large ecosystems of our planet, such as rainforests, wetlands, and marine ecosystems. By now it is generally accepted that biodiversity is an essential precondition for the stability of the current earth system.

Even though the earth crises are very complex, conventional solutions focus on a reduction in emissions by suggesting the sustainable use of resources. However, solutions to the earth crises do not only require efficient and non-polluting technology, but also fundamental changes in our life style. Whereas immediate changes are theoretically possible, the practical implementation of those changes throughout society is very difficult to achieve. Some examples:

(1) Use and development of efficient technologies that minimize emissions in all sectors. Unfortunately, so far the global use of such sustainable technology is not well implemented.

For example, dirty water or trash still undergoes no treatment in most parts of the world, polluting soils, rivers, and oceans.

(2) Implementation of specific laws. Such laws prescribe the needed solutions, such as the European Environmental Laws (WFD, REACH, and others). These laws are available immediately and globally, and can be developed further, if needed. There are a large number of technical and administrative solutions available, especially for dealing with water and food shortages, ranging from the protection of single water bodies to complex management strategies as practiced by the integrated water resource management. In practice water management worldwide is often weak.

(3) Socio-economic and corporate approaches, such as the call for a social market organization (Marx), a sustainable economy (Hauff), or a fairer global distribution of resources (Radermacher). Great potential for individual and social changes lie within the sectors of mobility, food, and distribution of goods. Also here the reality shows, that for changes there is still a long way to go.

The goal is a sustainable „Earth System Engineering“ that is based on tested and safe methods, and that will untighten the screw that is squeezing our earth system. But there might be an alternative to such traditional solutions: the use of „progressive-unconventional“ technologies, such as (also see figure 2):

(1) Increasing the shade on earth through the induction of sulfates into the high atmosphere or with the help of other methods.

(2) Increasing the albedo by whitening the landscape (roofs, fields) or by generating clouds.

(3) Ocean fertilization with iron, nitrates, or phosphate, and thus increasing the assimilation of CO_2 by algae; speeding up the turnover of the oceans with the help of large pumps that either bring deep water towards the surface or move surface towards the bottom. Such water movement increases the CO_2 uptake as surface water is more saturated with CO_2 than deep water.

When discussing the use of geoengineering, we also need to consider potential collateral damages and risks to future generations - risks that we consider acceptable in order to continue our lifestyle we are accustomed to. Therefore, risk assessments need to describe not only the technical feasibility of solutions but also their ethical consequences. IESP believes that there are massive challenges both in terms of risk and ethics associated with geoengineering which should render its use unlikely.

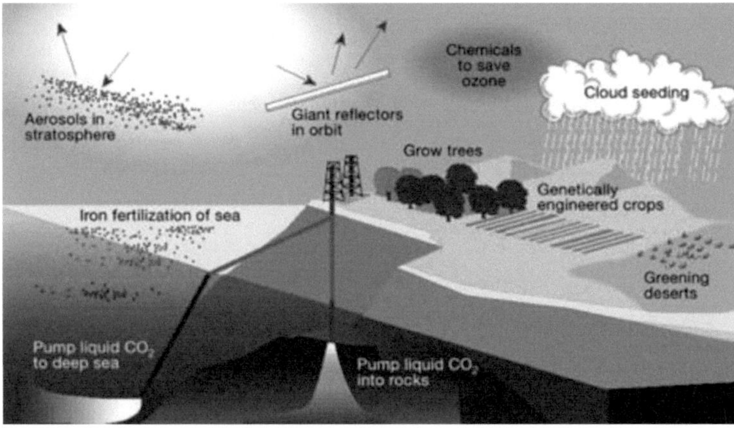

Figure 2: Brave new world? Overview over the methods used in geoengineering. (Source: DK. Smith /LLNL, in [10])

Without going into further detail, I will point out the following consequences for IWRM:

1. The search for „true sustainability" (i.e., the „warranty for the permanence of true human life", H. Jonas) does not have a counter measure once the risks are assessed.
2. The feasibility of the IWRM concept of integration still has to proof its usefulness in solving the four earth crises.
3. A safe supply of water and food is one of the most serious challenges for humanity today.
4. Water is normal managed locally. However, truly sustainable integrated solutions need to take global aspects into account.

To 1: Sustainability is well defined in the water sector. The technical as well as the organizational aspects of water management and water usage in general have to be tested based on efficiency and sufficiency. Research, development, and technical implementation still require major steps forward to reach Weizsäcker's factor 5.

To 2: It would be a mistake to focus on climate and energy only. Energy technologies that endanger water balance, biodiversiy, or global justice are useless. This is true for deep-sea drilling as well as for the ecologically questionable production of bioenergy. Such conflicts among different aims can be found frequently. They must not be ignored, but need to be discussed openly. At the same time, the use of huge amounts of energy for desalination and processing to obtain drinking water are quite questionable.

To 3: The global water sector has problems, because the technical and political developments in water management are too slow to adapt to the required changes. The millennium development goals will likely not be reached, especially not the goals for clean water. More and more risky substances are being discharged into the water, changing the water quality both inland and offshore. In many countries the water management seems to be too fragmented into separate sectors, and thus too weak for integrated problem solving. The continuing growth in human population size causes enormous challenges in the areas of irrigation and drinking water supply; at the same time, flooding and drought reduce the amount of arable land, or pushes that land towards the polar regions.

To 4: At first sight, water management might appear to be focused on local matters only. Such a perception is caused by tasks such as managing a certain stretch of river. But actually, water management needs to consider many global factors: The first important concept is that of virtual water which is considered in the area of global water volume management. The second important concept - global water quality - is discussed in the context of environmental norms and environmental dumpings. And finally, water management is involved in global attempts of “climate engineering”, which can affect our entire planet.

3 Prospect

Expectations that studies such as those associated with the IPCC will lead to effective solutions are high. But only interdisciplinary questions, a deepened understanding of the complexity and interconnectedness of the crises, and effective action will likely be able to step up to the enormous challenges ahead. The call for a better understanding of our earth system and for sustainable solutions in all sectors – especially the water sector – addresses the economy, development and everybody else who has responsibility in politics and society. We have been practicing earth system engineering since a long time. But today we realize that such actions are only permissible if they are sustainable.

The earth crises demand a transformation in our behavior towards our earth. And this request addresses each single one of us.

Literature

[1] Flannery, T.: Wir Wettermacher, Frankfurt am Main 2006.

[2] Grambow, M.: Integriertes Wasser-Ressourcenmanagement als Antwort auf drängende Wasserfragen – die Nachhaltigkeit als Dreh- und Angelpunkt einer globalen zukunftsfähigen Entwicklung, uwf-17 Nr. 3 Springer, Berlin Heidelberg 2009.

[3] Grambow, M. **Wei** M.**, Weiler** R, **Wilderer,** P. **:** Verantwortung für globale Herausforderungen übernehmen, Ergebnisse und Interpretation des „Earth System Engineering"-Kongresses , Korrespondenz Wasserwirtschaft, Heft 5 2009, DOI: 10.3243 / kwe 2009.10.005

[4] Jonas, H.: Das Prinzip Verantwortung - Versuch einer Ethik für die technologische Zivilisation, Frankfurt am Main 1979.

[5] Leggewie C, **Welzer**, H. Das Ende der Welt, wie wir sie kannten, Frankfurt am Main 2009.

[6] Lovelock J.: The revenge of Gaia: Why the Earth is fighting back – and how we can still have humanity. Penguin Books 2007.

[7] Marx, R.: Das Kapital - Ein Plädoyer für den Menschen, München 2008.

[8] Radermacher F.J.: Welt mit Zukunft - Überleben im 21. Jahrhundert, Hamburg 2007.

[9] Sloterdijk P. : Du mußt dein Leben ändern, Frankfurt a.M. 2009.

[10] The Royal Society,: Geoengineering the Climate, Science, governance and uncertainty, London 2009.

[11] Weizsäcker, E.U. et. al.: Faktor Fünf - Die Formel für nachhaltiges Wachstum, München 2010.

Dynamic downscaling of global climate projections for modelling the future regional water balance and water quality in Ukraine

Dirk Pavlik
Dennis Söhl
Technische Universität Dresden, Institute of Hydrology and Meteorology, Chair of Meteorology, 01067 Dresden, Germany, E-mail: dirk.pavlik@tu-dresden.de, Telefon: +49(0)351/46331316

Abstract

To overcome some of the world's most pressing water problems in the fields of water supply and sanitation, water and agriculture, ecosystem services and extreme events and –processes, the International Water Research Alliance Saxony (IWAS) was funded by the German Federal Ministry of Education and Research (BMBF) in 2008. Based on the approach of Integrated Water Resources Management (IWRM), the IWAS framework works on specific research topics in five model regions simultaneously: Eastern Europe (Ukraine/ Poland), Latin America (Brazil), Central Asia (Mongolia), South East Asia (Vietnam) and Middle East (Oman).

To asses the quality and quantity of natural water resources for implementing the IWRM concept in the model regions, the current conditions as well as the variations under a changing climate have to be taken into account. Accurate regional climate projections, based on global circulation model (GCM) simulations for specific scenarios, are necessary for the evaluation of future water availability in natural and anthropogenic systems. Therefore, the Climate Local Model (CLM) of the Consortium for Small Scale Modelling (COSMO) (CCLM) was initially implemented for the model region Ukraine/ Poland.

To adjust the CCLM to the physical and climatic conditions in the model region, a control run driven by ERA40 reanalysis data was carried out for the Period 1961 – 1990. The results of the control run (2m air temperature and precipitation) were evaluated with observed data. After evaluation of the model, global climate projections (ECHAM5/ MPI-OM) based on the Intergovernmental Panel on Climate Change (IPCC)-Special Report on Emissions Scenarios (SRES) were downscaled. The scenario results were compared with the control run to identify the climate change signals of 2m air temperature and precipitation.

Regarding to IWRM, the validated results of the regional climate projections are intended to be used for impact studies about possible changes in water availability and quality of surface water bodies.

Keywords

Eastern Europe, Climate Change, Regional Climate Model, CCLM, Dynamic Downscaling

1 Introduction

The protection and sustainable use of water resources are of utmost importance. The poor quality of water bodies is one of the main problems that can affect social welfare, economic development and environmental quality in developed countries and especially in the expanding industrial regions of transition countries. Climate change, land use change and increasing population pressure will aggravate this water problems.

A sustainable management of water resources requires the design of appropriate strategies, approaches, and measures to achieve an optimal distribution and use of water.

The climate is one of the key factors influencing the quantity and quality of water resources. Reliable predictions of the future climate are necessary to develop sustainable management concepts. Global circulation models (GCM) provide the most reliable and stable method to assess the response of the climate system to changes in the atmosphere. Nevertheless, GCMs are subject to some restrictions, whereby the most serious one in respect to regional impacts is the low spatial resolution. To bridge the large gap between the global and the regional scale a downscaling is necessary [1] [2]. The two main downscaling approaches are dynamical and statistical downscaling [1]. Due to insufficient data situations for most of the IWAS regions the dynamical approach was chosen. As a dynamical Regional Climate Model (RCM) the CCLM [3] was applied.

The objective is to model the future climatic conditions and to provide reliable data of regional climate scenarios for hydrological model applications.

2 Investigation Area

The Western Bug River rises in south western Ukraine and forms in its middle reach the eastern border of the European Union. Near Warsaw in Poland, the Western Bug River runs into the Vistula River, which drains into the Baltic Sea. The catchment has a total area of 40,000 square kilometres, of which about 11,000 square kilometres are located in Ukraine.

With annual averages of 2m air temperature and precipitation of about 7.4°C and 700 to 800 mm, respectively, the Western Bug River catchment has a temperate and humid climate, tending to be more continental towards the east [4].

The water quality of the Western Bug River is influenced by sewage disposal from domestic and industrial areas as well as from diffuse sources from cultivated areas. High biological and chemical loads occur due to out-dated or malfunctioning sewage treatment plants, nutrient leaching, mining activities, and pesticide deposits [5].

3 Data and Methods

The CCLM is a non-hydrostatic numerical Regional Climate Model (RCM), using a regular latitude/ longitude grid with a rotated pole and a terrain following height coordinate [3] [6]. It is based on the primitive thermo-hydrodynamic equations describing compressible flow in a moist atmosphere and was developed for applications on the meso-ß (20–200 km) and meso-γ scale (2-20 km).

The dynamical downscaling for the Western Bug River catchment ranged from a global scale of about 200 km down to a regional scale of about 7 km. In order to bridge the large scale ratio of about 29, a double-nesting approach was applied. In the first step, output from a global model was downscaled for a central part of Europe with 0.44 degrees (approx. 50 km) horizontal resolution (Fig. 1), i.e. with a downscaling factor of about four. Similarly, the output from the first nesting was downscaled for the region Belarus, Poland, the Western-Ukraine, and the Carpathian Mountains with 0.0625 degrees horizontal resolution (approx. 7 km) (Fig. 1), i.e. with a downscaling factor of about seven.

A control run based on ERA40 reanalysis data [7] was carried out for the period 1961-1990. The output variables 2m air temperature and precipitation were evaluated by comparing them to regionalized monthly station data. As reference data for the first nesting, the gridded CRU Dataset [8] with a spatial resolution of 0.5° was used. Due to the higher spatial resolution of the second nesting an appropriate resolved reference data set was necessary. This was derived from interpolated station data of the NOAA and ECA databases [9] [10]. The data set was quality checked and bias corrected at the Institute of Hydrology and Meteorology at the Technische Universität Dresden. Because of large gaps in the station data the evaluation period of the second nesting was reduced to 1973 - 1990. Model results and reference data were compared by difference calculation and statistical measures.

For the downscaling of climate the change scenarios A2 and B1 [11], the output of the GCM ECHAM5/ MPI-OM (run 1) [12] was used as forcing. The regionalized scenario results for the Western Bug River catchment were compared with the control run to identify the climate change signal of 2m air temperature and precipitation. The CCLM scenario runs were performed for the period 2010 to 2050 and evaluated for the four decades.

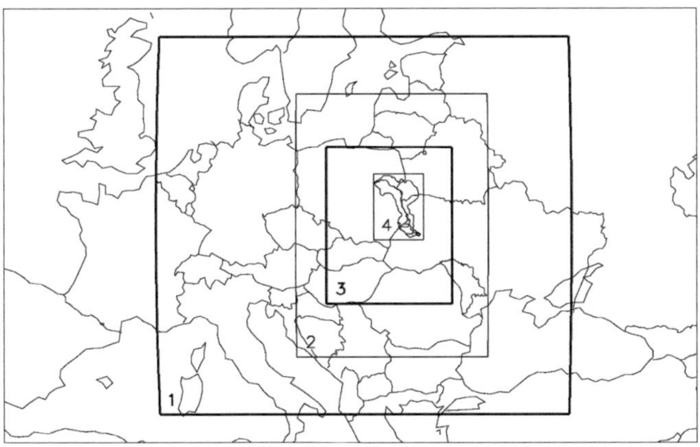

Fig. 1: 1-model domain[1] first nesting; 2-evaluation area[2] first nesting; 3-model domain second nesting; 4-evaluation area second nesting with Western Bug River catchment

4 Results

The first nesting of the control run was evaluated for the period 1961 to 1990 with the CRU dataset. The long term area mean of precipitation was overestimated by about +52 mm/year. More precisely, the low-land areas in the northern and central part of the study area have rather positive biases, whereas the high-mountain areas in the southern part show a higher variability in the differences of precipitation. The bias of the long-term area mean of 2m air temperature is about +0.27 K and has spatial variations between

[1] model domain – the total model area

[2] evaluation area – area without influence of the forcing at the boundaries, all presented results refer to the evaluation areas

-2.14 K and 1.9 K. The highest deviations occur in the southern part of the study area (Fig. 2), whereas the highest variability was found in the high mountain areas.

The results of the second nesting overestimate the long-term area mean of precipitation by about 43 mm/year. The spatial precipitation pattern was poorly reproduced. Overestimations of the precipitation occur especially in the hilly source area of the Western Bug River. Underestimations were found mainly in the south-western part of the study area (Fig. 2). The long-term monthly precipitation means were overestimated in spring and summer months with a maximum deviation of about 26 mm/month in May. Autumn and winter months show only small underestimations up to -9 mm/month in December. Positive deviations were found in October and March (Fig. 2).

The 2m air temperature was well reproduced by the model. The bias of the long-term area mean is about 0.0 K for the catchment area. Positive biases of more than 0.5 K were found in the western part of the study area outside of the catchment and negative biases of less than -0.5 K occur in the north east and the south west (Fig. 2). Autumn and winter temperatures were underestimated up to -0.5 K in January. Spring and summer temperatures showed a slight warm bias up to 0.4 K in May.

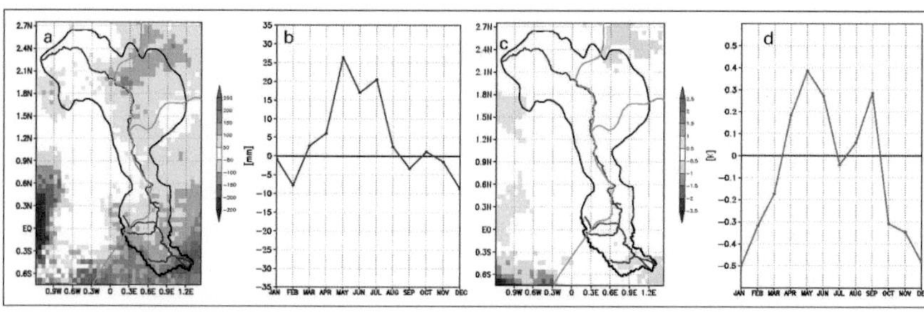

Fig. 2: bias of the second nesting: a) and b) precipitation; c) and d) 2m air temperature

The differences between the regional climate projections, and the control run are shown in Tab. 1. Both scenarios show an increase of 2m air temperature and precipitation in every decade compared with the reference period. As expected the changes of the scenario B1 are smaller than those of the scenario A2. Obviously, both scenarios have in common an increase of the mean 2m air temperatures in the last decade (2040-2050).

	Δ precipitation [mm]		Δ 2m air temperature [K]	
period	**A2**	**B1**	**A2**	**B1**
2010 – 2020	223	197	0.13	0.62
2020 – 2030	191	237	0.01	0.37
2030 – 2040	196	199	0.55	0.25
2040 – 2050	226	207	1.47	1.29

Tab. 1: Changes of long-term means of precipitation and 2m air temperature for the regional projections of the IPCC-SRES scenarios A2 and B1

5 Conclusions

The objective of the presented study is the application of a robust downscaling approach of global climate scenarios to the regional scale for hydrological applications. The downscaling was performed up to a spatial resolution of approximately 7 km. However, this scale is partially not appropriate for the modelling of hydrological processes. Furthermore, the results show that the model has not the ability to sufficiently reproduce the spatial distribution and the quantity of long-term precipitation. Maraun et al. [13] described this as "gap between dynamical models and the end user". Related to hydrological impact studies there is a need for the application of robust transfer approaches to bridge this gap.

Regarding to the estimated model errors, the 2m air temperature change signal is more robust than the precipitation change signal. Christensen and Christensen [14] found with a model-inter-comparison of different regional climate models, that the range of modelled climate change signals between the models is smaller than the range of the observed biases. This indicates, despite systematic errors in the model results, regional climate models are able to reproduce reliable climate change scenarios.

The presented results for the Western Bug River catchment show a smaller bias than the climate change signal for long-term means in all four decades. This suggests that the chosen downscaling approach can be applied for climate change studies on a regional scale. But for the usage of the results as input for hydrological impact studies it is necessary to improve the model results by application of an adequate transfer approach.

References

[1] J. Christensen, M. Hulme, H. von Storch, P. Whetton, R. Jones, L. Mearns, C. Fu. Regional Climate Information – Evaluation and Projections. In: J. Houghton, Y. Ding, D. Griggs, M. Noguer, P. van der Linden, X. Dai, K. Maskell, C. Johnson (eds.). Climate Change 2001: The Scientific Basis. Contribution of Working Group I to the Third Assessment Report of the Intergovernmental Panel on Climate Change, Chap. 10, Cambridge University Press, Cambridge, United Kingdom, New York, NY, USA. 2001.

[2] C. Hewitson, R., G. Crane. Climate downscaling: techniques and application, Climate Research (7). 1996.

[3] B. Rockel, A. Will, A. Hense. The Regional Climate Model COSMO-CLM (CCLM), Meteorologische Zeitschrift (17). 2008.

[4] M. B. Botyloch (Ed.). Environmental Atlas of Lviv (in Ukrainian). 2007.

[5] European Union. Management of the Bug, Latoritza and Uzh River Basins Ukraine, Report on Project Component 3 Upper Tisza / Western Bug, Challenges and management requirements, EuropeAid/114957/C/SV/UA. 2006.

[6] U. Böhm, M. Kücken, W. Ahrens, A. Block, D. Hauffe, K. Keuler, B. Rockel, A. Will. CLM - the climate version of LM: brief description and long-term applications. COSMO Newsl. 6. 2006.

[7] S. M. Uppala, P. W. Kållberg, A. J. Simmons, U. Andrae, V. Da Costa Bechtold, M. Fiorino, J. K. Gibson, J. Haseler, A. Hernandez, G. A. Kelly, X. Li, K. Onogi, S. Saarinen, N. Sokka, R. P. Allan, E. Andersson, K. Arpe, M. A. Balmaseda, A. C. M. Beljaars, L. Vande Berg, J. Bidlot, N. Bormann, S. Caires, F. Chevallier, A. Dethof, M. Dragosavac, M. Fisher, M. Fuentes, S. Hagemann, E. Hòlm, B. J. Hoskins, L. Isaksen, P. A. E. M. Janssen, R. Jenne, A. P.Mcnally, J.-F. Mahfouf, J.-J. Morcrette, N. A. Rayner, R. W. Saunders, P. Simon, A. Sterl, K. E. Trenberth10, A. Untch, D. Vasiljevic, P. Viterbo, J. Woollen et al. The Era-40 Re-Analysis, Quarterly Journal Of The Royal Meteorological Society (131). 2005.

[8] M. New, M. Hulme, P. Jones. Representing Twentieth-Century Space–Time Climate Variability. Part II: Development of 1901–96 Monthly Grids of Terrestrial Surface Climate. Journal of Climate (13). 2000.

[9] http://www.ncdc.noaa.gov/

[10] A. M. G. Klein Tank, and Co authors. Daily dataset of 20th-century surface air temperature and precipitation series for the European Climate Assessment. International Journal of Climatology, (22). 2002. (Data and metadata available at http://eca.knmi.nl)

[11] N. Nakicenovic, R. Swart (Eds.). Special Report on Emissions Scenarios. IPCC, Cambridge University Press, UK. 2000.

[12] E. Roeckner, G. Bäuml, L. Bonaventura, R. Brokopf, M. Esch, M. Giorgetta, S. Hagemann, L. Kornblueh, U. Schlese, U. Schulzweida, I. Kirchner, E. Manzini, A. Rhodin, A. Tompkins. The atmospheric general circulation model ECHAM5, Report No. 349, Max Planck Institute for Meteorology, Hamburg. 2003.

[13] D. Maraun, F. Wetterhall, A. M. Ireson, R. E. Chandler, E. J. Kendon, M. Widmann, S. Brienen, H. W. Rust, T. Sauter, M. Theme, V. K. C. Venema, K. P. Chun, C. M. Goodess, R. G. Jones, C. Onof, M. Vrac, I. Thiele-Eich. Precipitation Downscaling Under Climate Change. Recent Developments to Bridge the Gap between Dynamical Models and the End User. Reviews of Geophysics (in press). 2010.

[14] J. H. Christensen, O. B. Christensen. A summary of the PRUDENCE model projections of changes in European climate by the end of this century, Climatic Change (81). 2007.

CONTINUOUS SIMULATION OF CLIMATE CHANGE EFFECTS ON THE PORSUK WATERSHED HYDROLOGICAL PROCESSES

Albek, E.
Albek, M., Göncü, S., Yiğit, Z., Güngör, Ö., Şimşek, B.
Anadolu University, Environmental Engineering Department
İki Eylul Campus, Eskişehir, Turkey
e-mail: ealbek@anadolu.edu.tr, Telefon: +902223350580

Abstract

This study deals with the modeling of the effects of climate change on the hydrological processes in a watershed. The model is HSPF, used widely for simulating hydrology and water quality in watersheds. The time series supplied to the model to drive the hydrological processes have been prepared to reflect the expected climate change in the 21st century by downscaling the results of General Circulation Models to represent local climatic conditions. Two IPCC scenarios (A2 and B2) have been considered, with the A2 being the more pessimistic one showing large increases in temperature and decreases in precipitation. Simulations have shown that the watershed output, either snowpack yield or streamflow, will decrease in the future if the A2 scenario is considered in which case the sustainability of the water resources will no longer be maintained, especially on a seasonal basis. The B2 scenario also shows decreases in snowpack but significant changes in total streamflow are not observed. The simulations produce valuable information on how to plan for the sustainability of the water resources in the future.

Catchwords

Climate change, HSPF, Watershed modeling, Watershed planning

1 Introduction

Global warming and consequently climate change can be considered to be the most severe environmental problem the world is facing today. This problem, unlike other environmental problems encountered till now, will not be eliminated but civilization will be forced to adapt to the new conditions. Therefore it is of paramount importance to be able to produce reliable estimates about what climate change will bring in the future in order to find and implement the best adaptation methods based on scientific approaches.

One of the key issues in the adaptation process is the sustainability of water resources. Todays civilization depends on water resources much more than previous cultures did. Though each day methods are invented and applied to reduce water consumption, industrialization, agriculture and urbanization put large stresses which are likely to continue in the future on water resources.

This study focuses on two points in the adaptation process mentioned above. The first one involves the forecasting of climatic conditions in the 21st century based on scenarios likely to occur in the future. The second point is about the modeling of watershed hydrology. The study involves the Lower Porsuk Stream Watershed in western Inner Anatolia, Turkey where agriculture, industrial activities and urbanization put a significant pressure on the environment. The Porsuk Stream is heavily burdened by pollution and the problems which are encountered today will probably be aggravated in the future due to climate change in addition to other pressures.

2 Study Area and Method

The Lower Porsuk Stream Watershed encompasses the named stream beginning at the Porsuk Reservoir and ending at its confluence with the Sakarya River. The watershed has a surface area of 5800 square kilometers at an average altitude of 800 meters above sea level. A large city, Eskisehir, is located within the watershed and is the source of a heavy pollution burden. A second source of pollution is the Porsuk Reservoir which is highly eutrophic and supplies water to the stream. Fig. 1 shows the watershed together with the meteorological and streamflow and water quality monitoring stations whose data have been used in the modeling process. The figure also shows a small subbasin of the watershed from where results are presented in this paper.

The modeling environment is BASINS and the modeling program is HSPF (Hydrological Simulation Program - Fortran) (BASINS, 2007; Bicknel et.al., 1993). HSPF is a powerful modular program which is used widely for the simulation of the hydrological processes in watersheds and also can be utilized effectively to predict into the future using climate change scenarios as meteorological input (Göncü and Albek, 2010).

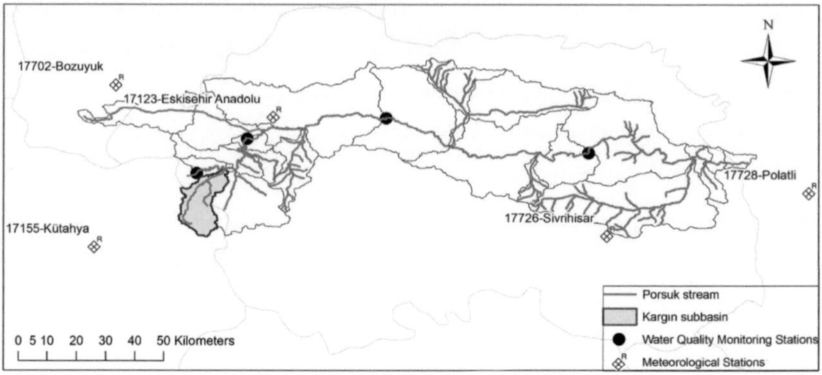

Fig.1, The Lower Porsuk Stream Watershed

Statistical downscaling has been applied to meteorological time series obtained from General Circulation Models (GCM). Two SRES (Special Report Emission Scenarios) from IPCC have been considered, namely A2 and B2 (IPCC-TGCIA, 1999). The scenario projections have been downloaded from the British Atmospheric Data Centre as the HadCM3 model outputs (BADC, 2009). Both scenarios assume a heterogeneous world but the B2 scenario is environmentally oriented while the A2 scenario puts economic development over environmental concerns.

The climatic conditions on the watershed are characterized by five meteorological stations located in and around the watershed as shown in Fig. 1. Results of statistical downscaling are shown in Fig. 2 for temperature and precipitation in the 17155 Kütahya meteorological station from 1961 till 2099 as this station affects the Kargın subbasin. Observations in the period from 1975 to 1990 have been used for calibration. Validation has been done with data from the 1990 – 2003 period. The Statistical Downscaling program SDSM has been used (Wilby et. al. , 2002; Wilby et. al. , 2007).

As observed in the temperature projections, the two scenarios display a similar course at the beginning (reaching the 12 degrees Celsius mark roughly in 2025). After around 2050 the A2 scenario diverges and takes on higher values as expected. For precipitation the scenarios behave similarly till 2050 (showing a small decrease). Afterwards the A2 scenario decreases while the B2 scenario shows a small increase. The curves shown are LOWESS (Locally Weighted Scatterplot Smoothing) smooths with a window length of 11 years (Cleveland, 1999). In both plots the 30 year averages at the beginning and at the end of the period are added to show the changes.

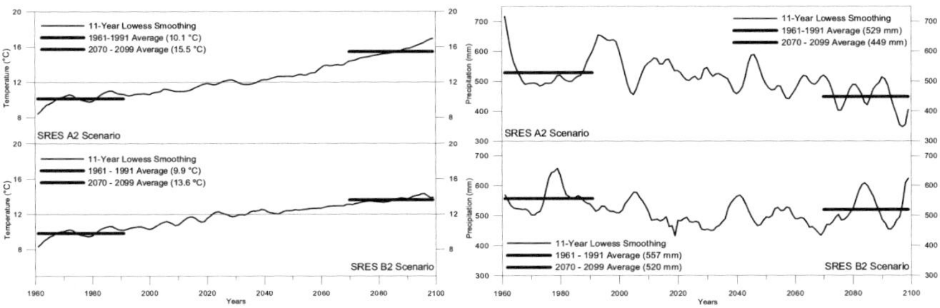

Fig.2, Temperature and precipitation projections for the Kütahya meteorological station

HSPF requires as input five more meteorological time series besides temperature and precipitation, namely cloudiness, dewpoint temperature, wind velocity, potential evapotranspiration (PE) and solar radiation. All except PE have been downscaled from GCM projections. PE has been calculated using the Penman Pan method.

The HSPF program has been used in the simulations with a time interval of one day. This increment is not enough to resolve events of short duration like flash floods but most other hydrological processes are simulated adequately. The choice of the time increment is also limited by data availability.

The model has been calibrated and validated by using streamflow and snow depth observations. The data sets for these are separated into two periods and the two portions are used separately for calibration and validation. As of to the preparation of this manuscript, two subbasins have been calibrated and validated. Good results have been obtained within an agreement of 10% between simulations and observations. In the following sections results belonging to the Kargın subbasin will be presented. This subbasin has been indicated in Fig. 1. It has a surface area of 168 square kilometers and is covered predominantly by mixed forests with sparse coverage (77%). Agricultural areas and pasture account for 10% and the rest is barren land.

Fig. 3 shows how the water yield from the snowpack covering the subbasin changes over time. For both scenarios, as clearly observable from the smooths, the snowpack yields are decreasing, mainly due to less snowfall with increasing temperatures. The decrease in the A2 scenario is also enhanced by decreasing precipitation. There are also completely snowfree periods in the later years.

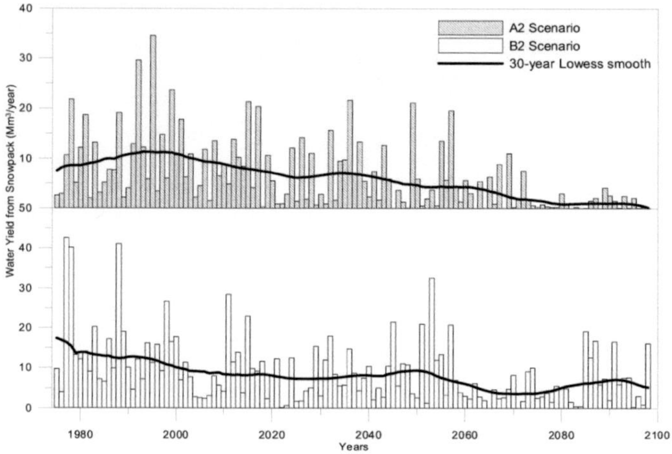

Fig.3, Water yield from snowpack in the Kargın subbasin

Fig. 4 shows how the Kargın stream draining the watershed behaves in the winter and summer months (represented by normalized February and August flows) in the A2 scenario. The summer flows are very low, as expected, and they decrease at a faster rate than winter flows. This result shows a feature which cannot be observed when aggregate values (i.e. yearly streamflow) are examined. The climate change behaves differently on a seasonal basis and at least in the subbasin modeled the summer effects are larger. The decrease in water availability in the irrigation season combined with reductions in water yield from snowpacks can have devastating effects, especially at sites where storage opportunities are restricted.

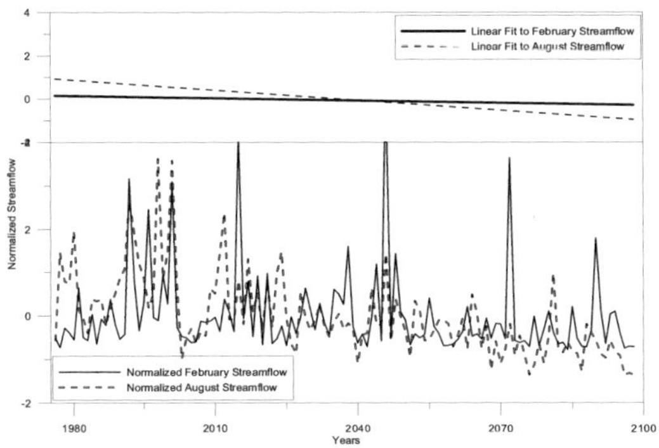

Fig.4, Streamflow leaving the watershed

3 Conclusion and Forecast

The foregoing results have shown that water resources will be in risk and hard to sustain in the future if the anticipated climate change will proceed as predicted by the scenarios. The B2 scenario does not reduce flows because there are no negative trends in precipitation as the A2 scenario. However, in each scenario the snowpack decreases steadily in the future. This will have important consequences as to the timing of melt events and ecological effects like the loss of the sheltering effect of snow cover.

The modeling study incorporates the meteorological aspects of climate change but changes in watershed properties (like infiltration rate, interception capacity, etc.) are not taken into account. Most watershed properties which are parameterized in the models will surely change with climate change. This will be incorporated into the models in future simulations. HSPF allows changes in parameters during the simulations.

Continuous simulation of climate change effects has the advantage of observing the response of watersheds to many different meteorological conditions possible. This can be utilized to design the future of watersheds with more insight and precision. Modeling in this respect stands out as the only possible tool in watershed planning.

This study has been supported as a research project by TÜBİTAK, the Scientific & Technological Research Council of Turkey, and Anadolu University.

List of Illustrations

List of Literature

BASINS. http://www.epa.gov/waterscience/basins/, U.S. EPA. 2007.

Bicknel B.R., Imhoff, J.C., Kittle, J.L., Donigian A.S. and Johanson R.C. Hydrological Simulation Program- Fortran User's Manual for Release 10, Environmental Research Laboratory Office of Research and Development U.S. Environmental Protection Agency, Athens, GA., EPA/600/R-93/174. 1993.

Carter, T.R., M. Hulme, and M. Lal. Guidelines on the Use of Scenario Data for Climate Impact and Adaptation Assessment. Version 1, IPCC-TGCIA. 1999

Cleveland, W.S. Robust Locally Weighted Regression and Smoothing Scatterplots, J.Am.Stat.Assoc., 1979

DAI CGCM3 Predictors. Sets of predictor variables derived from CGCM3 T47 and NCEP/NCAR Reanalysis. Montreal, QC, Canada, 15 pp. Available at: http://loki.qc.ec.gc.ca/DAI/doc/technique/DAI_Predictors_CGCM3andNCEP_January2010_EN.pdf. 2010

Göncü, S., Albek, E. Modeling Climate Change Effects on Streams and Reservoirs with HSPF, Water Resources Management. 2010

UK Meteorological Office, Hadley Centre. HadCM3 Model Data, [Internet]. British Atmospheric Data Centre, 2006-, 12.05.2009. Available from http://badc.nerc.ac.uk/data/hadcm3

Wilby, R.L., Dawson, C.W. and Barrow, E.M. SDSM - a decision support tool for the assessment of regional climate change impacts, Environmental Modelling Software. 2002,

Wilby, R.L., Dawson, C.W. SDSM4.2-A Decision Support Tool for the Assessment of Regional Climate Change Impacts , User Manual. 2007